Role of Seismic Testing Facilities
in Performance-Based Earthquake Engineering

GEOTECHNICAL, GEOLOGICAL AND EARTHQUAKE ENGINEERING

Volume 22

Series Editor

Atilla Ansal, *Kandilli Observatory and Earthquake Research Institute, Boğaziçi University, Istanbul, Turkey*

Editorial Advisory Board

Julian Bommer, *Imperial College London, U.K.*
Jonathan D. Bray, *University of California, Berkeley, U.S.A.*
Kyriazis Pitilakis, *Aristotle University of Thessaloniki, Greece*
Susumu Yasuda, *Tokyo Denki University, Japan*

For further volumes:
http://www.springer.com/series/6011

Michael N. Fardis • Zoran T. Rakicevic
Editors

Role of Seismic Testing Facilities in Performance-Based Earthquake Engineering

SERIES Workshop

Editors

Michael N. Fardis
Department of Civil Engineering
University of Patras, P.O. Box 1424
26504 Patras
Greece
fardis@upatras.gr

Zoran T. Rakicevic
Dynamic Testing Laboratory
& Informatics
Institute of Earthquake Engineering
and Engineering Seismology (IZIIS)
Salvador Aljende 73
1000 Skopje
Republic of Macedonia
zoran_r@pluto.iziis.ukim.edu.mk

ISSN 1573-6059
ISBN 978-94-007-1976-7   e-ISBN 978-94-007-1977-4
DOI 10.1007/978-94-007-1977-4
Springer Dordrecht Heidelberg London New York

Library of Congress Control Number: 2011939568

© Springer Science+Business Media B.V. 2012
No part of this work may be reproduced, stored in a retrieval system, or transmitted in any form or by any means, electronic, mechanical, photocopying, microfilming, recording or otherwise, without written permission from the Publisher, with the exception of any material supplied specifically for the purpose of being entered and executed on a computer system, for exclusive use by the purchaser of the work.

Printed on acid-free paper

Springer is part of Springer Science+Business Media (www.springer.com)

Preface

The Tohoku earthquake that devastated the North-eastern coast of Japan on March 11, 2011 and the smaller, yet also catastrophic, Christchurch (NZ) shock that hit the opposite end of the Pacific Rim less than 3 weeks before, were stark reminders of the weakness of humans and their works against the force of nature. These events came almost exactly 1 year after the extraordinary Bio-Bio earthquake in Chile and less than 2 years after the L'Aquila (IT) earthquake, the strongest and deadliest in Europe for almost 10 years. Within the overall picture of disaster left by these events, one can find bright spots. For example, the performance of buildings and civil infrastructures in the Tohoku earthquake looks like a success story, in view of the magnitude of the shock and of the disastrous effects of the 1995 Great Hanshin earthquake on buildings and civil infrastructures in Kobe. By contrast, the collapse of several RC wall buildings in Viña del Mar and Concepción in Chile shook our confidence to this time-and-again proven type of earthquake resistant construction and our knowledge of it.

In fact, every new earthquake demonstrates many still dark areas and gives the impression, to public and policy makers alike, that our Research and Technological Development (RTD) community of earthquake engineering has achieved little progress in over 40 years of intensive work. We have to bear in mind, though, that, unlike other engineering disciplines, earthquake engineering cannot give RTD answers by testing and qualifying few and small typical components, or even single large prototypes. The extreme diversity of Civil Engineering structures does not allow drawing general conclusions from few tests. Moreover, our structures are large in size and their response and performance under strong earthquakes cannot be meaningfully tested in an ordinary lab or in the field. Testing has to be of large-scale, if possible of real-scale specimens. It often has to be combined with heavy advanced computations, integrated with the large-scale experiments to complement them and extend their scope, even by coupling two different but simultaneous tests. So, nowadays earthquake engineering research infrastructures are much larger and expensive than their counterparts in other fields of Science and Technology. Moreover, their staff have to be very resourceful and imaginative, devising intelligent ways to carry out simultaneously different tests and advanced computations (often at different labs), in order to draw meaningful conclusions. This means that research infrastructures have to pool their resources and activities and either do their research together or jointly develop advanced testing and instrumentation techniques that will maximize their testing capabilities and increase the output and value of their tests.

Because the US and Japan have the largest overall seismicity among all developed regions of the world, and indeed concentrated mainly in some of their most populous, prosperous and economically important areas, it is natural for them to develop into the two world leaders in earthquake engineering RTD. They have established their leadership by investing heavily in a network of large and expensive, but complementary and well integrated experimental RTD infrastructures.

Europe as a whole has overall seismicity and exposure to earthquakes comparable to that of the US or Japan. Among the about 580 million of the inhabitants of Europe (the present EU, Turkey and the Western Balkans) 41 million (7% of the total) live in high seismicity areas and another 64 million (11% of the total) in moderate seismicity ones. In the last two decades of the twentieth century and in the first one of the 21st, earthquakes caused about 5,300 casualties in the EU (4,800 of them in Italy) and over 19,000 in Turkey alone. The earthquake disasters of 1980 in Irpinia (IT), 1989 in Spitak (Armenia), 1999 in Kocaeli (TR) and Athens (GR), as well as the recent one in L'Aquila (IT), are among the most costly ones in history. The effects of future seismic events in Europe may be even larger, especially in view of the ongoing urbanisation in some of Europe's most seismic areas. To mitigate the overall medium and long term impact of earthquakes, Europe should gradually reduce the seismic vulnerability of its new and existing structures by improving knowledge in earthquake engineering and by disseminating RTD results to its engineering and scientific and technical (S/T) community. This is especially needed in the most seismic but less developed European countries, which also happen to go through an era of frantic construction and infrastructure development.

Nowadays, research in earthquake engineering is mainly experimental and of large-scale. Large, expensive and advanced experimental research infrastructures can only be built by few technologically advanced States (or by supranational entities, such as the European Union). Moreover, the necessary integration and exploitation of such resources can only be done over large geographical areas, e.g., at the European level. The example to follow is that of the "Network for Earthquake Engineering Simulation" (NEES) program of the US National Science Foundation (NSF), which has invested to the present day over $210 m to upgrade the material infrastructures in 15 earthquake engineering laboratories around the US, couple them electronically via the high performance computational Grid, and directly support access to them by other research teams throughout the US.

In order to be one of the world leaders in earthquake engineering research together with the US and Japan, Europe must transcend the extreme fragmentation of its research infrastructures. EU Member States, Associated and Candidate Countries, being central in the development and financing of infrastructures, have made individually major investments in large and expensive experimental RTD facilities in earthquake engineering: the total present value of investment in material resources in Europe's national earthquake engineering research or university

Preface vii

Institutes is well over €200 m. However, individually these facilities don't have yet the critical mass of people and the broad range of expertise needed for major breakthroughs in the State-of-the-Art of earthquake engineering. Their overall capacity and the associated human resources are used suboptimally, not only owing to fragmentation among different States, but also because they are not open to use by outside researchers even from the same country, let alone from others. Besides, most of the large European seismic research infrastructures are established in few, technologically more advanced but not so seismic EU Member States, while countries with high seismicity have in aggregate fewer and less advanced RTD infrastructures. So, there is a need for better integration of the wider European RTD community in earthquake engineering into the activity of Europe's large experimental infrastructures.

Moving in the direction of integration and of common research infrastructures, the European Commission established in 1992 the European Laboratory for Structural Assessment (ELSA) at the JRC. ELSA is one of the largest research infrastructures of the JRC and has become a world leader in Pseudo-dynamic testing with substructuring for earthquake simulation and a world reference centre for experimental earthquake research. Shaking Table capabilities and Centifuge Test facilities, vital for structural or geotechnical experimental earthquake engineering research, respectively, are also offered in national labs scattered around Europe. In its Seventh Framework Programme [FP7/2007–2013], the European Commission is supporting further integration of all these earthquake research infrastructures through project SERIES ("Seismic Engineering Research Infrastructures for European Synergies") of its Research Infrastructures Programme (under grant agreement n° 227887).

SERIES's 23-strong consortium comprises the key actors in Europe's seismic engineering research (including 3 industrial firms). Its scope covers all aspects of seismic engineering testing, from eight reaction wall pseudodynamic (PsD) facilities and ten shaking table labs, to EU's unique tester of bearings or isolators, its two major centrifuges and an instrumented site for wave propagation studies. It offers to Europe's S/T community of earthquake engineering Transnational Access to a portfolio of world class seismic research infrastructures: EU's largest PsD facility, four diverse shake tables and the two centrifuges. It fosters co-operation within the wider European S/T community of earthquake engineering, through networking. In Networking Activities SERIES is setting up a public distributed database of past, present and future test results, installs distributed testing capabilities at all PsD labs, fosters development of up-and-coming laboratories at Europe's most seismic regions, drafts and applies protocols for qualification of seismic research infrastructures and engages the entire European community of earthquake engineering via the best possible instances: the European Association of Earthquake Engineering, EU's seismic code makers and their national groups, the European construction industry, as well as all relevant S/T associations or networks. All labs in the SERIES Consortium are also engaged in Joint Research Activities, to explore and prototype novel actuators (combination of electro-dynamic and hydraulic ones) for better control of fast tests or special applications, to advance new sensing and instrumentation systems, data assimilation in equipment-specimen models for better test control and optimisation of testing campaigns, as well as experimental studies of soil-structure interaction at all types of testing facilities.

The main impact of strong earthquakes that have hit developed countries in recent times was the serious damage to property and the heavy economic losses they inflicted, rather than the casualties. To cover the impact of earthquakes to aspects beyond human life, "Performance-based earthquake engineering" has developed in recent years as the most compelling engineering concept for protection against earthquakes.

"Performance-based engineering" in general focuses on the ends, notably on the ability of the engineered product to fulfil its intended purposes, taking into account the consequences of a failure to meet them. By contrast, conventional structural design is process-oriented: design codes emphasize prescriptive, easy to apply, but opaque rules which are just means to achieve the goal of satisfactory performance, but they don't describe the goal itself in clear and quantitative terms. Traditional code rules may be convenient for safe-sided, economic solutions under common combinations of design parameters, but leave little room for judgement and creativity and do not provide a rational basis for innovative designs that benefit from recent advances in technology and materials.

The general framework of Performance-based earthquake engineering is now well established. Its objective is to maximize the utility from the use of a structure by minimising its expected total cost over its lifetime, including the expected losses in future earthquakes. As far as development of specific methods and tools, the progress already made is very significant. It has been based so far mainly on numerical and analytical work. It is now high time to put the power and the advanced techniques presently available at large seismic testing facilities to the full service of Performance-based earthquake engineering RTD. Hence, one of the goals of the SERIES project is to promote Performance-based earthquake engineering through its experimental activities. The SERIES Workshop on "Role of Research Infrastructures in Performance-Based Earthquake Engineering" took place in the beautiful and historic Macedonian lake-side town of Ohrid in early September 2010, to serve exactly that purpose. It attracted renowned experts from around the globe who presented about 20 invited contributions, and a fairly large audience.

The Ohrid Workshop and the publication and diffusion of these Proceedings are part of the Networking Activities of the SERIES project and have been made possible through its funding by the European Community.

Michael N. Fardis, co-ordinator of the SERIES project and co-editor of this volume, gratefully acknowledges the support of his co-workers at the University of Patras, Ms Vassia Vayenas and Dr. Dionysis Biskinis, for the preparation of these Proceedings.

Patras, GR   Michael N. Fardis

Skopje, MK   Zoran T. Rakicevic

Contents

1 How Can Experimental Testing Contribute to Performance-Based Earthquake Engineering 1
Fabio Taucer and Artur Pinto

2 Earthquake and Large Structures Testing at the Bristol Laboratory for Advanced Dynamics Engineering 21
Matt S. Dietz, Luiza Dihoru, Olafur Oddbjornsson, Mateusz Bocian, Mohammad M. Kashani, James A.P. Norman, Adam J. Crewe, John H.G. Macdonald, and Colin A. Taylor

3 Structural and Behaviour Constraints of Shaking Table Experiments 43
Zoran T. Rakicevic, Aleksandra Bogdanovic, and Dimitar Jurukovski

4 Eucentre TREES Lab: Laboratory for Training and Research in Earthquake Engineering and Seismology 65
Simone Peloso, Alberto Pavese, and Chiara Casarotti

5 Cross-Facility Validation of Dynamic Centrifuge Testing 83
Ulas Cilingir, Stuart Haigh, Charles Heron, Gopal Madabhushi, Jean-Louis Chazelas, and Sandra Escoffier

6 Towards a European High Capacity Facility for Advanced Seismic Testing 99
Francesco Marazzi, Ioannis Politopoulos, and Alberto Pavese

7 Performance Requirements of Actuation Systems for Dynamic Testing in the European Earthquake Engineering Laboratories 119
Luiza Dihoru, Matt S. Dietz, Adam J. Crewe, and Colin A. Taylor

8 Model Container Design for Soil-Structure Interaction Studies 135
Subhamoy Bhattacharya, Domenico Lombardi, Luiza Dihoru,
Matt S. Dietz, Adam J. Crewe, and Colin A. Taylor

9 Computer Vision System for Monitoring in Dynamic Structural Testing 159
Francesco Lunghi, Alberto Pavese, Simone Peloso,
Igor Lanese, and Davide Silvestri

10 Quality Needs of IT Infrastructure in Modern Earthquake Engineering Laboratories 177
Mihai H. Zaharia and Gabriela M. Atanasiu

11 Use of Large Numerical Models and High Performance Computers in Geographically Distributed Seismic Tests 199
Ferran Obón Santacana and Uwe E. Dorka

12 Shaking Table Testing of Models of Historic Buildings and Monuments – IZIIS' Experience 221
Veronika Shendova, Zoran T. Rakicevic, Lidija Krstevska,
Ljubomir Tashkov, and Predrag Gavrilovic

13 Dynamic Behaviour of Reinforced Soils – Theoretical Modelling and Shaking Table Experiments 247
Jean Soubestre, Claude Boutin, Matt S. Dietz, Luiza Dihoru,
Stéphane Hans, Erdin Ibraim, and Colin A. Taylor

14 Evaluation and Impact of Qualification of Experimental Facilities in Europe 265
Maurizio Zola and Colin A. Taylor

15 Qualification of Large Testing Facilities in Earthquake Engineering Research 287
Özgür Kurç, Haluk Sucuoğlu, Marco Molinari,
and Gabriele Zanon

16 Performance Based Seismic Qualification of Large-Class Building Equipment: An Implementation Perspective 305
Jeffrey Gatscher

17 Experimental Evaluation of the Seismic Performance of Steel Buildings with Passive Dampers Using Real-Time Hybrid Simulation 323
Theodore L. Karavasilis, James M. Ricles,
Richard Sause, and Cheng Chen

| Contents | xi |

18 Experimental Investigation of the Seismic Behaviour of Precast Structures with Pinned Beam-to-Column Connections 345
Ioannis N. Psycharis, Haralambos P. Mouzakis, and Panayotis G. Carydis

19 Experimental Investigation of the Progressive Collapse of a Steel Special Moment-Resisting Frame and a Post-tensioned Energy-Dissipating Frame 367
Antonios Tsitos and Gilberto Mosqueda

Index 383

Contributors

Gabriela M. Atanasiu Multidisciplinary Center of Structural Engineering & Risk Management, "Gheorghe Asachi" Technical University of Iasi, Bdul. Dimitrie Mangeron, 43, Iasi 700050, Romania, gabriela.atanasiu@gmail.com

Subhamoy Bhattacharya Department of Civil Engineering, University of Bristol, Queen's Building, University Walk, Bristol BS8 1TR, UK, s.bhattacharya@bristol.ac.uk

Mateusz Bocian Department of Civil Engineering, University of Bristol, Queen's Building, University Walk, Bristol BS8 1TR, UK, Mateusz.Bocian@bristol.ac.uk

Aleksandra Bogdanovic Institute of Earthquake Engineering and Engineering Seismology, IZIIS, SS Cyril and Methodius University, Salvador Aljende 73, P.O. BOX 101, 1000 Skopje, Republic of Macedonia, saska@pluto.iziis.ukim.edu.mk

Claude Boutin Ecole Nationale des Travaux Publics de l'Etat (ENTPE), Université de Lyon, FRE 3237 CNRS, 3, rue Maurice Audin, 69120, Vaulx-en-Velin, France, claude.boutin@entpe.fr

Panayotis G. Carydis Laboratory for Earthquake Engineering, Department of Civil Engineering, National Technical University of Athens, Iroon Polytechniou 5 Zografou, 15780 Athens, Greece, pcarydis@central.ntua.gr

Chiara Casarotti Eucentre – European Centre for Training and Research in Earthquake Engineering, Via Ferrata 1, 27100 Pavia, Italy, chiara.casarotti@eucentre.it

Jean-Louis Chazelas Division Reconnaissance et Mecanique des Sols, IFSTTAR, Route de Bouaye, BP4129, Bouguenais 44341, France, jean-louis.chazelas@ifsttar.fr

Cheng Chen School of Engineering, San Francisco State University, San Francisco, CA 94132, USA, chcsfsu@sfsu.edu

Ulas Cilingir Department of Engineering, University of Cambridge, High Cross, Madingley Road, Cambridge CB3 0EL, UK, u.cilingir@sheffield.ac.uk

Adam J. Crewe Department of Civil Engineering, University of Bristol, Queen's Building, University Walk, Bristol BS8 1TR, UK, a.j.crewe@bristol.ac.uk

Matt S. Dietz Department of Civil Engineering, University of Bristol, Queen's Building, University Walk, Bristol BS8 1TR, UK, M.Dietz@bristol.ac.uk

Luiza Dihoru Department of Civil Engineering, University of Bristol, Queen's Building, University Walk, Bristol BS8 1TR, UK, Luiza.Dihoru@bristol.ac.uk

Uwe E. Dorka Steel and Composite Section, Department of Civil and Environmental Engineering, University of Kassel, Kassel, Germany, uwe.dorka@uni-kassel.de

Sandra Escoffier Division Reconnaissance et Mecanique des Sols, IFSTTAR, Route de Bouaye, BP4129, Bouguenais 44341, France, sandra.escoffier@ifsttar.fr

Jeffrey Gatscher Fellow Engineer, Schneider Electric, Nashville, TN 37217, USA, jeff.gatscher@schneider-electric.com

Predrag Gavrilovic Institute of Earthquake Engineering and Engineering Seismology, IZIIS, SS Cyril and Methodius University, Salvador Aljende 73, P.O. BOX 101, 1000 Skopje, Republic of Macedonia, gavrilovicpredrag@yahoo.com

Stuart Haigh Department of Engineering, University of Cambridge, High Cross, Madingley Road, Cambridge CB3 0EL, UK, skh20@cam.ac.uk

Stéphane Hans Ecole Nationale des Travaux Publics de l'Etat (ENTPE), Université de Lyon, FRE 3237 CNRS, 3, rue Maurice Audin, 69120 Vaulx-en-Velin, France, stephane.hans@entpe.fr

Charles Heron Department of Engineering, University of Cambridge, High Cross, Madingley Road, Cambridge CB3 0EL, UK, cmh78@cam.ac.uk

Erdin Ibraim Department of Civil Engineering, University of Bristol, Queen's Building, University Walk, Bristol BS8 1TR, UK, Erdin.Ibraim@bristol.ac.uk

Dimitar Jurukovski Institute of Earthquake Engineering and Engineering Seismology, IZIIS, SS Cyril and Methodius University, Salvador Aljende 73, P.O. BOX 101, 1000 Skopje, Republic of Macedonia, jurudim@yahoo.com

Theodore L. Karavasilis Department of Engineering Science, University of Oxford, Oxford OX1 3PJ, UK, theodore.karavasilis@eng.ox.ac.uk

Mohammad M. Kashani Department of Civil Engineering, University of Bristol, Queen's Building, University Walk, Bristol BS8 1TR, UK, Mehdi.Kashani@bristol.ac.uk

Lidija Krstevska Institute of Earthquake Engineering and Engineering Seismology, IZIIS, SS Cyril and Methodius University, Salvador Aljende 73, P.O. BOX 101, 1000 Skopje, Republic of Macedonia, lidija@pluto.iziis.ukim.edu.mk

Özgür Kurç Department of Civil Engineering, Middle East Technical University, Inonu Bulvari, Campus, Ankara 06531, Turkey, kurc@metu.edu.tr

Igor Lanese Eucentre – European Centre for Training and Research in Earthquake Engineering, Via Ferrata 1, 27100 Pavia, Italy, igor.lanese@eucentre.it

Domenico Lombardi Department of Civil Engineering, University of Bristol, Queen's Building, University Walk, Bristol BS8 1TR, UK, domenico.lombardi@bristol.ac.uk

Francesco Lunghi Eucentre – European Centre for Training and Research in Earthquake Engineering, Via Ferrata 1, 27100 Pavia, Italy, francesco.lunghi@eucentre.it

John H.G. Macdonald Department of Civil Engineering, University of Bristol, Queen's Building, University Walk, Bristol BS8 1TR, UK, John.Macdonald@bristol.ac.uk

Gopal Madabhushi Department of Engineering, University of Cambridge, High Cross, Madingley Road, Cambridge CB3 0EL, UK, mspg1@cam.ac.uk

Francesco Marazzi European Commission, Joint Research Centre, European Laboratory for Structural Assessment (ELSA), Institute for the Protection and Security of the Citizen (IPSC), via Enrico Fermi 2749, TP 480, 21027 Ispra (VA), Italy, francesco.marazzi@jrc.ec.europa.eu

Marco Molinari Department of Mechanical and Structural Engineering, University of Trento, Via Mesiano 77, 38100 Trento, Italy, marco.molinari@ing.unitn.it

Gilberto Mosqueda Department of Civil, Structural & Environmental Engineering, University at Buffalo – The State University of New York, Buffalo, NY 14260, USA, mosqueda@buffalo.edu

Haralambos P. Mouzakis Laboratory for Earthquake Engineering, Department of Civil Engineering, National Technical University of Athens, Iroon Polytechniou 5 Zografou, 15780 Athens, Greece, harrismo@central.ntua.gr

James A.P. Norman Department of Civil Engineering, University of Bristol, Queen's Building, University Walk, Bristol BS8 1TR, UK, james.norman@bristol.ac.uk

Ferran Obón Santacana Steel and Composite Section, Department of Civil and Environmental Engineering, University of Kassel, Kassel, Germany, ferran.obon@uni-kassel.de

Olafur Oddbjornsson Department of Civil Engineering, University of Bristol, Queen's Building, University Walk, Bristol BS8 1TR, UK, O.Oddbjornsson@bristol.ac.uk

Alberto Pavese Eucentre – European Centre for Training and Research in Earthquake Engineering, Via Ferrata 1, 27100 Pavia, Italy, alberto.pavese@eucentre.it

Department of Structural Mechanics, University of Pavia, Via Ferrata 1, 27100 Pavia, Italy, a.pavese@unipv.it

Simone Peloso Eucentre – European Centre for Training and Research in Earthquake Engineering, Via Ferrata 1, 27100 Pavia, Italy, simone.peloso@eucentre.it

Artur Pinto European Laboratory for Structural Assessment, Institute for the Protection and Security of the Citizen, Joint Research Centre, European Commission, I-21027 Ispra (VA), Italy, artur.pinto@jrc.ec.europa.eu

Ioannis Politopoulos Commissariat à l'Énergie Atomique, DEN/DANS/DM2S/SEMT, Bâtiment 603, CEA Saclay, 91191 Gif-sur-Yvette Cedex, France, ioannis.politopoulos@cea.fr

Ioannis N. Psycharis Laboratory for Earthquake Engineering, Department of Civil Engineering, National Technical University of Athens, Iroon Polytechniou 5 Zografou, 15780 Athens, Greece, ipsych@central.ntua.gr

Zoran T. Rakicevic Institute of Earthquake Engineering and Engineering Seismology, IZIIS, SS Cyril and Methodius University, Salvador Aljende 73, P.O. BOX 101, 1000 Skopje, Republic of Macedonia, zoran_r@pluto.iziis.ukim.edu.mk

James M. Ricles Department of Civil and Environmental Engineering, ATLSS Engineering Research Center, Lehigh University, Bethlehem, PA 18015, USA, jmr5@lehigh.edu

Richard Sause Department of Civil and Environmental Engineering, ATLSS Engineering Research Center, Lehigh University, Bethlehem, PA 18015, USA, rc0c@lehigh.edu

Veronika Shendova Institute of Earthquake Engineering and Engineering Seismology, IZIIS, SS Cyril and Methodius University, Salvador Aljende 73, P.O. BOX 101, 1000 Skopje, Republic of Macedonia, veronika@pluto.iziis.ukim.edu.mk

Davide Silvestri Eucentre – European Centre for Training and Research in Earthquake Engineering, Via Ferrata 1, 27100 Pavia, Italy, davide.silvestri@eucentre.it

Jean Soubestre Ecole Nationale des Travaux Publics de l'Etat (ENTPE), Université de Lyon, FRE 3237 CNRS, 3, rue Maurice Audin, 69120 Vaulx-en-Velin, France, jean.soubestre@entpe.fr

Haluk Sucuoğlu Department of Civil Engineering, Middle East Technical University, Inonu Bulvari, Campus, Ankara 06531, Turkey, sucuoglu@ce.metu.edu.tr

Ljubomir Tashkov Institute of Earthquake Engineering and Engineering Seismology, IZIIS, SS Cyril and Methodius University, Salvador Aljende 73, P.O. BOX 101, 1000 Skopje, Republic of Macedonia, tashkov@pluto.iziis.ukim.edu.mk

Fabio Taucer European Laboratory for Structural Assessment, Institute for the Protection and Security of the Citizen, Joint Research Centre, European Commission, I-21027 Ispra (VA), Italy, fabio.taucer@jrc.ec.europa.eu

Contributors xvii

Colin A. Taylor Department of Civil Engineering, University of Bristol, Queen's Building, University Walk, Bristol BS8 1TR, UK, colin.taylor@bristol.ac.uk

Antonios Tsitos Department of Civil Engineering, University of Patras, Rion 26500, Greece, atsitos@upatras.gr

Mihai H. Zaharia Multidisciplinary Center of Structural Engineering & Risk Management, "Gheorghe Asachi" Technical University of Iasi, Bdul. Dimitrie Mangeron, 43, Iasi 700050, Romania, mike@cs.tuiasi.ro

Gabriele Zanon Department of Mechanical and Structural Engineering, University of Trento, Via Mesiano 77, 38100 Trento, Italy, gabriele.zanon@ing.unitn.it

Maurizio Zola Mechanical Testing Department, Consultants of P&P LMC, Via Pastrengo 9, 24068 Seriate, BG, Italy, maurizio.zola@gmail.com

Chapter 1
How Can Experimental Testing Contribute to Performance-Based Earthquake Engineering

Fabio Taucer and Artur Pinto

Abstract Performance-based earthquake engineering (PBEE) entails the design, assessment, and construction of structures that perform under the action of frequent and extreme loads to the needs and objectives of its owners-users and society. PBEE implies that performance can be predicted and evaluated with sufficient accuracy. This requires the development of adequate tools for analysis, assessment and design, with a higher level of knowledge than what has been used in the past. In order to achieve this it is necessary to carry out research involving different actors and stake-holders, by means of analytical work, field investigations, and experimental testing. It may be argued that experimental testing has traditionally addressed the issues of performance, albeit fully underpinning the concerns of PBEE. This paper shows that although some levels of PBEE have been present in past and current experimental test practices, it is necessary to move a step forward by introducing the concepts of performance definition, engineering performance assessment and technological innovation performance enhancement, in order to produce results that can be fully used within the context of PBEE. This is first done by revisiting a set of experimental test campaigns carried out by the European Laboratory for Structural Assessment (ELSA) and indentifying the elements addressing PBEE in these tests. These elements are then extended to include all the necessary concepts contributing to PBEE and the steps needed to enhance test planning and collection of data from experimental campaigns within the framework of PBEE, thus setting the basis towards further development of PBEE in experimental testing.

F. Taucer (✉) • A. Pinto
European Laboratory for Structural Assessment, Institute for the Protection and Security of the Citizen, Joint Research Centre, European Commission, I-21027 Ispra (VA), Italy
e-mail: fabio.taucer@jrc.ec.europa.eu; artur.pinto@jrc.ec.europa.eu

M.N. Fardis and Z.T. Rakicevic (eds.), *Role of Seismic Testing Facilities in Performance-Based Earthquake Engineering: SERIES Workshop*, Geotechnical, Geological and Earthquake Engineering 22, DOI 10.1007/978-94-007-1977-4_1,
© Springer Science+Business Media B.V. 2012

1.1 Introduction

New and existing structures are required to be designed and assessed to perform under the action of frequent to extreme loads responding to the needs and objectives of owners and society. For this it is necessary that performance can be predicted with sufficient accuracy through the development of adequate tools for analysis, assessment and design, which in turn requires high levels of knowledge. The knowledge needed for the development and calibration of these tools has been gathered in the past through two main sources: field surveys and recording of structural performance during service life, and experimental testing.

Field surveys play an important role in verifying the performance of structures when subjected to various types of loading actions and for different limit states. For the serviceability limit state performance can be checked and registered through monitoring at fixed intervals in time or through constant recording of structural response. Examples are recording of accelerations due to ambient vibration and traffic (and pedestrian) loads, or monitoring of the evolution of crack opening in infill walls or masonry structures. For the case of earthquake loading, which is the main concern of the present paper, performance can be recorded in structures instrumented with accelerometers, providing valuable information for frequent, low level seismic actions, or through damage assessment following large earthquake events.

Post-earthquake field reconnaissance missions carried out with the objective of assessing damage to infrastructure and building stock are and have been a very important source for recording the performance of existing structures. This has allowed to confirm the vulnerability of certain classes of buildings and to test the validity of current design procedures and construction standards on newly designed ones. By having knowledge on the distribution of seismic excitations in the area affected by the earthquake event, and by recording the level of damage of a large number of buildings, it is possible to derive empirical fragility curves for different classes of buildings.

The collection of data following earthquake events suffers from several shortcomings: there is large uncertainty in quantifying the seismic excitation, a very large number of assessed buildings is necessary – although this is changing due to the advent of remote sensing (see PDNA Haiti 2010) – and there is difficulty in identifying uniform building structural classes. In particular, post-earthquake field data does not allow for the calibration of design procedures, nor for identifying the collapse mechanisms and the seismic demand associated to them.

Experimental testing, coupled with numerical modeling, allows to record the response of a structure in a controlled environment, thus allowing to calibrate procedures for the assessment of existing structures and the design of new ones. The present paper will focus on testing of large scale structures, as it is through the use of real- or near to full-scale models that it is possible to quantify the performance of structures. Moreover, the work will focus mainly on the pseudo-dynamic test method, as it allows testing of large models. Nevertheless, the concepts discussed in the paper are applicable to component testing as well as to shaking table tests.

Although a vast wealth of experimental data has been produced worldwide by many laboratories, most of the data has been recorded and processed with the aim of calibrating numerical models, design procedures and construction codes. Since most construction standards address only a few limit states, such as the Eurocodes, that consider the serviceability and life safety performance levels, experimental test campaigns have been planned with a limited scope often not registering all the data needed for a full performance based earthquake design.

The premise of this work is to propose a framework for planning a test campaign in order to gather the necessary data to complete the full cycle of Performance-Based Earthquake Engineering (PBEE). It takes as starting point the conclusions and recommendations of the working group session on harmonization of experimental and analytical simulations of the Bled workshop on "Performance-based seismic design concepts and implementation" (Fajfar and Krawinkler 2004; Fardis 2004), in particular in what regards the issue of 'testing procedures'.

The work will first introduce the concepts PBEE, followed by the introduction of several examples of tests carried out at the European Laboratory of Structural Assessment (ELSA) with the aim of describing the type of data recorded and identifying shortcomings within a PBEE framework. Based on this it will discuss the planning of a test campaign, and the instrumentation and post-processing needed to fulfill the needs of PBEE.

1.2 Performance-Based Earthquake Engineering Design

Seismic design has experienced a substantial evolution in the last 50 years, achieving the fundamental objective of life safety and accepting/incorporating solutions and technologies enabling critical facilities to remain operational after major seismic events. Current seismic design standards state clear objectives in terms of life safety (strength and ductility requirements), and state also objectives in terms of damage control that are typically checked indirectly based on the values calculated from ultimate limit states.

As economical aspects are also becoming overriding objectives in our societies, measurable consequences of earthquakes, such as structural and non-structural damage (e.g. repair costs) in earthquake events, as well as other economical consequences (e.g. loss of operation/revenue) and 'non-measurable' consequences, such as social impacts (quality of life), should also be considered in the planning and design of our infrastructures, living and production facilities. In fact, the economic losses resulting from the last major events in USA (Northridge 1994) and Japan (Kobe 1995), and as demonstrated more recently in Chile (2010), can be considered as the motivation for PBEE, which is deemed to provide an appropriate platform to achieve safer and more economic constructions.

The conceptual frameworks proposed in the USA for PBEE (Krawinkler 1999), such as Vision 2000, are a step forward to a more rational seismic design and assessment/redesign of engineered facilities. In fact, explicit consideration of

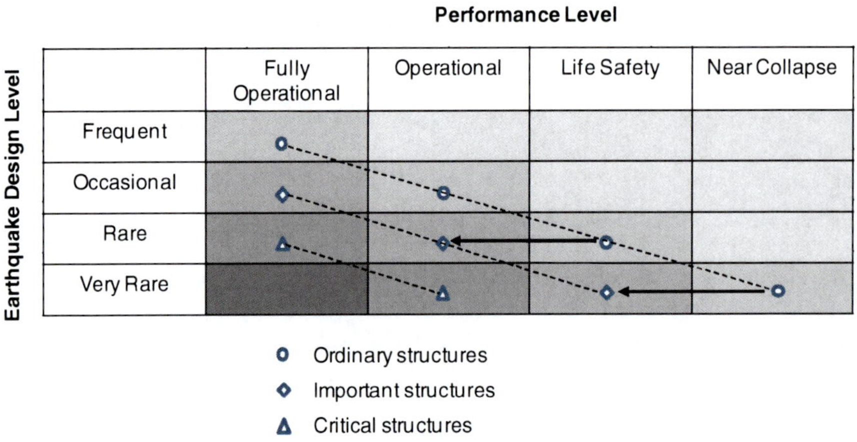

Performance Level

		Fully Operational	Operational	Life Safety	Near Collapse
	Frequent	o			
	Occasional	◇	o		
	Rare	▲	◇	o	
	Very Rare		▲	◇	o

Earthquake Design Level

o Ordinary structures

◇ Important structures

▲ Critical structures

Fig. 1.1 Vision 2000 matrix (multi-level performance objectives with corresponding demands) and increased performance expected by modern societies

multi-level performance objectives together with specific seismic intensities leads to a more controllable/predictable seismic performance (see Fig. 1.1). This represents a significant improvement relatively to the single-level explicit approach of current design codes because it requires explicit consideration and check of key performance objectives, and conveys to the designer that a structure is likely to be subjected to different seismic intensities during its life, including severe ones with low probability of occurrence.

Experience from recent earthquakes in Economically More Developed Countries (EMDC) indicate that there is increased expectation from modern societies concerning the performance of structures. It is claimed that also ordinary structures should remain operational after rare events, which implies a shift in the multi-level design procedure as illustrated in Fig. 1.1 (Pampanin 2009).

The minimal scope of structural seismic tests has been to check the performance of a model when subjected to the loading considered in its design and to check also its ultimate capacity in order to evaluate safety margins. In fact, the present limit-state based design codes explicitly consider one or two limit-states (safety and serviceability) and implicitly assume that the structure should be able to withstand (without collapse but with important/severe damage) earthquake intensities much higher than the design ones, which is achieved through capacity design (preferential-stable dissipation mechanisms) and requirements on ductility capacity. Explicit quantification of the seismic intensities associated to limit states other than safety is not given, nor is performance required to be checked. Therefore, one relies on prescriptive design procedures and on intended performances, which require verification and/or calibration. This implies that current practice in experimental testing does not provide sufficient data for the calibration of design procedures satisfying the performance requirements indicated by the Vision 2000 matrix. For example,

1 How Can Experimental Testing Contribute to Performance-Based Earthquake... 5

the operational performance level requires knowledge from the behavior of non-structural elements, which is often not recorded or included in experiments. Likewise, decisions on the extent of rehabilitating existing structures depend on an estimation of the cost of repair, which is also not addressed by most test campaigns.

In order to have a broad view of the variables that play a determinant role in the decisions for estimating the desired performance of a structure it is useful to examine the triple integral that is an application of the total probability theorem as proposed by PEER (Pacific Earthquake Engineering Research Centre):

$$\lambda(DV) = \iiint G[DV \mid DM]dG[DM \mid EDP]dG[EDP \mid IM]d\lambda[IM] \quad (1.1)$$

In Eq. 1.1 the decision variable, DV, such as cost, downtime and human lives, is calculated based on a pair-wise sequence of random variables representing the ground motion or intensity measure, IM, the structural response or engineering demand parameter, EDP, and the damage measure, DM.

In most experimental tests conducted to date only the first two variables of Eq. 1.1, IM and EDP, are recorded, together with a non systematic way of recording DM values, thus not allowing for a full estimation of DV. As a matter of fact, only a limited number of EDP parameters is registered depending on the scope of a particular test, i.e. identification of a given limit state and failure mode for a limited set of IM values. Where there is a true gap to enable full estimation of PBEE is in the recording and reporting during tests of DM variables and their transformation from qualitative to quantitative values, their relation with EDP values, and, wherever possible, with DV values.

1.3 Large Scale Tests at ELSA

Experimental verification of the performance of structures subjected to earthquake loading can be made through either shaking table (dynamic) or reaction wall (pseudo-dynamic) tests. However, if strain rate effects are important and condensation to a reduced number of test degrees of freedom (DOFs) is not realistic, dynamic testing should be sought. On the other side, if large-full scale models should be considered, pseudo-dynamic testing (PSD) becomes the appropriate solution because complex non-linear phenomena are often accurately simulated only at full or large model scales. Furthermore, expansion of the time scale makes up for much more handy tests, in that the tests can be stopped at any critical event and be restarted if necessary. Furthermore, PSD testing allows hybrid (physical and numerical) online simulation of large structures and systems to be carried out by sub-structuring techniques already familiar to analysts.

During the last 20 years the ELSA laboratory has performed more than 50 large scale tests. Two examples of such tests performed in support of Eurocode 8 (EN 1998-1 2004) are given in this section. One is concerned with the assessment of

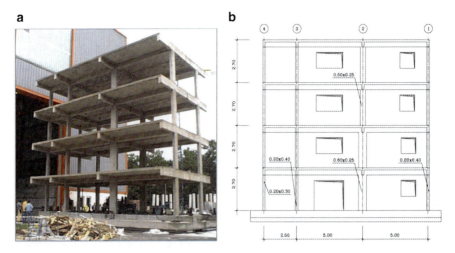

Fig. 1.2 Full-scale model of an existing RC frame structure: (**a**) Test set-up at ELSA; (**b**) Elevation layout with measures in metres

existing structures, for which a test protocol tailored for life-safety and for ultimate capacity was adopted, while the second example addresses testing of a full-scale reinforced concrete flat-slab building structure for evaluating life-safety and ultimate capacity limit states.

The first example concerns a series of pseudo-dynamic tests on two full-scale models of a 4-storey reinforced concrete frame (Fig. 1.2) representative of existing structures designed without specific seismic resisting characteristics (common practice of the 1960s in South European countries) (Pinto et al. 2002). Four testing campaigns were performed aiming at: (i) vulnerability assessment of a bare frame; (ii) assessment of a selective retrofit solution; (iii) earthquake assessment of an identical frame with infill masonry walls; and (iv) assessment of shotcrete retrofitting of the infill panels.

The tests on the model representing existing structures were focused on the behavior and performance for input motions corresponding to the design actions of new structures as well as on the assessment of their ultimate capacity. Therefore, an input motion corresponding to a 475 year return period (yrp) was adopted for the first test on the bare frame. The second test aimed at reaching ultimate capacity of the frames and was carried out with an input motion intensity corresponding to 975 yrp. The tests on the retrofitted structure and on the infilled frame structure adopted the same input intensities in order to allow for direct comparison with the original configuration. A subsequent PSD test with an intensity corresponding to 2,000 yrp was carried out.

The inter-storey shear against the inter-storey drift at the first storey and the respective envelope curves for the 475, 975 and 2,000 yrp earthquakes is given in Fig. 1.3a for the bare (BF) and infilled frames (IF). Both the inter-storey drift and inter-storey shear are *EDP* values, while the return period of the earthquake represents a parameter describing the *IM* value. In Fig. 1.3b the maximum inter-storey

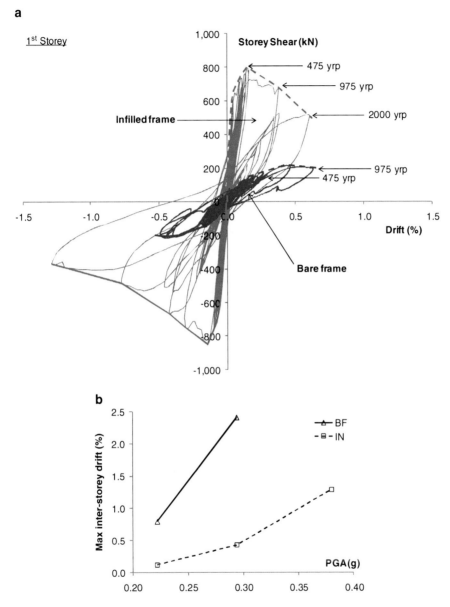

Fig. 1.3 Bare (BF) and Infilled (IN) frame tests: (**a**) 1st storey shear versus inter-storey drift and respective envelope curves for the 475, 975 and 2,000 yrp earthquakes; (**b**) maximum inter-storey drifts as a function of PGA

drifts are given as a function of peak ground acceleration (PGA), which is itself another way of representing the *IM* values. In Fig. 1.4 the maximum inter-storey drifts recorded for the 475, 975 and 2,000 yrp earthquakes are given at each of the four storeys for the two tested configurations BF and IF.

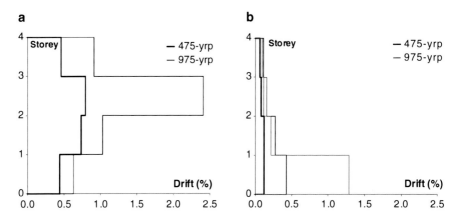

Fig. 1.4 Maximum inter-storey drift envelopes at each storey: (**a**) bare frames; (**b**) infilled frames

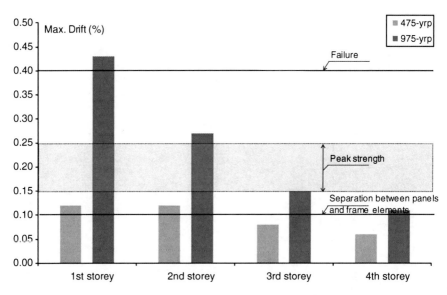

Fig. 1.5 Maximum inter-storey drift for the 475 and 975 year period PSD tests for the infilled frame: comparison with drift values corresponding to performance limit states (Pinto 2010)

Figure 1.5 represents the maximum *EDP* values corresponding to inter-storey drift for the infilled frame for the 475 and 975 yrp earthquakes, and attempts going a step further in correlating these *EDP* values with damage limit states for the infill panels: 0.10% drift for separation between panels and frame elements, between 0.15% and 0.25% drift for achieving peak strength and 0.40% drift for complete failure of the infill panels.

1 How Can Experimental Testing Contribute to Performance-Based Earthquake... 9

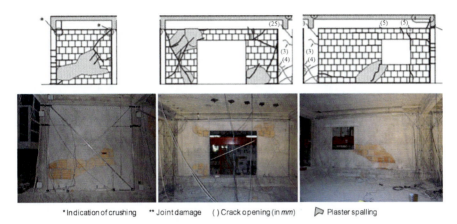

Fig. 1.6 Damage patterns for the infilled frame configuration after the 975-yrp pseudo-dynamic test (Adapted from Pinto 2010) and corresponding photographs at the first storey

Fig. 1.7 Types of damage in masonry infill panels (Adapted from Sortis et al. 1997)

In order to offer the possibility of determining the *DVs* of cost and downtime the actual damage of the infill panels was recorded during the tests, as shown in Fig. 1.6 for the infilled frame for the 975 yrp test. Figure 1.6 shows a sketch of the different types of damage patterns in terms of crushing, damage of the joint, crack opening and plaster spalling, together with photographic records at each of the three bays of the first storey at the end of the 975 yrp test.

Although sketches and photographs provide a complete picture of damage, these need to be translated into "damage levels" representing the *DM* values, which can then be used for estimating the *DV* values. For this there are many methodologies available in literature, such as the methodology proposed by Sortis et al. (1997) for evaluating the intensity of damage of infill panels. As shown in Fig. 1.7, five types of damage are considered. For each type of damage, the damage level is defined on the basis of the amplitude of the damage, namely crack opening and crushing (see Table 1.1). Following the classification proposed by Sortis et al. (1997), the damage observed in the infill walls of each storey at the end of each earthquake was classified. The results are presented in Table 1.2.

The methodology used for recording structural response and damage during the test of the existing reinforced concrete frame structure show that it is possible to calculate *DM* values as a function of *EDP* values for three *IMs* for two configurations (bare and infilled frame). Although not done as part of the test campaign, the

Table 1.1 Damage level evaluation in masonry infill panels (Sortis et al. 1997)

Damage level	Amplitude of observed damage (crack width/separation, mm)				
	Type 1	Type 2	Type 3	Type 4	Type 5
A = no damage	0	0	0	0	0
B = slight	≤2	≤2	≤1	0	0
C = medium	≤5	≤5	≤2	≤1	Crushing[a]
D = heavy	≤10	≤10	≤5	≤2	Crushing
E = very heavy	>10	>10	>5	>2	Crushing[b]
F = total	Total damage	Destruction	Part collapse	Extensive	–

[a] Indications of crushing
[b] Significant crushing

Table 1.2 Damage level evaluation in masonry in the infilled earthquake tests

Earthquake (yrp)	Storey			
	1	2	3	4
475	B	B	A	
975	E	C	B	
2,000	F	C		

DM values and type of damage given in Table 1.1 would allow to compute the *DV* values of cost and time of repair of the structure, thus completing the full cycle of PBEE.

The second example discussed in this paper and carried out at ELSA is the test campaign on a full-scale reinforced concrete flat-slab building structure. The presentation will focus on the choice of the instrumentation and how the results were used to interpret the performance of the structure, rather than showing the actual performance of the model during the test; for more details the reader may refer to the work of Zaharia et al. (2006).

The experimental program consisted in PSD earthquake tests on a three-storey reinforced concrete building structure (see Fig. 1.8a) designed according to the Portuguese code and representative of flat-slab buildings in European seismic regions. The experimental data and the results obtained from numerical simulations were used to assess the validity of several expressions proposed in literature, including Eurocode 2 (EN 1992-1-1, 2004), for computing the effective slab-width. The global load-displacement envelope of the tested structure was determined with reference to the drift corresponding to the maximum load-carrying capacity and to failure of the structure. Special consideration was given to second-order effects.

Two PSD tests were carried out using artificially generated input motions corresponding to a moderate-high European hazard scenario. For the first test, a 475 yrp was considered, while for the second test a 2,000 yrp input motion was computed by scaling the 475 yrp accelerogram by a factor of 1.73. The accelerograms were compatible with the seismic response spectrum of the Portuguese code.

1 How Can Experimental Testing Contribute to Performance-Based Earthquake...

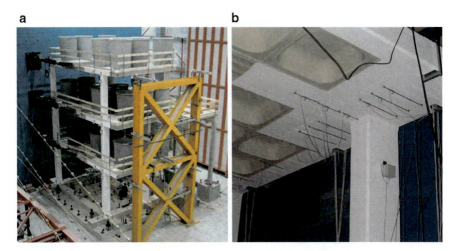

Fig. 1.8 Three storey reinforced concrete flat-slab building: (**a**) General view of the test set-up at ELSA; (**b**) Displacement transducers at the bottom of the first floor slab

The instrumentation used for recording the response of the structure was divided into two groups. The first group was reserved to measurements related to the pseudo-dynamic algorithm and comprises displacements and forces at the condensed – one per floor – degrees of freedom of the model. The second group includes all other measurements concerning rotations and relative displacements at several points in the structure.

The rotations of joints were measured by means of 52 digital inclinometers. Figure 1.9a shows the location of the inclinometers for the north frame. Four pairs of displacement potentiometer transducers were positioned at the bottom of each column, within a length equal to half the column depth (Fig. 1.9a), providing a second source of data with respect to the information given by the inclinometers at the base.

The effective width of the slab (slab participation) and the longitudinal deformation of the beams were measured by means of 40 displacement transducers positioned at the first floor level (see detail of Figs. 1.8b and 1.9b).

Figure 1.10 shows the cracking pattern of the slab at the bottom of the second floor, and its evolution from the 475 yrp to the 2,000 yrp earthquake test. These diagrams constitute a visual qualitative representation of *DMs*, which is an important part for recording the damage of the model. However, this information needs to be integrated with a quantitative measure of damage, which is provided by the displacements measured by the displacement transducers. More importantly, this allowed for computing the participation of the slab based on the rate of change of deformation between the slab and the column within the area of influence of the column (i.e. effective width). The results permitted to conclude that the current versions of Eurocode 2 and the Portuguese building code overestimate the slab width participation.

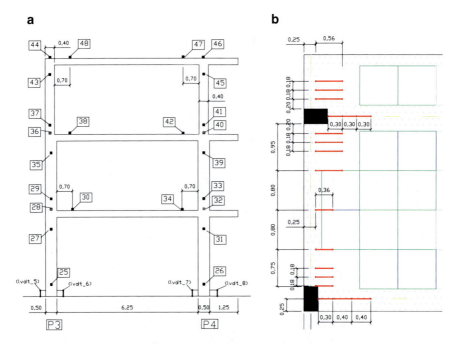

Fig. 1.9 Three storey reinforced concrete flat-slab building: (**a**) Location of inclinometers in the north frame; (**b**) location of displacement transducers to measure slab deformations

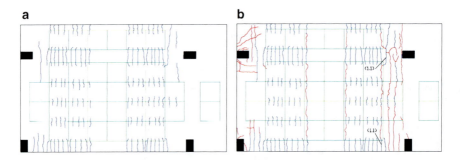

Fig. 1.10 Three storey reinforced concrete flat-slab building, cracking pattern below the second floor: (**a**) 475 yrp earthquake; (**b**) 2,000 yrp earthquake

The measurements given by displacement transducers, which provide information only at discrete points, was complemented by field vision systems. The methodology consists in positioning a set of reference points – below the first floor in the vicinity of the interior column – on the concrete surface, and to record, by means of a video camera, the evolution of displacements with respect to a reference point (i.e. the reaction floor of the laboratory). By post-processing the information recorded by

1 How Can Experimental Testing Contribute to Performance-Based Earthquake... 13

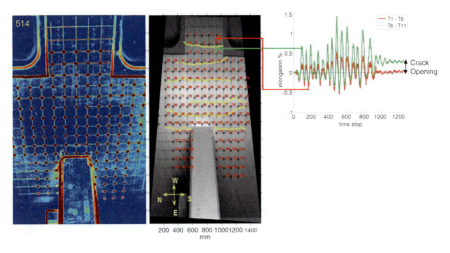

Fig. 1.11 Three storey reinforced concrete flat-slab building: field vision system at the bottom of the first floor at column P4; crack opening between target points 6 and 11

the camera it is possible to track the time history of displacements of the target points, enabling the computation of the opening and closing of cracks and the extent of deformations within the area near the column. The set-up of the system is shown in Fig. 1.11, whereby tracking the relative displacements of target points 6 and 11 it is possible to compute the magnitude of the crack opening between these two points. The ability of the system to track the evolution of cracking depends on the spacing of the target points, which need to be smaller than the minimum spacing expected between cracks. This method is an improvement with respect to the information given by displacement transducers that in general record deformations which are due to several cracks. Improvements on visual systems with respect to the choice of target points involve the use of a random texture, so that deformations can be recorded in an almost continuous way so that no previous estimation on the location of damage is needed.

The recording of crack widths through displacement transducers and vision systems may be considered as *EDP* values, as they give a direct quantifiable measure of the structure response, albeit at a very local level, thus enabling, with appropriate interpretation of damage, such as that presented in Fig. 1.7 and Table 1.1, a direct transformation into *DM* values.

Another set of response variables that is customary to record in pseudo-dynamic tests in moment resisting frame buildings are the rotations at nodes. By positioning inclinometers at the beam-column nodes at the center of the joint and at a distance equivalent to the theoretical plastic hinge length it is possible to obtain the rotation of the 'hinge' relative to the reference rotation measured at the center of the joint. Such a diagram of relative rotations is shown in Fig. 1.12b for the north frame, and corresponds to the maximum relative rotations for the 2,000 yrp earthquake test.

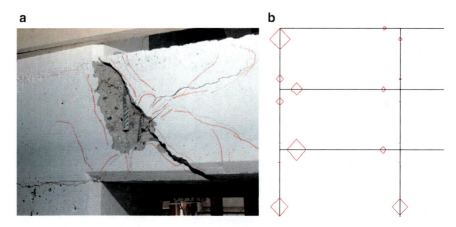

Fig. 1.12 Three storey reinforced concrete flat-slab building, north frame, 2,000 yrp earthquake: (**a**) Damage to the edge connection (column P3) at the second floor; (**b**) relative rotations of beams and columns

These rotations can also be considered as an *EDP* value, again at the local level, and allow, together with qualitative *DMs* given by photographic records of the damage of the joint as shown in Fig. 1.12a, to compute a quantitative measure of *DM*. Apart from providing an estimate of damage, joint rotations, together with storey displacements, were used to examine the importance of second order effects, which resulted to be determinant in the response of the model.

In spite of having recorded a wide range of structural responses, the tests on the flat-slab building stopped short in transforming *EDP* values into quantifiable *DM* values. Nevertheless, the *EDP*s were useful in determining the participation of the slab, in verifying the flexibility of the frame and in calibrating numerical models, which were the main objectives of the test campaign.

1.4 A Framework for PBEE Experimental Campaigns

The two examples proposed in the previous section provide an overview of current practice in experimental testing. As described in Sect. 1.2, experimental tests on full-scale structures using the PSD method have focused on several issues:

- Calibration of numerical models
- Calibration of design and assessment procedures
- Calibration of design standards such as the Eurocodes and determination of margins of safety
- Vulnerability assessment of existing structures
- Qualification of strengthening and retrofitting solutions

With the exception of the first, the remaining four issues have involved determining in some way the damage of the structure, and associating its performance to a given limit state. As mentioned in the previous sections, it has been customary to examine two limit states, serviceability and life safety, as required by most construction standards such as the Eurocodes. However, the issue of operationality and economical loss has become an important concern of modern societies, and at present is not fully addressed by current standards.

The use of controlled damage in selected members (e.g. formation of stable plastic hinges in beams) together with capacity design principles has allowed to take advantage of energy dissipation to contain the design base shear forces through the use of force reduction factors (q factor in the Eurocodes). This has led, especially in reinforced concrete moment resisting frames, to excessive deformations that without compromising the integrity of the main structural resisting system have lead to excessive damage of the infill wall panels, as observed in several cases following the 2,009 Aquila earthquake (MIC Report 2009). Newly four to five storey residential buildings designed according to the latest standards exhibited complete crushing of the infill walls at the lower stories (leading to damage of partitions, doors, windows and ceramic tiles in kitchen and bathrooms) and damage to furniture and falling of objects due to excessive deformations and accelerations at the top stories, forcing the dwellers to abandon the building for an extended period of time with important economic implications and loss comparable to the value of the reinforced concrete frame. These issues justify the consideration of damage to non-structural members at all limit-states so as to satisfy the performance matrix of Vision 2000 with consideration to the request of society of keeping buildings operational even for rare earthquakes, as shown in Fig. 1.1.

As demonstrated by the first example of Sect. 1.3, test campaigns have taken into account the interaction of building structures with infill walls. However, in order to assess the full economical impact of damage to non-structural elements it may be foreseen to include elements other than infill walls, such as door and window framings, ceramic tile surfaces in partitions and floors, glass facades and all other elements that can have a significant impact on operationality and economical loss. Another important issue that will need to be taken into account in the coming years is modeling the performance of non-structural elements used to fulfill the requirements of energy saving, insulation and sustainability. An example of the consequence of unforeseen poor performance of such systems was observed after the 2009 Aquila earthquake (Pinto 2010), where the external insulation material of the façade often failed out of plane due to insufficient support with the framing elements.

The issue of taking into consideration economic aspects requires to perform a larger number of tests at low levels of seismic excitation, and to adequately record *EDP* values that will allow to compute *DM* values. The tests described in the previous examples concentrated on recording *EDP* values, and, with the exception of the results of Table 1.2, the reported values fell short in providing a quantitative estimation of *DM*.

The transformation of *EDP* into *DM* values is a very important task that deserves more attention. Such transformation may be derived from post-earthquake field assessment data, from testing sub-assemblages, or more directly, by measuring during the test DM values through photographic/video documentation, recording of crack patters and vision tracking systems. However, since these DM values are in general qualitative, they need to be transformed into quantitative ones or into damage levels that can be used for the estimation of DV values. Examples of such work may be found in the work of Ghobarah (2004), which, based on analytical and experimental data, proposes damage levels in terms of drift for different structural systems.

In this sense, the collaboration with the construction industry is a very important one, as it allows to derive cost estimates as a function of the recorded DM values, notwithstanding the difference in cost between different regions in the world. An example of such work is found in Pagni and Lowes (2004), which relates damage states with the method of repair, providing precious information for assessing the replacement/repair cost. Likewise, interaction with stakeholders that are users of the tested structures is important to determine operational levels based on damage to non-structural components. Note that in this sense, especially for electro-mechanical equipment, shaking table tests can provide a very important contribution, which is already common in the nuclear and electric industry, but less common in other lifeline and critical infrastructure, such as hospitals and telecommunication services.

Performance, according to the Vision 2000 matrix (see Fig. 1.1), and in light of more stringent requirements that may be part of future construction standards, will need to be evaluated for several limit states at various levels of seismic excitation up to complete failure or collapse. Since full-scale testing, in general, can only make use of a single specimen, the number of tests is limited, and cyclic accumulation of damage may need to be taken into account. For the evaluation of serviceability and operational limit states this is less of a problem, as the accumulation of damage in structural elements that often remain in the elastic or near the yielding state plays a minor role in determining structural response. In fact, it may be possible to test various configurations of non-structural elements, as shown in previous tests at ELSA, using the same structure. For the life safety and near collapse limit states, which often involve considering only the structural system of resisting forces – with the exception of undamaged infill walls, generally in upper storeys – the number of tests is limited, as damage accumulation does play an important role. In this sense it is very important to make the best possible predictions of the response of the structure when sizing the seismic excitation for the test. In general, the number and type of tests will be a balance between the requirements of PBEE and the cost and feasibility of the test campaign.

Another requirement that is becoming an issue in experimental testing is the verification of the collapse of the structure and of its robustness, which numerical models have an inherent difficulty in estimating due to the role of second order effects and large displacements, the inherent larger deformation and residual strength of many materials and the change of structural configuration during collapse. In practice the estimation of collapse is generally derived from how far the structure is from its peak strength in terms of loss of maximum strength and from the slope of the tangent stiffness in the softening branch. A more precise estimation of collapse and robustness may be obtained by controlled demolition of the structure, using innovative vision systems to document the response and damage of the structure.

Although much has been said about *EDP*, *DM* and *DV* values, *IM* values deserve as much attention, especially when planning a test campaign with a limited number of sequences of tests. The *IM* values are chosen depending on the objectives of the test campaign as proposed in the beginning of the present section. The tests may be either pseudo-dynamic representing earthquake loading, or cyclic. For cyclic tests a choice must be made on the number and amplitude of cycles as well as on the distribution of storey forces, which may be triangular, rectangular, as well as other geometries, including adaptive pushover, which could be transposed from numerical analysis. Cyclic tests will privilege the calibration of numerical models as well as the determination of *DM* as a function of *EDP* values. Pseudo-dynamic tests are more suitable for checking performance, but require appropriate representation of the frequency content and number of cycles of the *IM* above a given threshold. Accelerograms derived from real earthquakes, in spite of being more realistic, have the disadvantage in not being able to excite all the modes of the structure. On the other hand, artificial accelerograms excite a broad range of frequency values compatible with a response spectrum of reference, but the response of the structure and accumulation of damage have to be interpreted carefully by taking into account the higher number of cycles at large amplitudes with respect to the use of real earthquakes. The work carried out by Pagni and Lowes (2004) goes in this direction by normalizing test results with respect to the number and amplitude of cycles.

Other aspects that need to be taken into account in determining the *IM* and that have a direct influence on the performance of structures are multi-axial excitations and asynchronous motions. Earthquake excitations in two directions on the horizontal plane are important in determining the response of torsionally unbalanced structures, such as the full-scale plan-wise irregular 3-storey frame structure model tested as part of the research project SPEAR (Seismic PErformance Assessment and Rehabilitation of existing buildings). In this project the longitudinal and transverse components of the earthquake were simultaneously applied to the structure. In fact, three DOFs per storey were taken into account: two translations and one rotation along the vertical axis, as opposed to the single DOF per storey that is usually taken into account in unidirectional PSD testing. The structure was subjected to two tests (with PGA of 0.15 and 0.20 g) each test with one accelerogram in each direction. Detailed analysis of the test results and test set-up can be found elsewhere (Negro et al. 2004). Asynchronous motions on the other hand are important for assessing the response of long structures, such as the six-pier bridge model tested at ELSA as part of the Vulnerability Assessment of Bridges (VAB) project (Pinto et al. 2004), where physical testing of two piers, including an isolated configuration, and on-line simulation of the remaining piers (non-linear numerical models) and deck (linear numerical model) was carried out via non-linear sub-structuring.

Lastly, test campaigns must represent as best as possible the prototype structure and the real conditions of loading. For this it is important to have a proper representation of the boundary conditions, which is especially important for sub-structured and hybrid simulation tests, including tests accounting for soil-(Foundation)-Structure

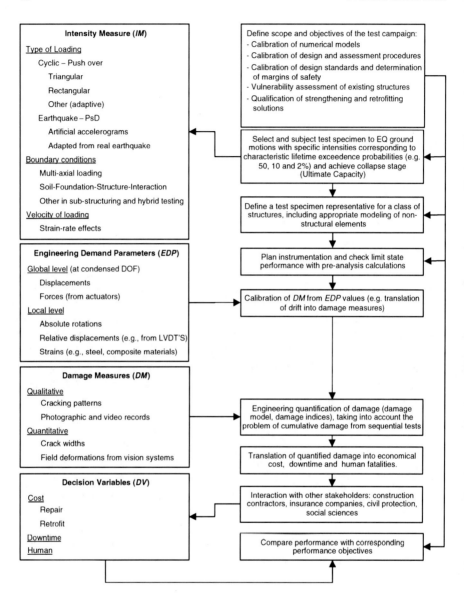

Fig. 1.13 A schematic representation of a framework for PBEE experimental PSD test campaigns

Interaction (S(F)SI) systems. Likewise, the execution of tests should consider the velocity of loading, which is especially important in PSD tests with strain-rate materials, such as energy dissipation devices and isolators (see Molina et al. 2003).

The concepts proposed in this section are summarized in Fig. 1.13, by providing a list of the *IM*, *EDP*, *DM* and *DV* values that need to be taken into account in a large scale – PSD – test and the sequence of steps in programming an experiment, as adapted from Pinto (2010), for full consideration of PBEE.

1.5 Conclusions and Recommendations

The issues concerning the aspects that need to be taken into account for the execution of experimental test campaigns within a framework of PBEE have been discussed and presented. The total probability theorem proposed by PEER is used to expose the variables that need to be considered as part of any test campaign addressing PBEE issues, namely: Decision Variables, Damage Measures, Engineering Demand Parameters and Intensity Measures.

Experimental test campaigns carried out at ELSA have shown that the collection of engineering demand parameters values has been adequately addressed, including the collection of qualitative information concerning damage measures. However, this information is not sufficient and more is needed to close the PBEE cycle. For example, more attention needs to be devoted into translating engineering demand parameters into damage measures, as well as in transforming qualitative into quantitative damage measures. For this, it is of outmost importance to properly plan the instrumentation so that the performance objectives defined for the test are met. This requires, in addition to the use of photographic and video records and visual interpretation of damage, capturing local deformations and using vision systems for recording field deformations.

As shown by the Vision 2000 matrix, more importance needs to be devoted to operational limit states. This calls for physical representation of all non-structural elements that have a role in determining cost and downtime, and for the involvement of all concerned stakeholders to derive these values from damage measures.

The selection of the intensity measure needs to be tailored as a function of the objectives of the test and is in general the result of a compromise between the number of tests needed for assessing performance and the limits of the model in not accumulating excessive damage. In this respect, accurate analytical predictions prior to the test are very important to select the intensity of the seismic input with respect to the desired performance of the model. Also, the boundary conditions of the model should represent as much as possible the real conditions of the prototype, such as multi-axial loading and asynchronous motion.

It is recommended that test campaigns should be planned with the objective of generating fragility curves addressing different performance levels, whereby an analytical model calibrated with the test results can be used to carry out Monte Carlo simulations using the prototype structure to represent a class of structures.

Lastly, data from tests should be documented in databases adequately addressing the recording of qualitative and quantitative damage measures, including the deformations obtained from field vision systems; efforts in this direction are being made for the setting up of a distributed database of experimental results in earthquake engineering in Europe as part of the SERIES project (Seismic Engineering Research Infrastructures for European Synergies) financed by the European Commission.

References

EN 1992-1-1 (2004) Eurocode 2: design of concrete structures – part 1-1: general rules and rules for buildings. European Committee for Standardization, Brussels

EN 1998–1 (2004) Eurocode 8: design of structures for earthquake resistance – part 1: general rules, seismic actions and rules for buildings. European Committee for Standardization, Brussels

Fajfar P, Krawinkler H (2004) Performance-based seismic design concepts and implementation. In: Proceedings of an international Workshop, Bled, Slovenia, 28 June–1 July. PEER report 2004/05, Berkeley, ISBN 0-9762060-0-5

Fardis MN (2004) A European perspective to performance-based seismic design, assessment and retrofitting. In: Proceedings of an international Workshop, Bled, Slovenia, 28 June–1 July. PEER report 2004/05, Berkeley, ISBN 0-9762060-0-5

Ghobarah A (2004) On drift limits associated with different damage levels. In: Proceedings of an international Workshop. Bled, Slovenia, 28 June–1 July. PEER report 2004/05, Berkeley, ISBN 0-9762060-0-5

Haiti Earthquake PDNA (2010) Assessment of damage, losses, general and sectoral needs. Annex to the action plan for national recovery and development of Haiti. Government of the Republic of Haiti with support from the International Community. Published by "United Nations Development Programme, Haiti (http://www.ht.undp.org/public/publicationdetails.php?idpublication=60&PHPSESSID=2d2e3e122375dc55fc42689bc09abbc1)

Krawinkler H (1999) Challenges and progress in performance-based earthquake engineering. International seminar on seismic engineering for tomorrow – in honour of Professor Hiroshi Akiyama, Tokyo, 26 Nov 1999

MIC Report (2009) Assessment mission, Italy earthquake 2009. Community civil protection mechanism, technical report, European Commission, Brussels

Molina J, Magonette G, Viaccoz B, Sorace S, Terenzi G (2003) Pseudo-dynamic tests on two buildings retrofitted with damped braces JRC scientific and technical reports, EUR20993EN, Ispra, Italy

Negro P, Mola E, Molina F, Magonette GE (2004) Full-scale PSD testing of a torsionally unbalanced three-storey non-seismic RC frame. In: Proceedings of the 13th WCEE, Vancouver

Pagni CA, Lowes LN A (2004) Tools to enable prediction of the economic impact of earthquake damage in older RC beam-column joints. In: Proceedings of an international workshop, Bled, Slovenia, 28 June–1 July. PEER report 2004/05, Berkeley, ISBN 0-9762060-0-5

Pampanin S (2009) Emerging solutions for damage-resisting precast concrete buildings: an up-date on New Zealand's practice and R&D. In: Proceedings of the SAFECAST workshop, Joint Research Centre, Ispra

Pinto A (2010) Large scale testing: achievements and future needs. In: Garevski M, Ansal A (eds) Earthquake engineering in Europe, vol 17, Geotechnical, Geological and Earthquake Engineering. Springer, Dordrecht/Heidelberg/London/New York, p. 359–381

Pinto AV, Varum H, Molina F (2002) Experimental assessment and retrofit of full-scale models of existing RC frames. In: Proceedings of the 12th European conference on earthquake engineering. Elsevier Science Ltd, London

Pinto AV, Pegon P, Magonette GE, Tsionis G (2004) Pseudo-dynamic testing of bridges using non-linear substructuring. Earthquake Eng Struct Dyn 33:1125–1146

Sortis A, Pasquale G, Nasini U (1997) Criteri di calcolo per la progettazione degli interventi – Terremoto in Umbria e Marche del 1997. Servizio Sismico Nazionale, Editrice Sallustiana, Roma, 1999 (in Italian)

Zaharia R, Taucer F, Pinto A, Molina J, Vidal V, Coelho, Candeias P (2006) Pseudodynamic earthquake tests on a full-scale RC flat-slab building structure, JRC scientific and technical reports, EUR 22192EN, Ispra, Italy

Chapter 2
Earthquake and Large Structures Testing at the Bristol Laboratory for Advanced Dynamics Engineering

Matt S. Dietz, Luiza Dihoru, Olafur Oddbjornsson, Mateusz Bocian, Mohammad M. Kashani, James A.P. Norman, Adam J. Crewe, John H.G. Macdonald, and Colin A. Taylor

Abstract Integrated within the Bristol Laboratory for Advanced Dynamics Engineering (BLADE) at the Faculty of Engineering of the University of Bristol, the Earthquake and Large Structures (EQUALS) Laboratory is the UK's largest dynamic test laboratory that specialises in earthquake engineering. The facilities contained include a six degree of freedom shaking table surrounded by a strong floor and adjacent strong walls. The capacity and capability of these facilities are described. The role and significance of the EQUALS infrastructure is demonstrated by discussion of a number of recent and ongoing projects. Subject areas encompassed include inclined cable dynamics, human-structure interaction, multiple support excitation, pile-soil interaction, non-linear self-aligning structures and corroded reinforced concrete.

2.1 Introduction

The dynamic performance of a civil engineering structure is a result of a series of interactions – between the structure's constituent parts and between it and the wider world – which can both alter the intrinsic behaviour and modify the imposed demand. Modelling is undertaken to increase the understanding of the mechanisms

---

M.S. Dietz (✉) • L. Dihoru • O. Oddbjornsson • M. Bocian • M.M. Kashani
• J.A.P. Norman • A.J. Crewe • J.H.G. Macdonald • C.A. Taylor
Department of Civil Engineering, University of Bristol, Queen's Building, University Walk, Bristol BS8 1TR, UK
e-mail: M.Dietz@bristol.ac.uk; Luiza.Dihoru@bristol.ac.uk; O.Oddbjornsson@bristol.ac.uk; Mateusz.Bocian@bristol.ac.uk; Mehdi.Kashani@bristol.ac.uk; james.norman@bristol.ac.uk; a.j.crewe@bristol.ac.uk; John.Macdonald@bristol.ac.uk; colin.taylor@bristol.ac.uk

M.N. Fardis and Z.T. Rakicevic (eds.), *Role of Seismic Testing Facilities in Performance-Based Earthquake Engineering: SERIES Workshop*, Geotechnical, Geological and Earthquake Engineering 22, DOI 10.1007/978-94-007-1977-4_2,
© Springer Science+Business Media B.V. 2012

occurring within such systems and in order to be able to predict the performance of those systems. Complex geometries, composite forms and convoluted loading impede the development and hinder the uptake of closed-form analytic techniques within design practice. Instead, performance predictions increasingly stem from numerical and particularly finite element simulation. The cost savings and sophistication enhancements brought by the continuing advancement of information technology seem set to continue this trend.

In the field of earthquake engineering, poor or unexpected performance generally occurs at the extremes of loading. Here, behaviour is governed by unorthodox mechanisms featuring material and geometrical nonlinearity. Furthermore, the pressure to reduce the excessive costs (see e.g. Erdik 1998) incurred by recent earthquakes drives rapid innovation within seismic resistant design practice. As such, performance predictions can be associated with significant extrapolation from known behaviour into an unknown domain. Extrapolation is associated with uncertainty; uncertainty in design practice requires careful management.

A physical model of a system provides a record of experimental observations to which a performance prediction for that system can be compared. High levels of correlation give confidence that the essential features of system response are being reproduced. This is not to say that physical models should by themselves be used for performance prediction. Like any model, the quality of a physical model is dependent on the validity and nature of its supporting assumptions. The sheer size of civil engineering structures makes full-scale modelling both expensive and impractical. The assumptions and approximations instilled by modelling at reduced scale, using dilated timescales or by employing hybrid techniques, attest that the imperative to demonstrate the viability of seismic resistant designs is best satisfied by a three-pronged attack using advanced analytical techniques in combination with numerical simulation and experimental testing.

Dynamic and seismic test laboratories contain the infrastructure and the expertise that allow such attacks to be orchestrated. The UK's largest such facility is an integral part of the Bristol Laboratory for Advanced Dynamics Engineering (BLADE) housed within the Faculty of Engineering at the University of Bristol.

2.2 Seismic Test Infrastructure at the University of Bristol

The BLADE Earthquake and Large Structures (EQUALS) Laboratory (Fig. 2.1) houses a 15 tonne capacity, six-degree-of-freedom earthquake shaking table surrounded by a strong floor and adjacent strong walls up to 15 m high. The shaking table is accompanied by a set of 40 servo-hydraulic actuators that can be configured to operate in conjunction with the shaking table, strong floor and reaction walls, providing a highly adaptable dynamic test facility that can be used for a variety of earthquake and dynamic load tests. Extensive instrumentation is available, including

Fig. 2.1 The Earthquake and Large Structures Laboratory

256 data acquisition channels. The facility is supported by an extensive manufacturing workshop equipped with numerically controlled machines.

The shaking table consists of a stiff 3 m by 3 m cast aluminium platform weighing 3.8 tonnes. The platform surface is an arrangement of five aluminium plates with a regular grid of M12 bolt holes for the mounting of specimens. The platform can accelerate horizontally up to 3.7 g with no payload and 1.6 g with a 10 tonne payload. Corresponding vertical accelerations are 5.6 and 1.2 g respectively. Peak velocities are 1 m/s in all translational axes, with peak displacements of ±0.15 m. Hydraulic power for the shaking table is provided by a set of six shared, variable volume hydraulic pumps, providing up to 900 l/min at a working pressure of 205 bar. The maximum flow capacity can be increased to around 1,200 l/min for up to 16 s at times of peak demand with the addition of extra hydraulic accumulators. The shaking table can be augmented by additional actuators to enable multiple-support excitation or travelling wave effects to be explored.

The EQUALS facility is supported by a multi-disciplinary group of academics and researchers specialising in advanced dynamics and materials from across the Civil, Aerospace, Mechanical Engineering, and Non-linear Dynamics fields. The technical workforce provides users with day-to-day support, specimen fabrication and manufacturing, as well as shaking table operation, electronics and instrumentation support. The multidisciplinary approach taken by the facility has engendered a special interest being directed towards the digital control systems of the facility which includes a 'hybrid test' capability in which part of the structural system of interest can be emulated by a numerical model embedded in the digital control system, while only a sub-component need be tested physically.

EQUALS is particularly suited to testing of small- to medium-sized specimens in order to investigate fundamental dynamic and seismic phenomena. Collaboration is viewed positively. Indeed, EQUALS was a founding member of the UK Network for Earthquake Engineering Simulation (UKNEES). Telepresence tools have been commissioned in order to share live or pre-recorded video and acquisition data with the wider world. Integration and collaboration has been further enhanced by the development of distributed hybrid testing, an approach that breaks a model into substructures that are tested in geographically-separated laboratories. Data transfer between the substructures takes place via either standard internet connections or UK-Light, a fast, dedicated fibre optic system. UKNEES have successfully performed three-site tests between Oxford, Cambridge and Bristol (Ojaghi et al. 2010a) and robust real-time two-site tests between Oxford and Bristol, which are believed to be a world first (Ojaghi et al. 2010b).

2.3 Recent and Ongoing Exemplar Projects

The role and significance of the EQUALS Laboratory is best demonstrated via the work that it undertakes. The EQUALS research portfolio covers topics as wide-ranging as the response of cables and cable-stayed structures, soil-structure interaction (the facility is equipped with two lamellar, flexible, shear boxes for geomechanics testing), the use of discrete damping elements in building structures, base isolation systems, torsional response of buildings, masonry structures, steel and concrete buildings, multiple-support excitation, travelling earthquake wave effects, non-linear self-aligning structures, dams, reservoir intake towers, retaining walls and strengthening systems with advanced composites.

The scope of the undertaking can be demonstrated by the ongoing Seismic Research Infrastructures for European Synergies (SERIES) project funded under the 7th Framework Programme of the European Commission. Herein, a distributed database to share past, present and future test results is to be implemented, the qualification of the test laboratory will be addressed, novel actuation, instrumentation and control technologies that enable better operation of tests will be explored and experimental studies of soil-structure interaction will be conducted. Moreover, access to the EQUALS facilities will be offered to groups of external researchers interested in the following areas of research:

- The seismic performance of cantilever retaining walls,
- The dynamic behaviour of soils reinforced by pile groups,
- High-performance composite-reinforced earthquake resistant buildings with self-aligning capabilities,
- Soil-pile group-structure interaction in the seismic domain,
- The seismic response of load-bearing masonry walls with acoustic barriers,
- Assessment of the seismic performance of flat-bottom silos.

Details of some other recent and ongoing exemplar projects are discussed in the following sections.

2.3.1 Cable Dynamics Under Multi-axis End Motion

The dynamic behaviour of cable structures such as cable-stayed bridges or guyed masts is complicated by the non-linear interactions occurring between the little-damped cables and the structural elements that they support. Large cable vibrations can result from small amplitude end motions potentially increasing in-service costs via fatigue. Two mechanisms in particular have been linked with damaging vibrations: firstly, direct excitation of the cable due to input motions close to one of its natural frequencies and, secondly, parametric excitation. The latter phenomenon is due to dynamic tension variations induced by the component of end motion along the axis of the cable, triggering a modal instability and cable vibrations at half of the excitation frequency.

Current design methodologies separate out the transverse and axial components of end motion neglecting that, in most cases, these excitations will coexist. Moreover, due to cable resonant frequencies resembling a pure harmonic series, the input frequencies required for parametric excitation also directly excite a mode close to the excitation frequency. The consequential cable motions alter the conditions under which parametric excitation occurs. By modelling this interplay, modified behaviour emerges.

2.3.1.1 Modelling the Dynamics of Inclined Cables

Non-linear mathematical modelling techniques were used to derive an analytical model of inclined cable dynamics (Macdonald et al. 2010a). Using the model, the location of the stability boundaries, in the presence of a directly excited mode, of modes excited either parametrically or via non-linear modal coupling were defined. The positions of the stability boundaries were confirmed by physical tests.

The physical model consisted of a 5.4 m length of piano wire (Fig. 2.2). Through the addition of lumped masses and winch-gear adjustment of the static tension, the non-dimensional cable parameters were made to match typical values for a 200 m long bridge cable. The lower end of the cable was anchored to and actuated via a simply-supported beam designed to carry the horizontal component of cable tension and to ensure that the imparted motion was constrained to be vertical.

A Linear Variable Differential Transformer (LVDT) was used to both monitor the vertical displacement of the lower anchorage and to provide feedback control for the servo-hydraulic actuator used to impart the excitation motion. Two multi-axis load cells were used to measure the reaction forces at the top and at the bottom of the cable. A video displacement measurement system was used to track both in-plane and out-of-plane cable motions.

2.3.1.2 Implications of Modelling for the Design of Cable Structures

Compared with previously defined simplified solutions, the presence of the directly excited response was found to significantly modify the position of the stability

Fig. 2.2 Physical model of inclined cable

boundaries. One consequence was that parametric excitation could occur over a wider frequency range than was previously thought. A simplified stability boundary was presented (Fig. 2.3) so that undesirable large amplitude cable vibrations can be avoided in practice (Macdonald et al. 2010b). Physical modelling also identified another non-linear mechanism capable of inducing significant cable vibrations at twice the excitation frequency. The mechanism was later captured by a refinement in the analytical model (Macdonald et al. 2010a). The hope is that the resulting physically-validated, simplified, and non-dimensional expressions can be used for improved analysis of cable vibrations in the design of cable structures.

2.3.2 Nonlinear Dynamics of Damage Resilient Buildings

To achieve life safety, current state-of-the-art elasto-plastic seismic design deliberately allows a structure to sustain damage (i.e. by permitting the formation of plastic hinges within its beams). By doing so, the number of earthquake fatalities has significantly reduced while the economical losses have grown vastly (see e.g. Erdik 1998).

To counteract the escalating financial cost of earthquakes the emphasis of earthquake engineering research has shifted towards more elaborate damage resilient

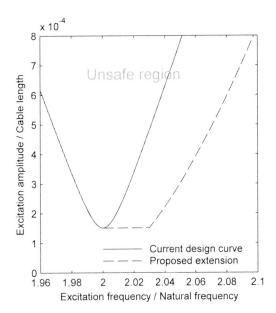

Fig. 2.3 Simplified design curve extension

buildings. Rather than having a moment-stiff connection between beam and column elements, these structures have spring loaded elastic joints between elements. These joints consist of a contact between the end of the beam and the face of the column together with a post-tensioning cable running between the beam and the column. The joints can open (rotate) through elastic elongation of the cable, allowing large deformations without permanent damage as well as giving the structure self centering capability. During joint opening, the cable tension force increases while the contact area shifts from being evenly distributed over the joint's height to being localised at the top/bottom of the joint. These characteristics of the joints give them a nonlinear moment-rotation relationship, which can induce complicated nonlinear dynamic response (Oddbjornsson 2009).

The dynamic response of this class of structures is relatively unknown and potentially complex. To explore and develop a comprehensive understanding of the fundamental mechanics, both at the local (individual joint) level as well as at the global (frame assembly) level, a research programme involving both numerical and physical modelling was conducted.

2.3.2.1 Numerical Simulation of Self Centring Frames

Distinct element numerical modelling is a cost effective way of exploring the fundamental mechanics of this class of structures. Thus, a generic single-story single-bay portal frame building was modelled employing the distinct element method

(Oddbjornsson et al. 2007). Quasi-static pushover tests on the numerical model confirmed the existence of a nonlinear relationship between moment and rotation of the joints, that the contact force relocates during loading, and that the cable tension does not remain constant.

Dynamic excitation of the numerical model revealed the existence of nonlinear dynamic phenomena. A nonlinear frequency response curve was observed with multiple coexisting solutions across a wide excitation frequency range. At distinct frequencies and amplitudes of excitation, the numerical model was observed to jump between different solutions on the non-linear frequency response curve. This behaviour bought with it sudden and significant increases or decreases in the magnitude of the response.

2.3.2.2 Physical Modelling of Self Centring Frames

Although the distinct element numerical model indicates that a structure of this type exhibits complicated nonlinear dynamic behaviour, a real structure made from real material with its imperfections and tolerances might not. To verify the existence of nonlinear response attributes within a real structure, a physical model is required. A physical model is also necessary to explore properties such as durability, robustness and repeatability. Furthermore, a physical model can reveal problems associated with construction, assembly and element detailing, that otherwise would go unnoticed in an abstract numerical model.

A quarter scale physical model (Fig. 2.4a), based on a typical light weight portal frame prototype building, was designed and constructed (Oddbjornsson et al. 2008). The model is single-bay, with a bay width of 2,100 mm (c/c), and single-story, with a height of 900 mm (to beam centre). Based on the artificial mass simulation rule, the model has an active seismic mass of 2 tonne. Following design guidelines (Stanton and Nakaki 2002), assuming 100 mm initial contact height, a tendon tension and cross section of 115 kN and 93 mm^2 for beam tendons and 64 kN and 52 mm^2 for column tendons were determined. Beam and column elements were made from $100 \times 100 \times 10$ mm steel square hollow section. The steel-on-steel contact area at each joint is 100×100 mm in size (Fig. 2.4b).

A quasi-static pushover test of the physical model confirmed its nonlinear softening stiffness characteristics (Fig. 2.5a). Repeat pushover testing confirmed the physical model's invariant stiffness characteristics. Snap-back tests were conducted to investigate the model's amplitude dependant response frequency, see (Fig. 2.5a).

The nonlinear resonance response curve of the physical model was investigated using the EQUALS shaking table. Coexisting solutions and jumps between its two branches were identified experimentally through increasing- and decreasing-frequency sine sweep testing over a range of excitation amplitudes (Fig. 2.5b). The role of damping was also explored by retrofitting the model with a metallic shear type dissipator (Dietz et al. 2010).

The rigorous test regime to which the physical model was subjected without significant damage demonstrated the remarkable robustness and durability of this

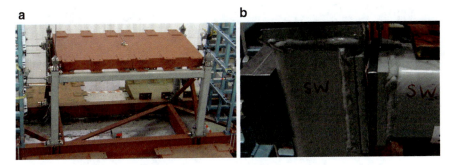

Fig. 2.4 The quarter scale physical model: (**a**) overview and (**b**) a beam-column joint

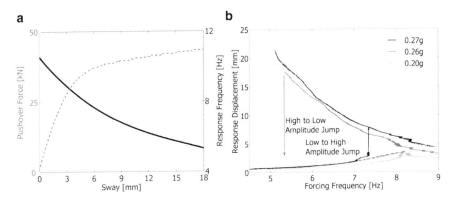

Fig. 2.5 (**a**) Stiffness and response frequency characteristics and (**b**) nonlinear resonance response curve of the physical model

new class of damage resilient structures. Indeed, the physical model survived the equivalent of more than 2,000 high amplitude earthquakes. It was demonstrated that while these structures exhibit complicated dynamic response, their performance space is invariant and well defined allowing them to be designed to operate within a safe domain.

2.3.3 Pile-Soil Kinematic Interaction

It is well established that the shearing stress-strain behaviour of soils affects the dynamics of soil-structure interaction at small, medium and large strains and that it may play a significant role on magnitude and patterns of seismic damage to buildings. Post-earthquake reconnaissance work has shown that pile supported buildings built on loose to medium dense sands or layered soils may suffer settlement and tilting under seismic excitation. In some cases (Mexico City 1985, Kobe 1995), damage to piles has been observed close to interfaces separating soil layers with very different shear

Fig. 2.6 Experimental investigation of pile bending within a layered soil deposit

moduli. While there is a large body of literature on soil-structure interaction for homogeneous deposits, experimental data on stratified deposits is scarce. An investigation was launched into soil structure interaction aspects and free field behaviour for layered deposits exhibiting stiffness contrast between the top and the bottom layers.

2.3.3.1 Physical Modelling of Pile-Soil Interaction

The model configuration (Fig. 2.6) featured a model pile and soil contained within a lamellar shear box known as the shear stack (Dietz and Muir Wood 2007). Monolayer and bi-layer deposits of granular materials with various particle characteristics were formed by pluviation. Tests with various boundary conditions for the top of the pile were carried out, i.e. with a free head capable of both translation and rotation, with a non-rotating head capable of translation only, and with a single degree of freedom oscillating mass to model a superstructure. The model was excited using a range of waveforms such as pulses, white noise, harmonic waves and scaled seismic records.

The bending response of the model pile was measured via strain gauges and the free field response was measured via accelerometers embedded in soil. The influences brought by the granular material characteristics, the stiffness ratio between the top and bottom layer in the deposit, the dynamic input parameters and the inherent design features of the shear stack were explored (Dihoru et al. 2010a). Aspects of soil-stack coupling and soil stiffness degradation were also investigated. The role

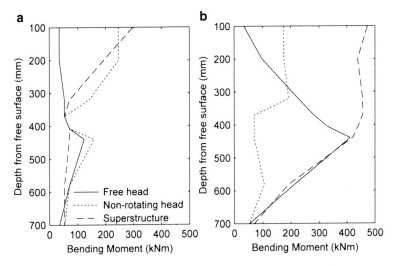

Fig. 2.7 Envelopes of maximum bending response for various pile end conditions with (**a**) 1.75 and (**b**) 80 ratio of shear modulus between soil layers (Reproduced by permission of Taylor & Francis Group)

played by the dynamic impedance of the deposit, the applied lateral force and the inertial force on the pile response was analysed (Dihoru et al. 2010b).

2.3.3.2 Pile Moment Patterns

The experimental results show that the pile-soil kinematic interaction is strongly influenced by the soil deposit configuration, in particular by the stiffness ratio between the layers (Dihoru et al. 2009). Figure 2.7 shows bending moment diagrams for two stiffness ratios between the bottom and the top layers and for three boundary conditions for the top of the pile. For the free end pile the maximum bending moment is recorded at the interface between the two granular layers. The fixing of the end of the pile and the presence of the superstructure change the pattern of the bending moment diagram, with maximum moment migrating towards the top of the pile.

The shear stress-shear strain results for the free field compared well with commonly-used empirical curves for sands (i.e. Seed and Idriss 1970). The dynamic hysteretic loops obtained for the bi-layered deposits showed some memory of the initial deposit stiffness and a clear link to the frequency content of the seismic input. Higher frequency loading cycles resulted in higher values of measured dynamic stiffness in both the monolayer and the stratified deposits. The dynamic measurements of shearing stress and strain gave useful insight into how the strain magnitude and the earthquake frequency content affect the hysteretic response of the deposit (Dihoru et al. 2010c).

The experimental data compared well with predictions of pile response provided by two classic analytical models (Dobry and O'Rourke 1983; Mylonakis 2001). The physical validation of the analytical models helps pave the way for their adoption into design practice.

2.3.4 Multiple Support Excitation of Bridges

Eurocode 8 – Part 2 asks for spatially-distributed structures such as long-span bridges to be designed to withstand the effects of multiple support excitation (MSE). This phenomenon occurs when the ground excitation due to an earthquake is considered to be different at each of the support points of a structure. Spatial variations of ground-support motion can be caused by both the finite velocity of seismic waves and the interference between them. The former gives rise to a wave-passage effect wherein the waveforms imparted at each of the ground supports are asynchronous, identical in shape but shifted in time. The latter combined with the former gives rise to support motion incoherence where excitation waveforms differ in terms of both time and shape.

The focus of MSE research has predominantly been on numerical and analytical studies of input motion (e.g. Loh et al. 1982) and structural response (e.g. Hao 1998; Lupoi et al. 2005). The infrastructure of the EQUALS laboratory combined with the integrated approach to research offered at the University of Bristol has prompted the design, construction and accurate control of the first dedicated multiple support excitation experimental rig for testing scale models of long span bridges (Fig. 2.8).

2.3.4.1 The Multiple Shaking Table Test Bed and Model

The MSE test bed comprises of five independently controllable, single axis shaking tables (Norman and Crewe 2008). The shaking tables run on ball-bushes across five parallel rails held at 1 m centres by a steel frame. Each table is driven by a high performance dual-stage servo-hydraulic actuator which spans between a strong wall of the EQUALS laboratory and the shaking table. The actuators were specified to work at around 1% of their 50 kN capacity to enable a very high level of motion control.

Designed by Zapico et al. (2003) for synchronous input studies, the model structure consists of a 1:50 scale 200 m long bridge with three equally spaced piers. 470 kg of additional dead load is added to the 4 m long model bridge deck to allow for scaling effects. The two end abutments use a steel pin to allow the bridge deck to rotate in plan but not in elevation.

The deck is supported by three piers, fixed top and bottom. While it is possible to investigate the effects of asymmetry by using piers of different lengths to model different ground topographies, herein a symmetrical bridge is considered with piers of length 435 mm at all three locations. At prototype scale, the model falls within the Eurocode 8 – Part 2 size range for bridges of non-uniform soil type but not within the size range for bridges with uniform soil types.

Fig. 2.8 The multiple support excitation test configuration

Numerical simulations reveal the response to be dominated by the first three modes of vibration; while the first and third modes are symmetrical, the second mode is asymmetrical.

2.3.4.2 Implications of Testing for Design

As described by Norman et al. (2006), the response of the model to synchronous and asynchronous seismic waveforms compatible with the Eurocode 8 Pt 1 Soil Type 1 response spectra for firm ground is shown in Fig. 2.9. The time delay for the asynchronous case was 7 ms between each input, equivalent to a surface wave velocity of 1,000 m/s. To prevent nonlinearity, peak displacements were scaled to 1 mm.

It can be seen that in this case the response of the bridge is greater for the asynchronous than for the synchronous case. The reason for this is that the frequency at which the first mode operates is lower than the peak plateau in the design response spectra, hence there is more energy in the excitation waveform at the second mode. However, when the input is synchronous the second mode cannot be excited as the model is symmetrical and the second mode is asymmetrical. When an asynchronous input occurs the second mode can be excited, which in this case has lead to a greater overall response.

These typical results reveal that if we consider only synchronous excitations, the response may be underestimated and hence the design unsafe. Contrary to the findings of this work, at present the Eurocode 8 – Part 2 does not require bridges of this size to be analysed under MSE if the ground conditions are homogenous. The design recommendations need to be readdressed.

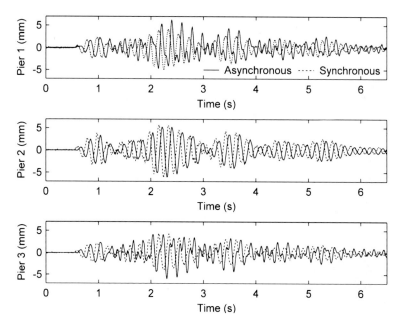

Fig. 2.9 The displacement response of Piers 1 – 3 to synchronous and asynchronous inputs

2.3.5 Interactions Between Pedestrians and Structures

Bridges, particularly footbridges, are vulnerable to pedestrian-induced vibrations and the recent trend for lighter and more slender structures has exacerbated this issue. Specifically, a disturbing number of bridges prone to human-induced lateral vibrations have been reported all over the world (see e.g. Dallard et al. 2001). The relatively large amplitudes of vibration of these bridges have been linked to a suggested phenomenon of pedestrians synchronising their steps with the bridge movement, locking in their phases such as to increase bridge motion. This has been thought to be especially valid for bridges having lateral natural frequencies which are close to the frequency of lateral walking cycles (i.e. from one step to the next step on the same foot).

2.3.5.1 Analytical Modelling of Pedestrian-Structure Interaction

Many models founded on the synchronisation notion have been proposed (Venuti and Bruno 2009; Racic et al. 2009). Often these models are based on uncertain forcing assumptions or have parameters acquired from back-calculation and/or set to fit

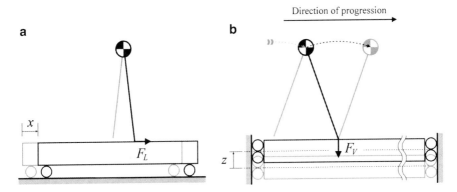

Fig. 2.10 Inverted pendulum pedestrian model in presence of: (**a**) – lateral bridge vibrations (direction of progression into page) and (**b**) – vertical bridge vibrations

the data. Moreover, measurements on two full-scale bridges seem to be inconsistent with the prevailing synchronisation hypothesis (Brownjohn et al. 2004; Macdonald 2008). Some other models considering modal interactions as the driving mechanism of excessive lateral vibrations cannot be easily generalised to cover all circumstances. The absence of a reliable loading model has arisen due to the scarcity of full-scale measurements on bridges and lack of a credible pedestrian model applicable to walking on an oscillating surface.

To address the latter issue a different modelling approach was adopted by Macdonald (2009), adopting an inverted pendulum model of human gait from the field of biomechanics (Winter 2005) and utilising fundamental concepts of kinematics and kinetics (Fig. 2.10a). Lateral balance is maintained by adjusting the position of lateral foot placement (in relation to the whole body centre of mass) (Hof et al. 2007), in contrast to changing the timing of foot placement as assumed by models of pedestrian synchronisation. As well as providing a reasonable approximation of the lateral forces of a pedestrian on stationary ground, in the presence of lateral bridge motion the model yields additional self-excited (or motion-induced) lateral walking forces (Macdonald 2009) which are consistent with the Arup negative damping model derived from back-calculation of the London Millennium Footbridge response (Dallard et al. 2001). The calculated self-excited component of pedestrian loading depends strongly on the bridge vibration frequency and can add negative or positive effective damping to the system in different frequency ranges. Furthermore, the model is able to explain shifts in vibration frequencies (from effective added mass, often negative) and simultaneous responses in two lateral modes, as measured on the Clifton Suspension Bridge (Macdonald 2008).

These findings have prompted a review of pedestrian vertical forcing which has, until now, been modelled as purely external loading to the structure, although there is growing evidence that self-excited forces may be relevant here too (Živanović et al. 2010). Therefore, ongoing research is considering a conceptually similar framework for a pedestrian walking on a vertically moving structure (Fig. 2.10b), revealing that this component of pedestrian loading also depends on the bi-directional pedestrian-structure interaction (Bocian et al. 2011). The model can be easily applied to other structures exposed to footfall-induced vibrations, such as floors and walkways. However, to validate and refine the proposed models, physical tests on people walking on oscillating surfaces are essential.

2.3.5.2 Experimental Identification of Pedestrian-Structure Interaction

Although some experimental attempts to quantify pedestrian forces while walking on vibrating surfaces have been made, most notably by Ingólfsson et al. (2011), who found results quite comparable with the inverted pendulum model (Macdonald 2009), there have been limitations to the experiments. Imposed behavioural restrictions, from insufficient walking areas on treadmills to imposed walking speeds, could bias results.

To truly capture the nature of the problem an extensive experimental campaign is required making provisions for such potential shortcomings. In particular, the effects of the surrounding environment (e.g. visual field) need to be accounted for and the variability of pedestrian walking parameters measured at wide range of bridge vibration amplitudes and frequencies. Especially important is a free-walking time history of pedestrian loading on oscillating ground making provisions for narrow-banded characteristics of this process. For that purpose an instrumented treadmill is being constructed, offering a generous walking surface and equipped with a feedback control system which allows for a rapid speed adjustment of the belt in response to the pedestrian behaviour. This means that more nuisances of gait will be captured and temporal variability permitted, allowing more realistic measurements of the pedestrian forces on moving structures.

A unique opportunity to study these underlying fundamental pedestrian relations exists by making use of the state-of-the-art facilities available at the EQUALS Laboratory. A motion capture system will be used to record the motion of body parts in space, while the treadmill is mounted on a shaking table, allowing great control of the vibrations of the supporting surface. The ongoing research is hoped to gain better understanding of the mechanisms of human-structure interaction and their consequences for the design of modern engineering structures.

2.3.6 Corrosion of Seismically Designed RC Bridge Piers

Corrosion of reinforcing steel is the single most dominant reason for the premature deterioration of reinforced concrete (RC) structures. Corrosion leads to a reduction in cross section of the steel and weakening of the bond and anchorage between

2 Earthquake and Large Structures Testing at the Bristol Laboratory...

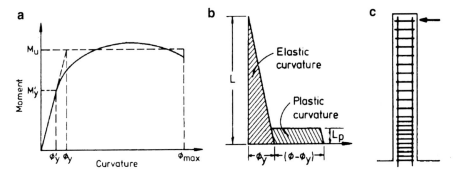

Fig. 2.11 Elasto-plastic approximation of moment-curvature diagram after Priestley and Park (1987): (**a**) moment-curvature relationship, (**b**) curvature distribution, and (**c**) cantilever column (Reproduced with permission of the American Concrete Institute)

concrete and reinforcement (Kashani et al. 2010). This directly affects the serviceability and ultimate strength of concrete elements within a structure. While much research has been conducted to study the effect of corrosion on the mechanical behavior of RC members, the seismic performance of corroded RC structures is not well understood. If maximum acceptable levels of corrosion can be determined while maintaining adequate levels of structural performance this will improve the level of safety and enable bridge owners and managers to employ an optimized maintenance strategy for bridge stock.

2.3.6.1 Analytical Modelling of Corroded Columns

Because seismic design philosophies rely on energy dissipation by post-elastic deformation it is important to characterize the ductility of corroded RC columns when subjected to seismic loading. The available ductility of flexural members can be estimated using a moment-curvature diagrams (see e.g. Fig. 2.11). The relationship between curvature ductility and displacement ductility for cantilever columns is then based on a plastic hinge analysis method which relies on the assumption of elasto-plastic response and an equivalent plastic hinge length.

Such an analysis taking into account the inelastic behaviour of the corroded column sections has been developed based on the classical fibre section method. An unconfined concrete model has been used for cover concrete, and a confined concrete model is used for core concrete. The variation in mechanical properties and cross sectional area of longitudinal reinforcement and the effect of corrosion induced cracking of cover concrete in compression zone have also been considered.

Fig. 2.12 Typical theoretical moment-curvature curves for a corroded RC section

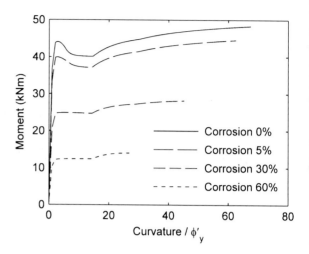

This type of analytical approach has been used to study various RC columns in both uncorroded and corroded conditions. Typical results are shown in Fig. 2.12 where the effect of corrosion on the curvature ductility and displacement ductility can be seen. It can be seen that the corrosion of longitudinal reinforcement dramatically reduces the ductility of reinforcement with low axial force. It is also observed that the effective stiffness of column is affected by corrosion of the longitudinal bars.

2.3.6.2 Experimental Modelling of Corroded Columns

To validate these results the first of four identical 250 mm square reinforced concrete columns 3,100 mm in height has been manufactured. The column has been designed to EC2 and detailed for seismic load according to EC8 criteria and has been subjected to 6 months of accelerated corrosion ready for dynamic testing (Fig. 2.13). A cycling lateral load of up to 50 kN will be applied using a hydraulic actuator and lateral deflections at the top of the column, rotation at the base, strains in the concrete and in the rebar, and crack widths in concrete will be measured.

The use of experimental testing for this type of work is essential because of the complex interaction between the corroded rebar and the concrete which makes any purely analytical modelling challenging. Some examples of the effects being verified by this experimental programme are the change the mechanical properties of steel reinforcement due to the corrosion and the bond-slip behavior of corroded structural elements (Du et al. 2005). In due course it is hoped to develop new finite element formulation using a multi-mechanical fibre-element technique with validated constitutive material models that can effectively predict the seismic performance of corroded concrete sections.

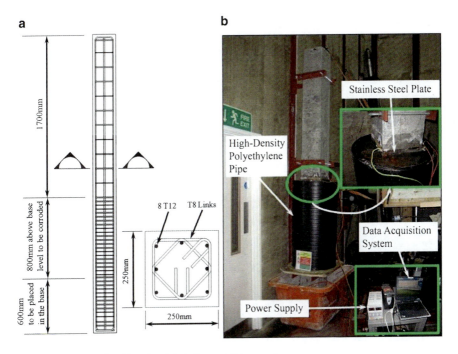

Fig. 2.13 Model RC bridge pier: (**a**) design and (**b**) corrosion

2.4 Summary

The subtleties in seismic performance of civil engineering structures demand the viability of designs to be assessed using a compound analysis featuring numerical simulation, analytical techniques and experimental testing. The infrastructure and expertise required for such analyses are housed within testing laboratories. The UK's largest such facility is the Earthquake and Large Structures (EQUALS) laboratory of the Bristol Laboratory for Advanced Dynamic Engineering (BLADE). The recent and ongoing exemplar projects discussed herein demonstrate the significance of the EQUALS infrastructure and the important role that it has in the advancement of structural dynamics and earthquake disaster mitigation research throughout the world.

Acknowledgments The research leading to these results has received funding from the European Community's Seventh Framework Programme [FP7/2007-2013] under grant agreement n° 227887. The authors acknowledge also the support of the following institutions: the Engineering and Physical Sciences Research Council (under grants GR/R51261/01, GR/R99539/01, GR/R99539/01, GR/T28270/01, EP/D080088/1, EP/D073944/1), Rannis (The Icelandic Centre for Research), Landsvirkjun (the Icelandic National Power Company), the RELUIS consortium, and URS/Scott Wilson.

References

Brownjohn JMW, Fok P, Roche M et al (2004) Long span steel pedestrian bridge at Singapore Changi Airport – part 2: crowd loading tests and vibration mitigation measures. Struct Eng 82(16):28–34

Bocian M, Macdonald JHG, Burn JF (2011) Modelling of self-excited vertical forces on structures due to walking pedestrians. In: Proceedings of the 8th international conference on structural dynamics (Eurodyn 2011), Leuven, Belgium, 1110–1117

Dallard P, Fitzpatrick AJ, Flint A et al (2001) The London Millennium footbridge. Struct Eng 79(22):17–33

Dietz M, Muir Wood D (2007) Shaking table evaluation of dynamic soil properties. In: Proceedings of the 4th international conference on Earthquake Geotechnical Engineering, Thessaloniki, Greece, paper 1196

Dietz M, Oddbjornsson O, Taylor CA et al (2010) Shaking table testing of a post-tensioned tendon frame retrofitted with metallic shear panel dissipator. In: 9th US/10th Canadian conference on earthquake engineering, Toronto, paper 1185

Dihoru L, Bhattacharya S, Taylor CA et al (2009) Experimental modelling of kinematic bending moments of piles in layered soils. In: IS-Tokyo 2009 – international conference on performance-based design in earthquake geotechnical engineering, Tokyo, Japan

Dihoru L, Taylor CA, Bhattacharya S et al (2010a) Stiffness design for granular materials – a theoretical and experimental approach. In: Proceedings of the 7th international conference on physical modelling in geotechnics, Zurich, paper 013

Dihoru L, Taylor CA, Bhattacharya S et al (2010b) Physical modelling of kinematic pile-soil interaction under seismic conditions. In: Proceedings of the 7th international conference on physical modelling in geotechnics, Zurich, paper 012

Dihoru L, Taylor CA, Bhattacharya S et al (2010c) Shaking table testing of free field response in layered granular deposits. In: Proceedings of the 14th international conference of earthquake engineering, Ohrid, paper 166

Dobry R, O'Rourke MJ (1983) Discussion on seismic response of end-bearing piles. J Geotech Eng Div ASCE 109:778–781

Du YT, Clark LA, Chan A (2005) Effect of corrosion on ductility of reinforcing bars. Mag Concr Res 57(7):407–419

Erdik M (1998) Seismic vulnerability of megacities. In: Booth E (ed) Seismic design practice into the next century: research and application. Balkema, Rotterdam

Hao H (1998) A parametric study of the required seating length for bridge decks during earthquake. Earthq Eng Struct Dyn 27(1):91–103

Hof AL, van Bockel RM, Schoppen T et al (2007) Control of lateral balance in walking – experimental findings in normal subjects and above-knee amputees. Gait Posture 25(2):250–258

Ingólfsson ET, Georgakis CT, Ricciardelli F et al (2011) Experimental identification of pedestrian-induced lateral forces on footbridges. J Sound Vib 330(6):1265–1284

Kashani MM, Crewe AJ, Canisius TDG (2010) Ductility of corrosion damaged RC bridges in seismic assessment. In: Proceedings of the 5th international ASRANet conference, Edinburgh

Loh CH, Penzien J, Tsai YB (1982) Engineering analyses of SMART 1 array accelerograms. Earthq Eng Struct Dyn 10(4):575–591

Lupoi A, Franchin P, Pinto PE et al (2005) Seismic design of bridges accounting for spatial variability of ground motion. Earthq Eng Struct Dyn 34(4–5):327–348

Macdonald JHG (2008) Pedestrian-induced vibrations of the Clifton Suspension Bridge UK. In: Proc Inst Civ Eng – Bridge Eng 161(BE2):69–77

Macdonald JHG (2009) Lateral excitation of bridges by balancing pedestrians. Proc R Soc Lond A 465(2104):1055–1073

Macdonald JHG, Dietz MS, Neild SA et al (2010a) Generalised modal stability of inclined cables subject to support excitations. J Sound Vib 329(21):4515–4533

Macdonald JHG, Dietz MS, Neild SA (2010b) Dynamic excitation of cables by deck and/or tower motion. Proc Inst Civ Eng – Bridge Eng 163(2):101–112

Mylonakis G (2001) Simplified model for seismic pile bending at soil layer interfaces. Soil Found 41:47–58

Norman JAP, Crewe AJ (2008) Development and control of a novel test rig for performing multiple support testing of structures. In: Proceedings of the 14th world conference on earthquake engineering, Beijing, paper no. 02-02-0051

Norman JAP, Virden DW, Crewe AJ et al (2006) Modelling of bridges subject to multiple support excitation. Struct Eng 84(5):26–28

Oddbjornsson O (2009) Dynamics of nonlinear elastic moment resisting frames. PhD thesis, Univeristy of Bristol, Bristol

Oddbjornsson O, Alexander, NA, Taylor CA et al (2007) Computational analysis of precast concrete frames with post-tensioned tendons. In: Proceedings of the 11th international conference on civil, structural and environmental engineering computing. Civil-Comp Press, paper 167

Oddbjornsson O, Alexander, NA, Taylor CA et al (2008) Shaking table testing of nonlinear elastic moment resisting frames. In: Proceedings of the 14th world conference on earthquake engineering, Beijing, paper 11–0110

Ojaghi M, Lamata-Martinez I, Dietz M et al (2010a) UKNEES – distributed hybrid testing between Bristol, Cambridge and Oxford Universities. In: 9th US/10th Canadian conference on earthquake engineering, Toronto, paper 1024

Ojaghi M, Lamata-Martinez I, Dietz M et al (2010b) Real-time hybrid testing in geographically distributed laboratories. In: 14th European conference on earthquake engineering, Ohrid, paper 872

Priestley MJN, Park R (1987) Strength and ductility of bridge columns under seismic loading. Struct J Am Concr Inst 84(1):61–76

Racic V, Pavić A, Brownjohn JMW (2009) Experimental identification and analytical modelling of human walking forces: literature review. J Sound Vib 326(1–2):1–49

Seed HB, Idriss IM (1970) Soil moduli and damping factors for dynamic response analysis. Report no. EERC 70–10. Earthquake Engineering Research Center, Berkeley

Stanton J, Nakaki S (2002) Precast Seismic Structural Systems PRESSS Vol. 3–09: Design guidelines for precast concrete seismic structural systems, University of Washington, Seattle

Venuti F, Bruno L (2009) Crowd-structure interaction in lively footbridges under synchronous lateral excitation: a literature review. Phys Life Rev 6:176–206

Winter DA (2005) Biomechanics and motor control of human movement. Wiley, Hoboken

Zapico JL, Gonzalez MP, Friswell MI et al (2003) Finite element updating of a small scale bridge. J Sound Vib 268(5):993–1012

Živanović S, Pavić A, Ingólfsson ET (2010) Modelling spatially unrestricted pedestrian traffic on footbridges. J Struct Eng ASCE 136(10):1296–1308

Chapter 3
Structural and Behaviour Constraints of Shaking Table Experiments

Zoran T. Rakicevic, Aleksandra Bogdanovic, and Dimitar Jurukovski

Abstract Large scale experiments performed on seismic shaking tables in the last 40 years have significantly influenced the development of earthquake engineering in general. They provide very useful information on materials, joints, elements and the overall structural behaviour when subjected to different earthquake motions in controllable laboratory conditions. Based on this information great improvement has been made in the modelling and physical understanding of the structural behaviour of different structural systems. These improvements, in turn, have enhanced the seismic design code practice. Based on a large number of performed tests, the purpose of experimentation could be classified in the following categories: testing of physical models for better understanding of the dynamic behaviour of structures, development of technologies in earthquake engineering, and proof and seismic qualification tests of vital mechanical and electrical equipment. In this paper a critical analysis of tests performed in Dynamic Testing Laboratory at Institute of Earthquake Engineering and Engineering Seismology–IZIIS, Skopje, R. Macedonia, is done from the point of view of the constraints present in using shaking tables for different testing categories.

Z.T. Rakicevic (✉) • A. Bogdanovic • D. Jurukovski
Institute of Earthquake Engineering and Engineering Seismology, IZIIS,
SS Cyril and Methodius University, Salvador Aljende 73, PO BOX 101,
1000 Skopje, Republic of Macedonia
e-mail: zoran_r@pluto.iziis.ukim.edu.mk; saska@pluto.iziis.ukim.edu.mk;
jurudim@yahoo.com

M.N. Fardis and Z.T. Rakicevic (eds.), *Role of Seismic Testing Facilities in Performance-Based Earthquake Engineering: SERIES Workshop*, Geotechnical, Geological and Earthquake Engineering 22, DOI 10.1007/978-94-007-1977-4_3,
© Springer Science+Business Media B.V. 2012

3.1 Introduction

Shaking tables installed in laboratories in a number of countries worldwide have provided almost 40 years of practical experience gathered through experiments in earthquake engineering. Although the characteristics of shaking tables, such as their size and capability of simulating higher kinematic values and frequencies, are permanently enhanced, their application is limited due to two reasons. First of all, their technical application is limited, and secondly, the experimental tests on these systems are quite costly. Irrespective of this, during the last 40 years wide international experience and knowledge that deserves attention due to its great contribution to the development of earthquake engineering and the dynamics of the systems in general has been gathered. The Dynamic Testing Laboratory of the Institute of Earthquake Engineering and Engineering Seismology (IZIIS) at the "Ss. Cyril and Methodius" University, Skopje, R. Macedonia, marks 38 years of experience gathered in application of the single component programmable shaking table and 28 years of experience in application of the biaxial programmable shaking table with 5.0×5.0 m in plane and a payload of 40.0 ton. During this period, 170 shaking table projects for different purposes have been carried out. The authors of this paper would like to present this rich professional and practical experience to the international scientific community by elaborating the limiting factors of these tests that should be known and critically assessed. Naturally, these experimental tests are based on known laws in the field of dynamics and as such they are susceptible to limitations dictated by the theory, but there are also limitations resulting from other reasons, which are the subject of this paper. In other words, shaking table experiments are applied in certain fields where all known limiting factors for the specific experiment have to be explored depending on the problem involved.

Shaking table experiments basically represent a technically justified and economically sensible approach to gathering valid information for the development of earthquake engineering. If properly used, they represent a good substitution for information on the behaviour of structures obtained under the effect of actual earthquakes. The incorporation of knowledge for the purpose of raising the economically justified protection of structures and people in the standards and codes of design prior to occurrence of earthquakes means reduction of possible economic losses and loss of human lives.

The limiting factors of shaking tables arising from their characteristics are given in the second section of this paper through the analysis of about 110 systems installed in a big number of laboratories worldwide. For the last several years, these systems have been applied to perform a large number of investigations resulting in the publication of a large number of scientific papers in renowned world journals or in the proceedings of World and European conferences on earthquake engineering and structural control. More information along these lines is given in Sect. 4.3 of this paper.

The practical application of shaking tables is reduced to one of the following possible purposes: investigation of the dynamic behaviour of structures by means of physical models, performance of tests for the needs of development of new technologies of construction, or new devices by which the safety of structures against the effect of different dynamic loads is enhanced, and tests for proving the quality of vital elements, or components in the field of mechanical and electrical industry whose exposure to earthquake effect may cause extensive direct or indirect damage. Each of these categories of shaking table experiments has its own specificities that have a limiting effect on planning and performing these investigations. These limitations should be known and well analyzed by professional researchers in the laboratories.

The limiting factors during the performance of shaking table experiments for all the categories of tests are separately analyzed in this paper. The authors are aware of the existence of other arguments that limit or affect the extensive use of shaking tables in earthquake engineering for which some others may present different views. The intention of the present authors is to point out that these investigations are not without limitations, which should however be appropriately understood and analyzed by researchers.

3.2 Shaking Tables

In the beginning, the development of earthquake engineering as a scientific discipline was mainly based on information obtained on the basis of observations of dynamic behaviour of built structures during actual earthquake effects. The real earthquake as a natural experiment without limitations revealed the flaws made by the structural engineers in the process of design or construction and "punished" for these flaws. Similarly to other engineering disciplines, earthquake engineering scientists were permanently led by the idea of verification of the seismic stability of their products (the engineering structures) prior to their construction. Regardless the extent of similarity between different types of engineering (mechanical, electrical, civil engineering, etc.), there is a characteristic in the concept of verification of the quality of products that makes them essentially different. Namely, unlike other types of engineering, the products of civil engineering are unique from the aspect of their geometry and main mechanical properties. If one adds this fact, which is specific only for seismic excitation, then it is more than clear that the concept of seismic verification of the stability of structures should be based on another approach. This approach is based on the idea that investigations of the stability and dynamic behaviour of structures should be performed on physical models tested under simulated natural earthquake effects.

Although in the first half of the twentieth century efforts were made to build a laboratory system for simulation of earthquakes, the first types of earthquake simulators with programmable effect were produced and made available to the earthquake engineering scientists as late as the beginning of the 1970s due to the insufficient level of technological knowledge in the mechanical, electrical and electronic industry. Since then, more than 110 shaking tables with programmable characteristics have been installed in laboratories worldwide (Earthquake shaking table - Wikipedia).

If one looks at the list of all these shaking tables and compares their main characteristics, such as the size of the shaking table, the payload, the degrees of freedom, the kinematic quantities (stroke, velocity, acceleration) and frequency, one can see a big difference in these performance characteristics. Each of the characteristics has a different effect upon the quality of simulation and particularly the quantity of technical and scientific contribution made by the experiment. Also, each of the characteristics does not by itself say much about the quality of the possibilities, i.e. limitations in application. Integrally, all the characteristics define the spectrum of applicability for scientific and practical needs.

As to the size of the shaking tables, it basically ranges between 0.465 (the smallest) and 300 m^2 (the largest), while the given size of payload ranges from 1.0 to 2,000 t. The number of degrees of freedom that can be simulated on the shaking tables ranges from 1 to 6. The shaking tables with a single degree of freedom of motion usually provide the possibility of simulation of earthquake motion along a single axis (x or y or z), whereas those with six degrees of freedom provide the possibility of simulation of motion and rotation along the three axes. The spectrum of kinematic characteristics is different. So the stroke ranges between ±5 and ±244 cm, whereas the velocities range between 40 and 200 cm/s. The accelerations range from 0.5 to 10.0 g, while the working frequency is from 20 to 200 Hz. All these characteristics are mutually dependent and affect the total cost.

It can be estimated that the cost of an integral system from the list of existing shaking tables in the world ranges between 10^5 US \$ and 10^9 US \$. The cost of these systems affects the cost of the experiments as well as the maintenance of the systems.

Shaking tables that are used for testing the quality of equipment expected to be exposed to dynamic effects in serviceability conditions have more specific requirements. Namely, the equipment requires a higher level of acceleration and a higher frequency range. For some functional tests on equipment, a higher stroke and velocity are needed. Depending on the component which is tested, there is often a need for shaking tables with a larger payload (a few hundreds of tonnes).

The application of shaking tables in the development of new technologies may also result in the need for a system with specific characteristics. These tests are usually associated with tests in the field of structural control.

It is known that a system for simulation of earthquakes in laboratory conditions cannot thoroughly satisfy all the needs. In its design, one should endeavour covering as wide a spectrum of its application as possible depending on the budget available for procurement of equipment, laboratory, measuring equipment and other accessory devices.

While using shaking tables, one must take care of their regular maintenance and check the design parameters at the expiry of semi-annual and annual periods.

3.3 Published Testing Results in the Last Years

Since the design and installation of the first modern shaking table in the USA, at UC Berkeley in 1972, many shaking table experiments have been conducted in the world's well recognized research institutions and laboratories on all continents.

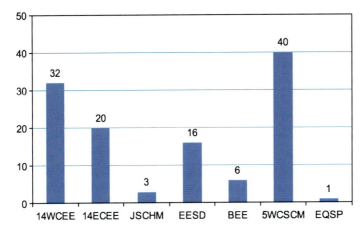

Fig. 3.1 Published results from tests done in the course of the last 3 years; *14WCEE* World Conference on Earthquake Engineering; *14ECEE* European Conference on Earthquake Engineering; *5WCSCM* World Conference on Structural Control and Monitoring; *EESD* Earthquake Engineering and Structural Dynamics, Wiley; *BEE* Bulletin of Earthquake Engineering, Springer; *JSCHM* Journal of Structural Control and Health Monitoring, Wiley

Laboratory reports have been written and published, testing results have been presented at World, European, regional and other international conferences as well as a huge number of papers have been published in conference proceedings and relevant scientific journals.

A brief survey of published results from shaking table tests carried out during the last 3 years, i.e. since 14WCEE, has been done. About 118 papers were published in the proceedings of the last 14th World and 14th European Conference on Earthquake Engineering, the 5th World Conference on Structural Control and Monitoring and in selected relevant scientific journals (Fig. 3.1).

Out of the 118 papers, 55 refer to shaking table tests performed for the needs of the development of new technologies (structural control devices, new materials and components of structural health monitoring systems), 52 involve tests on physical models for different purposes (structural behaviour and safety, improvement of codes for earthquake resistant design, etc.) and 11 are related to proof test experiments on various equipment.

3.4 Categories of Shaking Table Experiments

There is a big variety of shaking table experiments depending on what is tested, for which purpose, according to which protocols, methods and/or standards the test is performed, the mode of simulation of loading conditions, intensity and test sequence, uniaxial or multi-axial testing, type of physical model and material used for its construction, scaling factor, etc.

Taking into account the practice of the world scientific-research laboratories and based on their own experience, the authors of this paper consider that shaking table experiments can generally be divided into three main groups:

1. Experiments for testing structural behaviour and safety (physical models)
   This category involves experiments performed on physical models of hypothetical or actual structures constructed to a certain scale in accordance with the similitude laws from experimental mechanics.

   The purpose of these tests is to define the behaviour of different structural systems composed of different construction materials with or without non-structural elements and to define, at the same time, the level of safety under earthquake effects.

   Depending on the types of structures and material, one can define the failure pattern, starting from the occurrence of the first cracks and their propagation, or the yielding zone until the final loss of stability and failure.

   The testing programme should be conceptualized in such a way that as much as possible reliable information on the model is obtained, including 3D behaviour, evaluation of requirements and corresponding consequences such as type and development of damage. This is of a particular importance since the recorded experimental results represent the basis for verification and improvement of analytical models of structures, analytical procedures and computer codes (programmes) that are used in further phases of combined experimental-analytical investigations.

   On the other hand, the experiments from this group play an important role in the development of investigations and practices in earthquake engineering. In addition to being used for verification of analytical procedures and modelling, they extensively contribute to the calibration of standards for seismic design and the design of engineering structures in general, particularly when speaking about new structures that incorporate innovations, new technologies and materials.

2. Proof test experiments
   The proof test experiments are carried out on different types of equipment, devices and assemblages, either complex or regular, for electrical and industrial facilities, as well as various products in accordance with required procedures and for a particular requirement. The procedure has to follow certain standards regarding: nuclear power plant components, substation's equipment, electrical engineering, mechanical engineering and civil (earthquake) engineering.

   The testing methods can generally be divided into three main categories:

   - Specific proof tests. These are used to check the qualification of the equipment for a certain requirement. With these tests, the equipment is tested for the effect of response spectrum, time history or other parameters that are characteristic for the location where the equipment will be placed. Since the equipment is tested according to a specified performance requirement not with regard to its ultimate state, specification for the testing is usually given by its end user.

3 Structural and Behaviour Constraints of Shaking Table Experiments

- General proof tests. The objective of these tests is to evaluate the qualification of equipment for a wide spectrum of application. The testing specification involves most or all known requirements that, on the other hand, may result in the generation of a very severe test excitation.
- Fragility tests. These tests are used to define the ultimate state of the equipment. The information obtained from these tests can later be used to demonstrate the adequacy of the given requirements or applications.

The excitations used during the tests can be divided into two main groups:

- Single frequency excitations. These are mainly used to define the resonant frequencies and damping of the equipment. If it is shown that the equipment has only one resonant frequency, or clearly distinguished resonant frequencies without coupling, or if justified out of other reasons, these excitations can be used for complete qualification of the equipment. Characteristic tests that are the most frequently applied are: continuous-sine test, sine-beat test, decaying-sine test, sine-sweep test.
- Multiple frequency excitations. These excitations enable simulating broadband test motion which will cause participation of a greater number, or all modes in the response of tested equipment in which loss of function or failure can be caused by modal interaction. The multiple frequency motion tests enable closer and more realistic simulation of a typical seismic excitation. Characteristic tests that are the most frequently applied are: time history test, random-motion test, complex-motion tests, random motion with sine dwells, random motion with sine beats, combination of multiple sinusoids, combination of multiple sine beats, combination of decaying sinusoids.

Which test method will be selected depends on the nature of the expected vibrations in the environment (e.g. seismic hazard of the location) and the very nature of the equipment itself.

3. Experiments for the development of new technologies

Tests are conducted on various structural control devices and systems, components of structural health monitoring systems, seismic isolation devices as well as other devices and equipment based on new technologies for which there are no strictly defined standards and procedures for testing and qualification.

The testing standards and procedures have to be agreed between the laboratory and the manufacturer (or the one who has made an inquiry) in order to collect as much information as possible and test the device under as many simulations as possible. This means that there are no limitations as to how the tests will be done and in most of the cases, a combination of methods characteristic for the preceding two groups of shaking table experiments is used (1 and 2 above).

During these tests, it is of prime importance to provide conditions by which the functional characteristics of the devices/systems will be realistically simulated.

3.5 Structural and Behaviour Constraints

Shaking table tests performed on physical models in laboratory conditions realistically predict structural behaviour during an earthquake and provide the possibility of eliminating the unwanted structural flaws prior to the construction of the structures. Although the theoretical principles of application of physical models in earthquake engineering for predicting the dynamic behaviour of structures are simple, there is a big number of limiting factors in practice that belong to one of the following groups:

- Limitations arising from the characteristics of the shaking tables;
- Limitations arising from the impossibility of satisfying all the similitude relationships;
- Limitations based on phenomena that can hardly be experimentally simulated;
- Limitations arising from the categorization of the objectives of the tests.

The limiting factors should be well analyzed in all phases of planning of the experiments. Some of them should be accepted as facts that cannot be compensated for in the experiment, but some of them may be appropriately scientifically interpreted for the purpose of appropriate assessment of the experimental values (strain rate effect, size effect). The effects of limitations per the previously stated groups are presented in the subsequent text.

3.5.1 Shaking Table Constraints

As explained in Sect. 3.2, shaking tables have their own characteristics as size, payload, kinematic quantities, frequency range, overturning moments, etc., that must be respected and cannot be changed. The size of the shaking table basically determines the geometrical similitude (the scale) and this, as basic information, affects further the remaining scales regarding time, velocity, displacement, material properties, reproduction of dead loads, earthquake forces, etc. If the size of the shaking table is small in respect to the geometry of the structure, then the scale of quantities is large which leads to small scale physical models in which case the satisfying of all the similitude laws arising from the concept of the physical model design becomes very complicated. In the case of small scale models and depending on the material used to produce the physical models, serious problems may arise during fabrication and handling in the laboratory itself. As to simulation of time scaling, in the case of small scale models and true replica modelling, the acceleration ratio is $a_m/a_p = 1$, whereas the time ratio is $t_m = t_p l_r^{1/2}$ (l_r – length scale factor). In these relationships, it is obvious that the lesser l_r the shorter the time for true replica simulation, i.e. the higher the frequencies. If in the real structure (the prototype structure), the dynamic response ranges from 0.3 to 25 Hz (the usual range that models the dynamic behaviour of the structures) then the physical model on the

3 Structural and Behaviour Constraints of Shaking Table Experiments 51

shaking table will have to be reproduced for $l_r^{1/2}$ times higher frequency range. This means that, if the usual maximum working frequency of the shaking table is 50 Hz, then $l_r^{1/2}$ can be maximum 2, leading to a length scale ratio of 4. Geometrical scales of this order would surely be interpreted as scales for large (not small) scale models. From this simple simulation, it can be concluded that it is very important to know all the limiting factors referring to the shaking table and the investigator is obligated to make an unbiased assessment of these factors and propose solutions to minimize their effects.

As to the character of the test, importance is given to the kinematic limitations of the shaking table as the stroke, the velocity and the acceleration and particularly the frequencies at which their intersecting points occur. These characteristics are crucial as to the extent the physical model will be excited to the level of important nonlinear states of dynamic behaviour. These limitations that are dictated by the kinematic properties of the shaking table should be analyzed interactively with the payload capacity and overturning moment control. Models with big mass located high above the centre of the shaking table will use much of the energy controlling the platform to control overturning moments, rolling, pinching and twisting of the shaking table.

3.5.2 Constraints of Similitude Relations Fulfilment

To design any physical model, the Buckingham's theorem is applied to define the main similitude relations. For true replica reproduction, it is necessary to satisfy all the defined relationships, which is a very complex process in practice. The process of satisfying the similitude relationships will be demonstrated through a simple problem along with comment on the physical and practical limitations. Namely, if the problem of stress state in a frame structure for which a physical model is to be constructed is considered, the equation of the variables defining the stress state would be as follows:

$$\sigma = F(\vec{r}, m, c, k, l, t, \rho, E, a, g, \sigma_0, \vec{r}_0)$$

Where:

$\sigma_0, \vec{r}_0$	initial stress and displacement variables
m, c, k	mass, damping and stiffness matrices
E, ρ	Young's modulus and mass density
a, g	acceleration and gravity acceleration
l, t	structural geometry and time
$\sigma, \vec{r}$	stresses and deformations of the structure due to earthquake motion.

By application of the Buckingham's theorem, the above dimensional equation is transformed into the following dimensionless equation:

$$\frac{\sigma}{E} = f\left(\frac{\vec{r}}{l}, \frac{m}{Elt^2}, \frac{c}{Elt}, \frac{k}{El}, \frac{t}{l}\sqrt{\frac{E}{g}}, \frac{a}{g}, \frac{gl\rho}{E}, \frac{\sigma_0}{E}, \frac{\vec{r}_0}{l}\right)$$

For true replica modelling, the physical model needs to satisfy all the dimensionless relationships defined in the above equation. Each dimensionless product in the above equation has its own specificities in satisfying these relationships, but due to limited space, the limitations arising from the relationships between material properties, time and kinematic properties will be briefly commented.

If index r indicates the ratio between the variables of the physical model and those of the prototype, from the dimensionless relationship one can obtain that time scaling is $t_r = l_r^{1/2}$. This relationship shows that the time for simulation of the earthquake motion and consequently the responses of the model shall be lesser (shorter) for the values of the square root of the length scale. So, for models of smaller scales (application of small shaking tables), the simulation time will be considerably shorter, whereas for larger scale models, the simulation time will be somewhat shorter than the time of the real earthquake motion. This means that the frequencies of the physical model will be reciprocally changed in respect to time so that, due to the frequency limitation of the shaking table, distorted (modified) earthquake time history will be simulated in case of small scale models. From the relationships, it is also evident that accelerations will be simulated to a scale ratio of 1, namely the acceleration amplitudes of both the prototype and the model are the same, the difference being in the time shortening by a factor of $l_r^{1/2}$. Accordingly, the time histories of displacement will be $l_r^{1/2}$ times shorter, but their amplitudes will be l_r times smaller. If l_r is a small value (small scale model), then the measured time histories of the physical model will be small which aggravates the process of registration.

The other dimensionless relationship to be commented refers to the main material properties. Namely, the ratio between the Young's modulus of the model and the prototype is equal to the product of the corresponding relationships between material mass density and length scale ratio. This relationship is simple but is practically unfeasible because the Young's modulus affects the amount of strain, the mass density and the size of inertial forces. This relationship ($E_r = \rho_r.l_r$) requires special analysis in design of the physical models and selection of materials. Namely, if small scale models are at stake, then materials with high values of ρ (which will influence the inertial forces) and low values of E will be required. The latter will lead to selection of materials with low strength which will make the procedure of fabrication and instrumentation more complicated.

The remaining dimensionless products are characterized by other problems and generate other practical limitations. The analysis of all the constraints arising from dimensionless relationships in design is necessary for the purpose of testing a physical model that will enable the obtaining of realistic values. In case of performance of proof tests or tests for technological innovations, the analysis of the limitations should be done from the aspect of what is expected from the experiment. In other words, the experiment (including the physical model) should be designed such that one could obtain response to the key questions as to the purpose of the test, eliminating as much as possible any dilemmas that could lead into ambiguous interpretation of the results.

3.5.3 Constraints Related to Some Physical Properties

While selecting the material to be used for the physical model, it is endeavoured that this model has a similar σ-ε relationship as that of the material of the prototype. For nonlinear models, the σ-ε relationship should cover the entire range of expected nonlinearity, i.e. until the occurrence of failure of the material. In addition to the standard tests on the σ-ε relationship, for the model material, one should also analyze the effect of some other physical phenomena upon the expected simulation as are: the ageing problem, the size effect, the strain rate effect, the long term effect, the ductility, the fabrication problem and so on. Some of the mentioned effects are briefly discussed below.

The ageing problem is particularly important if physical models are used to explore the dynamic behaviour of structures that are centuries old. Almost all these structures are constructed of archaic materials that have suffered changes of their mechanical properties due to the ravages of time. Depending on the serviceability conditions, the concrete, i.e. reinforced concrete experiences change of its strength, i.e. σ-ε relationship in a way that, in some cases, the strength is increased, while in some other cases, it is decreased. There is no solution as to how to include this phenomenon in selecting or developing the material for the model. The researcher is obligated to provide a critical estimation of its effect upon the final results from the tests.

The size effect has considerably been explored whereat, for the main materials as concrete, micro-concrete and steel, it has been proved that the strength of the materials is increased with the decrease of proportions. Namely, the strength of the materials deteriorates if their proportions increase. The investigations have proved that this change is not linear and can hardly be analytically defined. If steel is taken into consideration for which the producers of material (bars, steel plates) usually provide the strength characteristics, for the materials of physical models which are based on concrete (micro concrete), these differences should be defined by the researcher and incorporated in the model computation. Although there are a number of investigations of this effect, the referent values should be taken informatively because the dimensions of the elements (beams, columns, plates) depend on the scale of the quantities.

The strain rate effect in different models is of concern to be analyzed for small scale models. The speed of change of deformations is defined by the $v_r = l_r^{1/2}$ relationship, which means by the square root of the length scale. According to the investigations performed for steel and concrete and the given analytical relationships, which are based on empirical knowledge, it can be considered that, for models to a scale higher than 1/20 (l_m/l_p), these effects of strength of materials are of the order of several percents. Taking into account the remaining factors of effect upon change of σ-ε, this effect of several percents can be neglected.

In the case of smaller scales (1/100), this effect should certainly be seriously considered and incorporated in the final results from the investigations.

The remaining effects as the long term effects, the ductility, the fabrication, the creeping, the shrinkage, etc., that don't depend on the remaining similitude factors but have effect upon the real model reproduction should be analyzed with particular attention in order to evaluate their individual effect upon the final results.

3.5.4 Constraints Based on the Category of Testing

For physical models intended for experimental investigation of the dynamic behaviour of special/important structures exposed to earthquake motion and phenomena that may occur during such an effect in the course of the serviceability period of the structures, as well as for the needs of development and improvement of the standards and analytical procedures that are used in practice, the analysis of the constraints for these models is given in the preceding Sects. 3.5.2 and 3.5.3. In this part, some specific requirements referring to categorization of the tests for development of technologies and performance of tests for the needs of proving the quality of products will be given.

The application of shaking tables in the development of technologies of construction of structures in seismic areas, or the development of structural devices, or structural control systems represents a unique technologically justified procedure for definition of the functioning and efficiency of the systems that provide the necessary seismic safety of structures and equipment. Considering that new technologies and new control systems or devices have known technical, technological and mechanical characteristics and as such should be incorporated into actual structures, it is logical to conclude that they should be tested to a full scale. Namely, irrespective of whether the principal structure is a model, a fragment of a hypothetical structure, the new technological system should be produced and incorporated into the model to the scale of 1:1. This means that it is not necessary to perform time scaling, but it is necessary that the model of a structure or a fragment has dynamic characteristics similar to those of actual structures. In such conditions of realization of an experiment, it can be said that the efficiency of the components or technical solutions is realized in conditions of actual functioning. To enable such an approach, based on our practical experience, the only limiting factor is the size of the shaking table at plan and its capacity. Such tests can hardly be realized on shaking tables with an area of less than 20.0 m^2 and payload capacity of less than 40.0 t for the planned level of seismic excitation higher than ±1.0 g.

The proof test investigations are usually intended for verification of the functioning of the mechanical or electrical equipment during earthquake effect when installed in technologically important systems whose failure is dangerous. These tests are realized on full scale equipment whereat the following should be taken into account:

- The experiment should be realized according to a procedure defined with a standard or according to a programme and a procedure defined by the end users of the equipment, the producers and the laboratory at which the test is realized;

3 Structural and Behaviour Constraints of Shaking Table Experiments 55

- The shaking table should enable simulation of the design floor response spectra or the expected earthquake ground motion in full scale time along with the entire range of frequency content;
- These tests are frequently realized in simulated serviceability conditions. Such requirements of the user of the equipment should be accepted with a lot of criticism, while the laboratory should provide conditions for the safety of the people and the equipment in case of average during the realization of the experiment.

If the laboratory can provide the above stated conditions, then it is qualified to realize such tests. Considering that, in practice, equipment with a mass of several kilograms to several hundreds of tons and with different size and frequency content is tested, it is clear that limitations arise, first of all, from the size of the shaking table, the payload and the frequency range.

3.6 IZIIS Experience and Examples

The IZIIS' Dynamic Testing Laboratory marks 38 years of experience gathered in application of the single component programmable shaking table and 28 years of experience in application of the biaxial programmable shaking table proportioned 5.0×5.0 and characterized by a payload of 40.0 ton. During this period, 170 experimental projects have been carried out for different purposes.

The tests that have so far been carried out involve tests on different physical models of actual and hypothetical buildings and engineering structures constructed of different materials (reinforced concrete, steel, wood, masonry) for different purposes as well as tests on buildings pertaining to cultural heritage within the frames of numerous domestic and international scientific research and applicative projects for the purpose of definition of dynamic behaviour, prediction and definition of failure model and safety level of different types of structures under the effect of actual earthquake excitations. The experience acquired with these tests has further been applied in the improvement of methods of analysis, design, repair and strengthening of structures and, at the same time, calibration of the national codes for seismic design (Jurukovski et al. 1989b, c, 1991a, b, 1992, 1993, 1994, 1996b; Rakicevic et al. 2007, 2009a; Krstevska et al. 2007; Tasshov et al. 1992a, b, 2007, 2008a, b; Gavrilovic et al. 1990, 1996, 2001; Bojadziev et al. 1993; Boschi et al. 1995).

For the last three decades of use of the seismic shaking table at IZIIS, more than 80 projects (tests for seismic qualification of different electrical and mechanical equipment for industrial facilities and nuclear power plants (NPP-s) of the type of different valves and electro-mechanical drives for NPP-s, instrument transformers, circuit breakers, switches, relays, and alike) have been realized (Rakicevic et al. 2008, 2009; Jurukovski et al. 1986a, b, 1987, 1988, 1989a, 1996a, 2005; Mamucevski et al. 1985a, b, 1986a, b, c, 1987a, 1988a, b, c, 1989, 1990, 1992; Tashkov et al. 1986, 1988, 1994, 2000, 2009).

Through the realization of 30 projects, IZIIS has extensively contributed to the development and experimental tests on systems based on new technologies – different systems for seismic isolation, passive systems for structural control, dampers for different purposes, application of new materials for repair and strengthening (Rakicevic et al. 2001, 2009b; Jurukovski et al. 1984a, b, c, 2004a, b).

Examples characteristic for the stated three basic groups of experiments performed on the IZIIS' shaking table are presented in the subsequent text.

3.6.1 Example – Physical Modelling

Experimental investigations presented through this example were a part of the research work carried out within the international project entitled "Basic and Applied Research Study for Seismic Modelling of Mixed Reinforced Concrete Masonry Buildings", accomplished by the Institute of "Scienza delle construzioni" and the Institute of "Tecnica delle construzioni" both from the University of Bologna and IZIIS (Jurukovski et al. 1988–1992). The project was financially supported by the government of Regione Emilia Romagna, Italy.

The main purpose of this project was to develop a methodology for a seismic design and strengthening of mixed reinforced concrete masonry buildings built in the mid 1950s as a part of the large seaside hotel complex in Rimini-Italy. The research program that was realized was composed of several phases:

1. In situ investigation including full-scale tests of a selected building "Villa Paola Hotel" by forced vibration and ambient vibration measurements tests.
2. Mathematical modelling of tested building, design of hypothetical building and physical model; shaking table test of reduced-scale models: non-linear shaking table test of non-strengthened model; Correlation of experimental and analytical model.
3. Design of strengthening methods and strengthening of the models; shaking table testing of strengthened models.

According to the size of the shaking table a reduced 1:3 scale model was chosen. The design of the model was performed on the basis of the Theory of Models. For the considered geometry scale (1/3), an artificial mass model type was used. According to the similarity principles additional masses for each floor were calculated and a reduced value for the modulus of elasticity (1/2) was adopted, both for the mortar and for the concrete of the frame.

The results obtained by testing of masonry specimens and brick and mortar samples extracted from the original prototype (Villa Paola Hotel), were considered as basic data in the design of the model. To prove that the designed properties are reached, shear-compression testing of both the model and the hypothetical prototype walls was undertaken.

The model was designed to provide a realistic picture of the nonlinear behaviour of the system. To simulate this particular requirement the scale of the model strain (ε_r) was taken to be equal to 1. The acceleration ratio (a_m/a_p) was also adopted to be equal to 1, while assuming the scaled modulus of elasticity $(E_r = 1/2)$, consequently the stress ratio, σ_r, was also 1/2.

3 Structural and Behaviour Constraints of Shaking Table Experiments

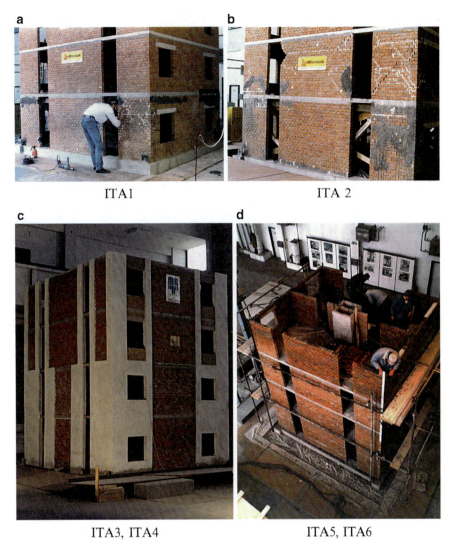

Fig. 3.2 Tested models on IZIIS' shaking table (Jurukovski et al. 1989b, c, 1991a, b, 1992). (**a**) ITA1, (**b**) ITA 2, (**c**) ITA3, ITA4, (**d**) ITA5, ITA6

Six models were tested in total (Fig. 3.2): ITA1–Original model; ITA2–Repaired original model by mortar injection; ITA3–Strengthened model by external reinforced concrete additional walls without connection of strengthening walls reinforcement with the foundation; ITA4–Strengthened model by external reinforced concrete additional walls with anchored of strengthening walls reinforcement with the foundation; ITA5–Strengthened model by central reinforced concrete core and ITA6–ITA5 model repaired by mortar injection.

Fig. 3.3 Oil-immersed current transformer AGU-525 kV mounted on IZIIS' shaking table (Rakicevic et al. 2009a)

The monitoring system that was used in the testing had a capacity of 32 channels, which provided a real time data acquisition of accelerations, displacements and strains during the tests. The instrumentation set-up of a total of 30 transducers was chosen to obtain the best description of the model response to seismic excitations.

One model was used both for linear and different non-linear test levels, so a procedure with gradual increase of the excitation intensity had been applied. Using several models for different non-linear tests would significantly increase the cost of testing. The other constraint was the size of the bricks, as well as the size and mechanical properties of the mortar between the bricks in the walls of the models. Namely, it is very difficult to simulate the σ-ε relationship of the prototype having in mind the size and ageing effect of the masonry.

3.6.2 Example – Seismic Qualification

Seismic qualification of the oil-Immersed current transformer AGU-525 kV (Fig. 3.3), manufactured by KONCAR Instrument transformers Inc, from Zagreb,

Croatia, was conducted by time history shaking table testing for Moderate Seismic Level, in accordance with the IEEE Std. 693–2005, IEEE Recommended Practice for Seismic Design of Substations (Rakicevic et al. 2009a).

Two test set-ups having different supporting conditions were considered. In the first one the AGU-525 kV was anchored on the shaking table, while for the second one the transformer was mounted on a steel support structure.

Two instrumentation schemes were developed in accordance with the IEEE 693-2005 requirements. One for the current transformer having 27 sensors located on 20 different positions and the second for the current transformer mounted on the support structure having 36 sensors located on 26 different locations. Three types of sensors were used: accelerometers, strain gauges and displacement transducers.

All sensors through appropriate signal conditioners have been connected to the data acquisition system, which has a capacity for simultaneous acquisition of 72 channels with a maximum sampling speed of 101.4 kS/s and 24 bit resolution. For the needs of these tests data were acquired at a sampling rate of 1,000 S/s.

The testing programme was prepared in appropriate order to be in compliance with IEEE Std.693-2005 standard, and fully realized. The current transformer AGU-525 kV, manufactured by KONCAR Instrument transformers Inc, from Zagreb, Croatia, without and with support met all of the requirements of the IEEE Std. 693-2005 regarding the Moderate Level of Seismic Qualification.

3.6.3 Example – New Technologies

In order to estimate the efficiency of the developed PDDs (Prestressed Damping Device), manufactured by GERB Schwingungsisolierungen GmbH & Co. KG, Germany, in controlling the structural response due to seismic excitations, and to develop further a procedure for optimal design and placement of these and similar devices in the process of earthquake resistant design of structures, an appropriate research program has been proposed for realization.

In the first part of the research related to experimental investigation an appropriate program has been considered and fully realized (Rakicevic et al. 2009b). Namely, a hypothetical steel frame structure has been designed in accordance with the latest Eurocode 3 and Eurocode 8 requirements, as full scale structure and has been tested on the 5 × 5 m MTS bi-axial shaking table at IZIIS' Dynamic Testing Laboratory, without and with PDDs under simulation of 17 different real recorded earthquake time histories. Five different configurations of the same structural model, one without and four with PDDs (Fig. 3.4) having different positions along the height of the frame structure have been tested. For each earthquake time history, several runs have been done by selecting appropriate scaling factors.

The instrumentation set-up for tested models was composed of accelerometers, linear potentiometers, LVDTs and strain gages. A total of 51, as well as 63 transducers,

Fig. 3.4 Set-up of MODEL02 and MODEL04 on the shaking table (Rakicevic et al. 2009b)

have been used for recording the structural response for the model without and with PDDs, respectively.

In all tested models the configuration mounting and functional condition for all PDDs were fully provided. The maximum recorded stroke of PDDs, during the most severe tests was within their operational range.

3.7 Conclusions and Recommendations

For the last 40 years, shaking tables have demonstrated their importance in the development of earthquake engineering. A big number of phenomena have been discovered during tests on physical models, important analytical models have been developed and codes for design of seismically resistant structures have been improved.

In addition, seismic shaking tables justified their application in the development of new technologies of construction in seismic areas as well as the performance of tests for control of the resistance of vital mechanical and electrical equipment to different dynamic effects.

Despite the evident valuable results achieved by the application of shaking tables, there are still limitations of their use. The use of shaking tales must therefore be analyzed carefully by users to avoid misinterpretation of results. It should be known, however, that limitations are different depending on the category of experimental tests for which shaking tables are used.

In this paper, based on long years of experience in this field, the authors provide a critical review of possible limitations which should be known to users of shaking tables.

Acknowledgments The authors express their gratitude to IZIIS and the Dynamic Testing Laboratory for making the publishing of this paper possible.

References

Bojadziev M, Tashkov Lj, Paskalov T (1993) Model testing of the seismic stability of the stone fill dam 'Kozjak' on Treska River. IZIIS report 93-027

Boschi E, Rovelli A, Funiciello R, Giufre A, Mihailov V, Tashkov Lj, Krstevska L, Mamucevski D, Stamatovska S (1995) Analysis of the seismic risk of existing column monuments in Rome by seismic shaking table testing of a model, vol 2. IZIIS report, Oct 1995

Bulletin of Earthquake Engineering (2008–2010) vols 6–8. Springer

Earthquake Engineering & Structural Dynamics (2008–2010) vols 37–39. Wiley

Earthquake Spectra (2008–2010) Volumes 24–26. EERI

Earthquake Shaking Table – A World List of Shaking Tables. Wikipedia http://en.wikipedia.org/wiki/Earthquake_shaking_table. Accessed 10 Jan 2011

Gavrilovic P, Gramatikov K, Tashkov Lj, Krstevska L, Mamucevski D (1990) Dynamic Analysis of Timber Truss Frame Systems Project JFP-760/40 AES-192 USDA. IZIIS Report 90-47

Gavrilovic P, Sendova V, Tashkov Lj, Krstevska L, Ginell W, Tolles L (1996) Shaking table tests of adobe structures. Report IZIIS 96-36

Gavrilovic P, Sendova V, Kelley S (2001) Earthquake protection of Byzantine churches using seismic isolation. Macedonian – US joint research. Report IZIIS 2001

Jurukovski D, Mamucevski D, Petkovski M (1987) Report on testing of the seismic resistance of the current transformer ISF 420/525 manufactured by Energoinvest – Sarajevo. IZIIS Report 87-22

Jurukovski D, Mamucevski D (1988) Report on testing of the seismic resistance of metal enclosed high voltage equipment type MOP-A3 manufactured by Energoinvest – Sarajevo. IZIIS report 88-109

Jurukovski D, Rakicevic Z (2004a) Shaking table testing of a frame model with GERB Base Control System (BCS). IZIIS report 2004-50

Jurukovski D, Rakicevic Z (2004b) Shaking table testing of a frame model with GERB Tuned Mass Control System (TMCS). IZIIS Report 2004-51

Jurukovski D, Rakicevic Z (2005) Proof tests on a shaking table of a ventilated facade system manufactured by SONICO, Italy. IZIIS report 2005-33

Jurukovski D, Mamucevski D, Tashkov Lj, Hadzi-Tosev N, Trajkovski V (1984a) Shaking table test of a five storey steel frame structure base isolated according to GERB's Spring-Dashpot System. IZIIS report 83-242

Jurukovski D, Mamucevski D, Tashkov Lj, Hadzi-Tosev N, Trajkovski V (1984b) Shaking table tests of a five storey steel frame model fixed at the base and isolated by rubber elements. IZIIS report 84-123

Jurukovski D, Mamucevski D, Tashkov Lj, Hadzi-Tosev N, Trajkovski V (1984c) Shaking table tests of a five storey steel frame model isolated by GERB vibration isolation elements. IZIIS report 84-128

Jurukovski D, Mamucevski D, Bojadziev M (1986a) Report on testing of seismic and dynamic stability of high voltage isolator type RVZ-420-1 manufactured by Rade Koncar – Zagreb. IZIIS report 86-141

Jurukovski D, Mamucevski D, Trajkovski V (1986b) Report on testing of seismic stability of high voltage circuit breaker type SFE-13 manufactured by Energoinvest – Sarajevo. IZIIS report 86-34

Jurukovski D, Mamucevski D, Bojadziev M, Krstevska L (1989a) Report on seismic withstand testing of high voltage isolator type RS-4202 manufactured by MINEL Belgrade. IZIIS report 89-29

Jurukovski D, Tashkov Lj, Petkovski M, Gavrilovic P (1989b) Basic and applied research study for seismic modelling of mixed reinforced concrete masonry buildings: shaking table test of reduced scale model. IZIIS report 89-66

Jurukovski D, Tashkov Lj, Petkovski M, Mamucevski D (1989c) Basic and applied research study for seismic modelling of mixed reinforced concrete masonry buildings: shaking table test of reduced scale repaired model. P IZIIS report 89-75/1

Jurukovski D, Tashkov Lj, Petkovski M, Krstevska L (1991a) Basic and applied research study for seismic modelling of mixed reinforced concrete masonry buildings: shaking table test of reduced scale model strengthened by Reinforced Concrete Central Core. IZIIS report 91-92

Jurukovski D, Tashkov Lj, Petkovski M, Mamucevski D (1991b) Basic and applied research study for seismic modelling of mixed reinforced concrete masonry buildings: shaking table test of reduced scale model strengthened by External Reinforced Concrete Walls. IZIIS report 91-01

Jurukovski D, Tashkov Lj, Petkovski M, Krstevska L (1992) Basic and applied research study for seismic modelling of mixed reinforced concrete masonry buildings – shaking table test of reduced scale model strengthened by Reinforced Concrete Central Core after repairing by Cement Mortar Injection. IZIIS report 92-27

Jurukovski D, Tashkov Lj, Bojadziev M, Garevski M, Mamucevski D (1993) Shaking table test of 1/30 scale model of Palasport in Bologna: model with wooden roof-model 1. IZIIS report 93-48

Jurukovski D, Tashkov Lj, Bojadziev M, Mamucevski D (1994) Shaking table test of 1/30 scale model of Palasport in Bologna: model with continuous shell roof – model 2. IZIIS report 94-17

Jurukovski D, Gavrilovic P, Tashkov Lj, Krstevska L, Rakicevic Z, Petrusevska R (1996a) Experimental investigation of seismic stability of prefabricated facade system with thin stone claddings. IZIIS report 96-76

Jurukovski D, Tashkov Lj, Bojadziev M (1996b) Dynamic testing of the Alma-Ata Bank Model in scale 1/11 on shaking table. IZIIS report 96-02

Krstevska L, Tashkov Lj, Gramatikov K (2007) FP6 project PROHITECH – shaking table testing of Mustafa-Pasa Mosque model. Final report. IZIIS March 2007

Mamucevski D, Bojadziev M (1992) Vibration tests of bimetallic relay TRM-12 manufactured by Rade Koncar – Skopje. IZIIS report 92-18

Mamucevski D, Jurukovski D (1987a) Seismic and dynamic testing of valves for nuclear power plants manufactured by Energoinvest – Sarajevo. IZIIS report 87-98

Mamucevski D, Jurukovski D (1987b) Experimental testing of elements for elastic foundation of facilities and devices produced by Jugoturbina – Karlovac. IZIIS report 87-92

Mamucevski D, Jurukovski D (1988) Experimental seismic and vibration tests of controlling drives produced by Energoinvest-Sarajevo. IZIIS report 88-104

Mamucevski D, Jurukovski D (1989) Experimental seismic investigations performed for electromechanical drives manufactured by Energoinvest – Sarajevo: experimental results for the EMPN 250-10-C-0/KK-0-G-1-9-0-0 drive. IZIIS report 89-44

Mamucevski D, Krstevska L (1988a) Seismic and dynamic testing of needle valves for nuclear power plants produced by 'Prva Iskra' – Baric. IZIIS report 88-44

Mamucevski D, Jurukovski D, Micajkov S (1985a) Dynamic experimental testing of vehicles produced by Rakovica in Belgrade using a biaxial shaking table. IZIIS report 85-98

Mamucevski D, Jurukovski D, Percinkov S (1985b) Seismic and dynamic testing of valves for nuclear power plants produced by Energoinvest – Sarajevo. IZIIS report 85-36

Mamucevski D, Bojadziev M, Jurukovski D (1986a) Experimental seismic and dynamic testing of the high-voltage separating contactor type RVZ-420-1 manufactured by Rade Koncar – Zagreb. IZIIS report 86-144

Mamucevski D, Jurukovski D, Percinkov S (1986b) Experimental determination of the dynamic characteristics of hydraulic dampers manufactured by GOSA. IZIIS report 86-105

Mamucevski D, Petkovski M, Jurukovski D (1986c) Experimental seismic and dynamic tests of a low voltage control panel type VM16 VMF6 produced by 'Rade Koncar' Zagreb. IZIIS report 86-96

Mamucevski D, Jurukovski D, Petkovski M (1988b) Experimental seismic and dynamic testing of the current transformer A64-420 produced by Rade Koncar-Zagreb. IZIIS report 86-49

Mamucevski D, Stamatovska S, Petkovski M, Jurukovski D (1988c) Experimental investigations of the pressure and differential pressure transmitter P 151 manufactured by ATM – Zagreb. IZIIS report 88-36

3 Structural and Behaviour Constraints of Shaking Table Experiments

Mamucevski D, Bojadziev M, Percinkov S (1990) Experimental investigation of the vibration effects on the distributing panel OR-1 manufactured by Rade Koncar – Skopje. IZIIS report 90-27

Proceedings of 14th European conference on Earthquake Engineering, 2010, Ohrid, Macedonia

Proceedings of 14th world conference on Earthquake Engineering, 2008, Beijing, China

Proceedings of 5th world conference on structural control and monitoring, 2010, Tokyo, Japan

Rakicevic Z (2007) Proof tests on a shaking table of a new brick laying system patented by PREXO SEALING, Norway. IZIIS report 2007-26

Rakicevic Z (2008) Seismic qualification of current transformer IST 123 kV manufactured by ENERGOINVEST-RAOP, Sarajevo by Time History Shaking Table Testing. IZIIS report 2008-58

Rakicevic Z, Jurukovski D (2001) Optimum design of passive controlled steel frame structures. IZIIS report 2001-59

Rakicevic Z, Matevski V (2009) Seismic qualification by Time History Shaking Table Testing of oil-immersed current transformer AGU-525, capacitor voltage transformer VCU–765, combined instrument transformer VAU–420 and voltage transformer VPU–525, manufactured by KONCAR instrument transformers Inc. IZIIS report 2009-42, 43, 59 and 60

Rakicevic Z, Zlateska A, Golubovski R, Matevski V (2009a) Shaking table testing of Wienerberger two storey masonry models without and with antiseismic devices. IZIIS report 2009-41

Rakicevic Z, Zlateska A, Jurukovski D (2009b) Shaking table effectiveness testing of GERB PDD (Prestressed Damping Device) Control System. IZIIS Report 2009-40

Ristic D, Tashkov Lj, Micov V, Bojadziev M, Popovski M (1992) Investigation of the dynamic behaviour of a bridge model with built-in neoprene bearings under actual earthquake effects simulated by a seismic shaking table. IZIIS report 92-44

Structural Control and Health Monitoring (2008-2010) vols 15–17. Wiley

Tashkov Lj, Krstevska L (2007) Shaking table test of a segment of brick-masonry wall at Beauharnois Powerhouse. IZIIS report 2007-51/1

Tashkov Lj, Krstevska L (2009) Seismic qualification of Lindner raised floors type NORTEC G30 ST Germany. IZIIS report 2009-20

Tashkov Lj, Petkovski M (1986) Experimental determination of the dynamic characteristics of the high voltage circuit breaker K3AT3 manufactured by Rade Koncar – Zagreb. IZIIS report 86-112

Tashkov Lj, Bojadziev M, Krstevska L (1988) Testing of damping efficiency of the damper type 297.04.000 for conductor 490/65 by using a two-component shaking table. IZIIS report 88-64

Tashkov Lj, Krstevska L, Gavrilovic P (1992a) Study for seismic strengthening conservation and restoration of churches dating from the Byzantine Period (9th–14th century) in Macedonia vol 9 shaking table test of models. IZIIS report 92-71/9

Tashkov Lj, Petkovski M, Krstevska L, Mamucevski D (1992b) Research study for evaluation of seismic resistance of the 105-storey Ryugyong Hotel in Pyongyang DPR Korea vol 4: shaking table test of 1/40 scale model of the building. IZIIS report 92-06

Tashkov Lj, Jurukovski D, Mamucevski D, Rakicevic Z, Stamatovska S (1994) Shaking table test of key lock systems products of Heerum – Germany under seismic and resonant conditions. IZIIS report 94-51/2

Tashkov Lj, Tomovski I, Rakicevic Z (2000) Seismic qualification on high voltage circuit breaker type SFL 13 three pole drive manufactured by ELOP-IRCE, Sarajevo. IZIIS report 2000-23

Tashkov Lj, Kokalanov G, Krstevska L, Aleksovska M (2008) FP6 project PROHITECH – seismic upgrading of Byzantine church by reversible innovative base isolation ALSC floating-sliding system. WP12 final report, IZIIS Aug 2008

Tashkov Lj, Krstevska L, Gramatikov K (2008a) FP6 project PROHITECH – shaking table testing of Fossanova model. Final report, IZIIS report March 2008

Tashkov Lj, Krstevska L, Gramatikov K (2008b) FP6 project PROHITECH – shaking table testing of Saint Nicholas model. Final report, IZIIS report July 2008

Chapter 4
Eucentre TREES Lab: Laboratory for Training and Research in Earthquake Engineering and Seismology

Simone Peloso, Alberto Pavese, and Chiara Casarotti

Abstract Italian awareness about the seismic vulnerability of its building stock dramatically increased after two earthquakes hit the nation: Umbria-Marche earthquake (1997) and Molise earthquake (2002). These two seismic events caused important losses in terms of human life as well as to the economic and artistic wealth. From here the decision to take important actions aiming to the reduction of the national seismic risk: creation of a new seismic zonation of Italy; adoption of a new seismic code; foundation of a research center on earthquake engineering. The paper reviews the development of Eucentre Foundation (European Centre for Training and Research in Earthquake Engineering) and its experimental laboratory TREES Lab (Laboratory for Training and Research in Earthquake Engineering and Seismology). A brief description of the experimental facilities at TREES Lab is reported, describing the principal characteristics of Shaking Table, Bearing Tester System, Reaction Wall-Strong Floor Structure and Mobile Unit. Furthermore, an introduction to some past and current research projects is given to explain what can be done exploiting the capabilities of the TREES Lab facility. Finally, the experimental activities within the SERIES (Seismic Engineering Research Infrastructures for European Synergies) project are described.

S. Peloso (✉) • C. Casarotti
Eucentre – European Centre for Training and Research in Earthquake Engineering,
Via Ferrata 1, 27100 Pavia, Italy
e-mail: simone.peloso@eucentre.it; chiara.casarotti@eucentre.it

A. Pavese
Eucentre – European Centre for Training and Research in Earthquake Engineering,
Via Ferrata 1, 27100 Pavia, Italy

Department of Structural Mechanics, University of Pavia, Via Ferrata 1, 27100 Pavia, Italy
e-mail: alberto.pavese@eucentre.it; a.pavese@unipv.it

M.N. Fardis and Z.T. Rakicevic (eds.), *Role of Seismic Testing Facilities in Performance-Based Earthquake Engineering: SERIES Workshop*, Geotechnical, Geological and Earthquake Engineering 22, DOI 10.1007/978-94-007-1977-4_4,
© Springer Science+Business Media B.V. 2012

4.1 Introduction

Italian public awareness about the seismic risks increased during the last decades as a consequence of a number of events across the nation. Seismic events such as the Umbria-Marche earthquake (1997) or the Molise earthquake (2002) strongly hit Italy pointing out the vulnerability of the building stock. Those two events strongly affected the nation under several points of view: causing important losses in terms of human life and serious damages to the artistic wealth and to the socio-economic structure of the hit areas (Casarotti et al. 2009b).

After those events the design code was revised, with particular emphasis on the seismic design and the seismic zonation of Italy. At the same time, it was decided to create an Italian research center which would focus on the seismic risk reduction: Eucentre, European Centre for Training and Research in Earthquake Engineering. Eucentre is a non-profit foundation located in Pavia, a city in the north of Italy. It was launched by the Dipartimento della Protezione Civile (DPC – Italian Civil Protection Department), the Istituto Nazionale di Geofisica e Vulcanologia (INGV – National Institute for Geophysics and Vulcanology), the Università degli Studi di Pavia (University of Pavia) and the Istituto Universitario di Studi Superiori di Pavia (IUSS – University Institute for Advanced Study of Pavia), with the aim of promoting, sustaining and overseeing training and research in the field of the reduction of seismic risk.

Essential part of Eucentre is the TREES Lab: Laboratory for Training and Research in Earthquake Engineering and Seismology. It was created pursuing the idea of building an experimental laboratory able to compete with the major US and Japanese laboratories. Today, TREES Lab is one of the most powerful laboratories within the European seismic research framework.

The Eucentre experimental facility includes a unidirectional Shake Table (ST), a Bearing Tester System (BTS) and Strong Walls-Strong Floor system. Since Eucentre was founded, its staff dealt with a number of projects including experimental activities: among them the risk assessment of existing reinforced concrete buildings and bridge piers and the development of retrofit strategies based on advanced materials. At the moment, several researches involve the use of the TREES Lab facilities, such as the characterization and development of bearing and isolation systems and the experimental assessment of the seismic response of modular panel structures.

Last but not least, Eucentre is partner of the SERIES (Seismic Engineering Research Infrastructures for European Synergies) project. As it will be described in the following, three research projects have already been selected for the execution of tests on the shaking table. These three projects respectively aim to investigate the seismic behavior of textile retrofitted unreinforced masonry structures, spray-concrete panel structures and mixed reinforced concrete – unreinforced masonry buildings.

4.2 Eucentre Foundation

The European Centre for Training and Research in Earthquake Engineering (Eucentre) was established in Pavia, Italy, in June 2003 by the Dipartimento della Protezione Civile (DPC), the Istituto Nazionale di Geofisica e Vulcanologia (INGV), the Università degli Studi di Pavia and the Istituto Universitario di Studi Superiori di Pavia (IUSS). Nowadays Eucentre is a non-profit foundation, the governing body of which still reflects the original founders: its main objective is promoting, sustaining and overseeing training and research in the field of the reduction of seismic risk. A view of the Eucentre complex is shown in Fig. 4.1.

Eucentre works for the development of applied research in the field of seismic engineering, oriented towards reaching concrete goals of evaluation and reduction of seismic vulnerability and risk. The staff of Eucentre is organized in several groups collaborating on common objectives but with different specializations. In more detail, Eucentre is made up of the following research areas: reinforced concrete structures, masonry structures and monuments, precast structures, experimental methods and techniques for the seismic vulnerability reduction, computational mechanics and advanced materials, structural analysis, geotechnical engineering and engineering seismology, telecommunications and remote sensing, technological innovation, design methods and seismic risk. It is worth mentioning that about 200 people, graduate students and researchers, work at Eucentre. Besides, Eucentre is supported by several practitioners and companies.

As part of an initiative to support the world of engineering practice, Eucentre actively collaborates with practitioners, with the aim of supporting them by sharing information through educational and cooperative activities, facilitating the exchange of information among members and others, fostering a sense of shared commitment

Fig. 4.1 European Centre for Training and Research on earthquake engineering

among the diverse communities dedicated to earthquake risk management, promoting research, speaking with a common voice to public forums and legislative bodies on behalf of the diverse risk management community. An important activity done at Eucentre is the training of engineers and technicians with scientific and professional capabilities in the field of seismic engineering. Short-courses oriented to professionals are frequently organized focusing on seismology, geology, geotechnics, behavior of materials and structures, design of new structures, evaluation and retrofit of existing structures, this last even in emergency situations. Retraining of professional engineers has a crucial role for the reduction of the seismic vulnerability of the nation: practitioners have the important duty of spreading the results of the research in the everyday practice. Nevertheless, at Eucentre, research is considered at least as important as practice toward the final goal of reduction of seismic risk. For this reason, the Eucentre education actions are not just oriented to professionals. Eucentre staff is involved in education at different levels from under-graduate (mainly at the University of Pavia in collaboration with the Structural Mechanics Department of the Faculty of Engineering) to graduate and post-doctoral. The biggest part of the educational activities is related to the graduate courses. Eucentre is strictly involved in the MSc program for the reduction of seismic risk and PhD program on earthquake engineering at the school for reduction of seismic risk (ROSE School – IUSS).

Moreover, Eucentre acts establishing temporary and permanent cooperation agreements with companies that operate in the different areas related to earthquake engineering. Main objective of these partnerships is to improve the interaction between research, practice and industrial production. For this reason, Eucentre collaborates with several companies and research centers carrying out scientific and technical consultancy at a national and international level.

Even more important is the collaboration with the DPC for the development and definition of specific lines of public action, guidelines and regulatory documents. The main advantage concerning these activities is the possibility of gathering together the different experiences of Eucentre's partners: the Civil Protection Department accustomed to face emergency management problems, the practitioners dealing with the technological related design problems, the academics and researchers often having a deeper knowledge of the state of the art at an international level.

4.3   Eucentre TREES Lab

Experimentation has always had a fundamental role for scientific research, being essential for the validation of numerical models and theories as well as for the performance definition of systems and components. In the structural and seismic engineering framework, experimentation is recently assuming a key role due to the development of new technologies. Improving both test execution and data acquisition and processing, it is possible to study the behavior of structures and materials in conditions very close to the real applications.

Fig. 4.2 General view of the Eucentre TREES Lab

TREES Lab (Laboratory for Training and Research in Earthquake Engineering and Seismology), the Eucentre experimental facility (see Fig. 4.2), has been specifically designed according to the most innovative technologies (Calvi et al. 2005). Its high performance equipment allows conducting dynamic, pseudo-dynamic and pseudo-static tests on full-scale prototypes, thus reducing the uncertainties of interpretation and correlation with actual structural conditions. Moreover, the Mobile Unit of Eucentre TREES Lab allows the execution of on-site tests.

TREES Lab features four experimental facilities: the high performance uniaxial Shaking Table (ST); the bi-axial Bearing Tester System (BTS) for testing of full-scale bearing and isolation devices, with high dynamic and force capabilities; the Strong Floor-Reaction Wall System for full-scale pseudo-static and pseudo-dynamic tests; the Mobile Unit (MU), with the most advanced tools for in situ non-destructive tests and for fast seismic vulnerability analyses. It is worth mentioning that besides being a mobile lab, the MU also works as a communication center able to keep a real time contact with the remote control room developed at Eucentre according to last generation technologies.

TREES Lab staff and facilities are involved in a number of national and European research and development projects, both in public and private partnership. In particular, Eucentre is Official Consultant of the Italian Department of Civil Protection and works in collaboration with the Official Laboratory of Materials and Structures of the Structural Mechanics Department of the University of Pavia.

In the following paragraph a brief review of the main characteristics of the four facilities of Eucentre TREES Lab is reported.

4.4 Shaking Table

One the most interesting ways to have information about the seismic behavior of structures is to let them undergo a seismic excitation while recording the displacements, deformations and accelerations at different points across the structure. This way of testing structures is currently the only possibility to account for seismic-induced rate-dependent effects, clearly lost when using pseudo-static or pseudo-dynamic testing strategies. The principal objective of the shaking table is the simulation of the seismic effects on large and full-scale structures or structural components.

When building a shake table, the first choice to be made is about the number of degrees-of-freedom of the table itself. Since the Eucentre TREES Lab table had to fit within the European research framework, it was chosen to create a very powerful uniaxial shaking table in the attempt to complete the European panorama. It has to be understood that having a limitation on the power-supply required to move the table (i.e. the capacity of the hydraulic system), a balance must be sought between the number of degrees-of-freedom of the available movements and the maximum payload and acceleration of the table. At Eucentre, it was chosen to build a uniaxial shake table (1 degree-of-freedom) able to displace a very high payload (the maximum is about 140 ton) with a high acceleration capacity, the peak acceleration is about 6 g. Maximum payload and acceleration are clearly counteracting each other, hence the maximum acceleration can be reached with the bare table. Increasing the mass of the specimen on the table, the peak acceleration decreases: nevertheless the TREES Lab Shaking Table is able to reach a maximum of about 2 g with a payload equal to 70 ton. Fortunately, natural earthquakes do not have this kind of accelerations, however when testing a substructure, potentially located at a certain height from the ground, the dynamic amplification given by the structure has to be accounted for, hence it is important to be able to reach such high accelerations. Table 4.1 shows some of the main characteristics of the Eucentre TREES Lab Shaking Table. Further details can be found on (Airouche et al. 2008).

Table 4.1 Main characteristics of the shaking table

Platen dimensions	5.6 m × 7.0 m
Peak displacement	±580 mm
Peak velocity	±2.2 m/s
Peak acceleration (bare table)	±6.0 g
Peak acceleration with rigid payload (70 ton)	±1.8 g
Flow rate	11,000 l/min
Maximum dynamic force	±1,700 kN
Maximum static force	±2,000 kN
Maximum rigid payload	140 ton
Maximum overturning moment	±4,000 kN m
First frequency of vibration of the table	84 Hz
Dissipation of the system	350 N
Control software	MTS adaptive

Fig. 4.3 Shaking table and servo-hydraulic actuator at Eucentre TREES Lab

Figure 4.3 is a picture of the servo-hydraulic actuator used for the movement of the Shaking Table.

4.5 Bearing Tester System

The Bearing Tester System (BTS) is a high performance multi-axial system for dynamic testing of full-scale bearing and isolation devices (see Fig. 4.4).

The BTS is one of the few large bearing testing machines in the world and the only one in Europe with dynamic capabilities. Exploiting the customized design of the actuation system and the control software (in cooperation with MTS Systems Corporation) the BTS can carry out biaxial tests on real scale bearings and seismic isolation devices in static and dynamic conditions. The combined control of seven servo-hydraulic actuators allows managing displacements and rotations corresponding to 5 degrees-of-freedom, in particular: longitudinal horizontal and vertical displacements and rotation about the vertical, longitudinal and transversal axes. The main performances of the BTS are reported in Table 4.2.

Since its construction the BTS has been used for several purposes including industrial needs and research on new types of materials and isolators. Additionally, qualification and acceptance tests following the main world testing protocols, as required by the seismic design codes, are usual activities performed using the TREES Lab BTS.

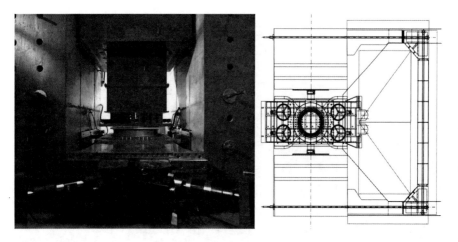

Fig. 4.4 Bearing tester system at Eucentre TREES Lab (front view and technical drawings)

Table 4.2 Main characteristics of the Bearing Tester System

Platen dimension	1,700 mm × 1,990 mm
Table mass	22.0 ton
Degrees of freedom	5 [longitudinal, vertical, roll, pitch, yaw]
Peak displacement	Horizontal: ±600 mm, vertical: ±75 mm
Peak velocity	Horizontal: 2,200 mm/s, vertical:±250 mm/s
Peak acceleration	Horizontal: ±2,000 mm/s^2
Maximum force	Horizontal: 2,000 kN, vertical: 40,000 kN ± 10,000 kN
Maximum overturning moment	20,000 kN m
Operational frequency range	0–20 Hz
Flow rate	27,000 l/min

4.6 Reaction Structure

One of the most common ways to assess the seismic behavior of structures and structural elements is the pseudo-static technique. This consists in a slow application of forces (or displacements) to the structure to be tested, typically imposing series of cycles at increasing levels of deformation. Through the application of such a loading history, it is possible to study the cyclic structural behavior, thoroughly investigating the stiffness and strength modifications due to the damage progression. Furthermore, the hysteretic behavior gives important information about the energy dissipation capacity of the tested structure.

An alternative testing procedure is the pseudo-dynamic technique in which the large- or real-scale structures are subjected to loading histories corresponding to the inertia forces caused by a seismic excitation. Although this kind of test is more complex on the computational point of view, the key point remains the application of forces or displacement at a certain location across the structure.

4 Eucentre TREES Lab

Fig. 4.5 Reaction structure at Eucentre TREES Lab

Essential for the realization of such type of tests is a reaction-structure able to resist the forces which are necessary to deform and seriously damage the tested prototype. Very important is that the reaction-structure does not experience significant deformations during the test. The reaction system at Eucentre TREES Lab (see Fig. 4.5) is composed of two strong-walls and a strong-floor: each of them has been built using precast concrete blocks post-tensioned together. These blocks have a cubic shape with 2.4 m side and an inner cavity of 1.2 m side, their weight is 259 kN. The two strong-walls are 12 m in height, hence enough to test a three-story real-scale structure, while their lengths equal 9.6 and 14.4 m. The strong-floor, extending below the walls to connect them in a unique structure is 12 m by 16.8 m with an open space of 9.6 m by 14.4 m to host test specimens. The walls are arranged in an L-shape configuration, i.e. they are disposed on two adjacent sides of the strong-floor, to allow for 3D pseudo-static or pseudo-dynamic testing.

Besides the reaction structure, to be able to perform pseudo-static and pseudo-dynamic tests, it is also crucial to have a number of actuators for the application of forces and displacements to the tested structures. TREES Lab equipment includes several servo-hydraulic actuators to be used for this kind of tests; their principal characteristics are reported in Table 4.3

It is worth mentioning that the facility is also available to external customers or research partners for performing demonstration and qualification tests on large-scale prototypes as well as for validation of innovative constructions or retrofitting

Table 4.3 Main characteristics of the actuators for pseudo-static and pseudo-dynamic testing

Model	Number	Maximum force [kN]	Maximum displacement [mm]
MTS 243.40T	3	+291; −496	±381
MTS 244.41	1	+445; −649	±254
MTS 243.90T	1	+2,000; −2,669	±254
MTS 243.45T	2	+445; −649	±508
MTS 243.4	2	+291; −496	±508

techniques. This offers a major opportunity to the European construction industry to enhance its competitive position in world-wide markets, especially in countries with significant seismic risk.

4.7 Mobile Unit

The Mobile Unit is an experimental and numerical laboratory and a communication centre. Designed for intervention during a post-earthquake emergency, the Mobile Unit is an effective tool for on-site structural assessment, structural evaluation consultancy and quick response intervention. It consists of an equipped vehicle able to transport instrumentation and digital acquisition system in order to perform on site experimental investigation (see Fig. 4.6). Initially designed to be used during post-earthquake emergency, the Mobile Unit enables a fast and comprehensive data collection, the storage of all the acquired data and their transmission through satellite internet connection.

Current experimental assessment techniques and numerical methods are implemented to be carried out with the Mobile Unit for fast and reliable assessment of buildings, bridges and monuments. The required apparatus for experimental testing has been implemented on board of the vehicle. In more detail, the experimental equipment of the Mobile Unit includes: thermo-camera, georadar, laser scanner, pachometer, sclerometer, sonic and ultrasonic device, accelerometers, geophones, inclinometers, flat-jack test equipment. Some of these instruments clearly need a data acquisition system, for this reason the Mobile Unit has been equipped with a 16 bit data acquisition system, based on National Instrument technology, able to record 500 ksample and support up to 200 channels. Some of these channels are for synchronized sampling, hence enabling structural identification tests.

One of the most advanced features implemented on the Mobile Unit is the database and the system for the management of the data flow, either numerical or multimedia. The Eucentre database structure has been developed in order to store all the data acquired from experimental tests both in lab and outdoors. To perform as data collecting unit, independently of the mutable connectivity conditions, the Mobile Unit has been provided with an internal database, local partial mirror of the main database at Eucentre.

Fig. 4.6 Eucentre mobile unit

Besides, the Mobile Unit is not just a laboratory but also works as communication centre. Connectivity toward third parties is quite robust allowing connections via Wi-Fi, 3G and satellite. In particular the wireless connection is designed to allow fast data transfer between the assessor or experimental teams and the Mobile Unit database. On the other side, 3G and satellite connections can be used to link the Mobile Unit with the central unit at Eucentre or with anybody able to connect to the videoconferencing system installed on the Mobile Unit. For further details about the Mobile Unit and its applications on field, see (Casarotti et al. 2009a).

4.8 Experimental Activities at Eucentre TREES Lab

Exploiting the experimental infrastructures described above, a number of research projects took place at Eucentre since its foundation. In the following a brief description of some concluded research projects will be reported as an example of the work done as research, of the activity in support to the Italian Civil Protection Department (DPC) and of the collaboration with the industries.

The very first large project involving the use of the Shaking Table was the "Experimental-Numerical verification of OPCM 3274 20/3/03 guidelines for the assessment of existing RC structures vulnerability". Founded by the DPC, this research work aimed to verify the guidelines included in the OPCM 3274 (Ordinanza del Presidente del Consiglio dei Ministri n°3274), the first step toward the new Italian seismic code for construction. For the verification of such guidelines applied to a typical irregular building of the 1950s–1960s, a 1:2 scale reinforced concrete (RC) building has been designed, for gravity loads only, in compliance with the old pre-seismic code and adopting technological solutions of the past such as smooth steel bars. In Fig. 4.2, it is possible to see the building on the Shaking Table at the end of the testing campaign. The building was tested on the Shaking Table to assess its behavior against the performances prevision of the current seismic codes such as NTC and Eurocode 8 (Pavese and Lanese 2010). Particular attention was given to the influence on the seismic response of the beam-column joints behavior and the frame-panel interaction. Furthermore, within the research project "Use of advanced materials for strengthening and repair of RC structures in seismic area" funded by Fondazione Cariplo, the tested building was then repaired and retrofitted using Fiber Reinforced Polymers (FRP). This allowed to verify the possibility of intervention on existing RC structures and to evaluate the effectiveness of the retrofit technique based on the application of FRP.

The Italian DPC is one of the founders of Eucentre, which explains their close collaboration. Eucentre is not only doing research for the DPC, but it is also giving support both on field and with laboratory activities. It is worth mentioning, as an example of the work done at Eucentre TREES Lab in support of the Italian DPC, the testing campaign on Friction Pendulum System (FPS) isolators. After the 2009 Abruzzi Earthquake an important reconstruction took place around the city of L'Aquila. A total of 185 isolated residential buildings have been built with an original and efficient solution adopting isolated RC slabs on the top of which the buildings have been realized (Bertolaso et al. 2010). This tremendous effort involved the installation of more than 7,000 FPS. According to the Italian seismic code, 5% of those isolators needed to be tested, hence more than 350 devices had to undergo seismic testing for acceptance (factory production control). Eucentre TREES Lab worked incessantly for this testing campaign allowing the reconstruction not to slow down. Although this campaign did not have a scientific aim, the results were not only useful for the DPC: as a matter of fact this testing campaign was probably the largest on this type of isolators resulting in a very large database that has been used for scientific research.

Besides the activities in cooperation with the DPC, Eucentre collaborates with a number of private companies for the development and testing of innovative construction solutions and for the extension of the application of existing technologies to construction in seismic areas. An example of this latter activity is the experimental and numerical investigation on the seismic behavior of walls realized using hollow blocks filled with concrete or using sandwich panels (panels made of polystyrene and spray-concrete on the two sides). Since these techniques are not currently included in the Italian seismic code, it was necessary to perform a complete investigation. Exploiting the TREES Lab Reaction-Structure, the campaign started from the quasi-static testing of single panels (see Fig. 4.7). During this phase several panels have been tested varying the height to length ratio, the acting axial load; panels were tested with and without openings. Then the experimental campaign moved to the testing of the connections between panels and between panel and slab. The tests allowed assessing the cyclic behavior of those structural elements giving information on the energy dissipation characteristics. Besides, cracking patterns were investigated understanding criticalities and typical collapse modes

Fig. 4.7 Cyclic testing of hollow blocks panel (*left*) and FEM modelling of building (*right*)

characterizing these kinds of panels (Bournas et al. 2010). The second step was the numerical study: a thorough analysis of the experimental data and several non-linear and linear analyses have been used to define the q-factor to be used in the design phase of buildings. Additionally, a simplified linear analysis method and the corresponding modeling solution (see Fig. 4.7) have been developed and implemented creating a useful tool that could be used by professional engineers (Peloso et al. 2010). This work allowed the preparation of a document for the companies: guide-line for the design of seismic resistant buildings realized with these construction techniques. This was also the first step toward the final goal of obtaining technical approval.

Finally, it is worth mentioning that the current project of collaboration with the European Union is for the Development of Rapid Highly-specialized Operative Units for Structural Evaluation (DRHOUSE), in which Eucentre is partner of the Italian Civil Protection Department and the Italian Department of Fire Brigades. The project is in line with the perspective of integrating the shortage of disaster-response capacity at the European level in the field of post-earthquake structural evaluation. The main objective is the development and implementation of a new Civil Protection Module able to ensure a rapid and effective response in the field of the post-earthquake damage and safety assessment targeted to enhance the European Rapid Response Capability (ERRC) within the European Community Mechanism for civil protection. In order to ensure the modularity of the action and the interoperability among modules, the capability has been conceived as composed by three sub-modules, which can either work together or independently:

- Basic Seismic Assessment module (BSA) managed by the DPC aiming at the usability assessment of common buildings;
- Advanced Seismic Assessment module (ASA) managed by Eucentre for the assessment of strategic and/or complex structures, with dedicated advanced instrumentation and procedures
- Short-term countermeasures module (STC) managed by the Italian Department of Fire Brigades for quick intervention on damaged structures and infrastructures.

Such an intervention structure is also able to ensure in the future the possibility of a modular implementation by the member states according to the available resources. From the perspective of implementing an effective post-earthquake assessment system, multiple levels of expertise and knowledge are required and need to be coordinated. The implementation of such module will be developed in the framework of the European perspective on improving the effectiveness of emergency response by enhancing the preparedness and awareness of civil protection professionals and volunteers.

4.9 Experimental Activities Within SERIES Project

As the title of the project explains "Seismic Engineering Research Infrastructures for European Synergies" (SERIES) is an EU funded project that tries to create synergies between the various parts of Europe for the improvement of research on earthquake engineering. Among a number of actions aiming at this final goal, a very interesting peculiarity of this project is the trans-national access to a number of European experimental facilities. Among these facilities is the Eucentre TREES Lab. External users have the possibility to submit an application to a User Selection Panel (USP) for the realization of experimental activities with essentially no cost for the accepted users.

Within the SERIES project, three experimental tests on Shaking Table have been selected for execution at Eucentre TREES Lab. These projects aim respectively to (*i*) evaluate the effectiveness of POLYfunctional technical textiles for the retrofit of MAsonry STructures (POLYMAST); (*ii*) study the SEismic behavior of structural SYstems composed of cast in situ COncrete WAlls (SE.SY.CO.WA); (*iii*) investigate the seismic behavior of mixed reinforced COncrete–unreinforced MAsonry WALL Structures (CoMa-WallS).

In more detail, the first project aimed to evaluate the effectiveness of a new material for the structural retrofit and strengthening developed in the framework of the FP7 project "Polyfunctional Technical Textiles against Natural Hazards" (POLYTECT). This material is a high resistance textile fabric, made of multidirectional glass fibers, including a number of sensors for the monitoring of structures. This project included the seismic testing of an unreinforced masonry building up to the achievement of its

Fig. 4.8 POLYMAST project: ST test of the retrofitted structure

ultimate limit state, the subsequent repair and retrofit of the tested structure and the retesting of the strengthened building. As shown in the following Fig. 4.8, the tested specimen was a two-story real scale structure. On the right of the figure it is possible to see the high-resolution cameras used for the optical monitoring of the structure during the test. The final report relative to this experimental campaign is currently under preparation.

Next to come is the SE.SY.CO.WA project, the specimen of which is currently under construction. This study involves the testing on the ST of a three-story structure made of sandwich panels. These panels are built spraying concrete on a small-diameter steel net attached on the two sides of a polystyrene panel. The scope of this research is the validation of the theoretically and partially-experimentally anticipated good seismic behavior of this structural system. After the testing of single panels and connection wall-to-wall and wall-to-slab, the ST testing will allow the investigation of the seismic effects, including torsion, on a building realized with this construction technique. The test will give important information both about the global and local behavior highlighting possible structural lacks at the connection between different elements, at the anchorages to the foundation and characterizing the structural details across the whole structure.

The third project will investigate the seismic behavior of mixed reinforced concrete (RC) and unreinforced masonry (URM) wall structures. Although such structures are very common in practice, they have not been tested in the past. In mixed wall structures, the system stiffness and overall structural behavior strongly depend on both types of structural elements since the stiffnesses of the walls are comparable. The test specimen will be a four-story building built at half scale. The structure will be representative of a modern, engineered residential building with RC slabs. It has two RC and six URM walls; the first two and four of the latter are parallel to the direction of motion of the ST. The test results will allow addressing open issues in European standardization concerning the seismic design of mixed structural systems. Among the open issues are the choice of force-reduction factors q, assumptions concerning the distribution of base shear forces between different types of structural elements, the importance and quantification of compatibility forces due to different deformation characteristics of structural elements.

4.10 Conclusion

The paper briefly reviews the development of Eucentre Foundation (European Centre for Training and Research in Earthquake Engineering) and Eucentre TREES Lab (Laboratory for Training and Research in Earthquake Engineering and Seismology). Founded after two major seismic events that strongly hit Italy in 1997 and 2002, Eucentre quickly developed by collaborating closely with the Italian Department of Civil Protection (DPC), one of the original founders. In time, a number of collaborations with private companies and professional engineers have been used to sustain the activities of the Foundation trying to increase their effectiveness toward the goal of reduction of seismic risk. Besides, Eucentre and TREES Lab participate in several research projects funded by the EU or by Italian private foundations (such as CARIPLO Foundation).

Currently Eucentre gets and gives support to the Italian Department of Civil Protection, private companies and practitioners. Linking these three categories, Eucentre does research using public and private funding. The findings of the researches are then used to help the funders. Besides, the research results are spread to all the categories since, although at different levels, each of them must work toward the final common goal of reducing seismic risk: DPC, developing national preparedness and prevention actions; private companies, improving their handcrafts and construction technologies and products; practitioners, spreading throughout the territory the knowledge derived by the research and supporting, when needed, the DPC actions, as was the case in 2009 during the last post-earthquake emergency.

A brief review of the major research projects has been presented to better explain the use of the facilities at TREES Lab: Shake Table and Bearing Tester System, Reaction Walls-Strong Floor structure and Mobile Unit.

Acknowledgements The authors acknowledge the precious support of the Italian Department of Civil Protection that allowed the birth and growth of the Eucentre Foundation and TREES Lab. The authors thank the European Community's Environment ECHO, financial supporter of the projects which allowed the implementation and use of the Mobile Unit. The authors recognize the financial support given by CARIPLO Foundation in the past years for the project regarding the use of advanced materials for strengthening and repair of RC structures in seismic areas and the project for the development and characterization of enhanced sliding polymer composites for seismic isolation. The authors acknowledge also the financial support received from the European Community's Seventh Framework Programme [FP7/2007-2013] under grant agreement n° 227887 for the SERIES Project.

References

Airouche AH, Casarotti C, Thoen BK, Dacarro F, Pavese A (2008) Numerical modeling and experimental identification of the EUCENTRE TREES Lab shake table. In: Proceedings of the 14th world conference on earthquake engineering, Beijing

Bertolaso G, Calvi GM, Gabrielli F, Stucchi M, Manfredi G, Dolce M (2010) L'Aquila, Il progetto CASE, Complessi Antisismici Sostenibili ed Ecocompatibili. IUSS Press, Pavia

Bournas D, Pavese A, Peloso S (2010) Seismic behavior of prefabricated concrete sandwich panels. In: Proceedings of the 4th international conference on structural engineering, Cape Town

Calvi GM, Pavese A, Ceresa P, Dacarro F, Lai CG, Beltrami C (2005) Design of a large-scale dynamic and pseudo-dynamic testing facility. IUSS Press, Pavia

Casarotti C, Dacarro F, Pavese A, Peloso S (2009a) Mobile unit for fast experimental post-earthquake vulnerability assessment. In: Proceedings of the Ingegneria Sismica in Italia, Bologna

Casarotti C, Pavese A, Peloso S (2009b) Seismic response of the San Salvatore Hospital of Coppito (L'Aquila) during the 6th April 2009 earthquake. Progettazione Sismica 3:159–172, Special Abruzzo

Pavese A, Lanese I (2010) Verification of EC8-based assessment approaches applied to a building designed for gravity-loads through the use of shaking table tests. In: Fardis MN (ed) Advances in performance-based earthquake engineering. Springer, Dordrecht/Heidelberg/New York/London. ISBN ISBN 978-90-481-8745-4

Peloso S, Zanardi A, Pavese A, Lanese I (2010) Innovative construction techniques for buildings in seismic areas: structural modeling and design issues. In: Proceedings of the 14th European conference on earthquake engineering, Ohrid

Chapter 5
Cross-Facility Validation of Dynamic Centrifuge Testing

Ulas Cilingir, Stuart Haigh, Charles Heron, Gopal Madabhushi, Jean-Louis Chazelas, and Sandra Escoffier

Abstract This paper compares the results of dynamic centrifuge tests on shallow foundations conducted at two different geotechnical facilities, IFSTTAR (Institut français des sciences et technologies des transports, de l'aménagement et des réseaux, formerly LCPC), France and Cambridge University, U.K. Both facilities ran tests on a single degree of freedom model structure with its shallow foundation located on dry sand and subjected to dynamic shaking. Measurements were taken at both facilities allowing direct comparisons to be made. Fundamentally the results obtained were found to agree well via comparison of soil amplification profiles and moment-rotation cycles. However, higher frequency components agreed less favourably. This variation is thought to be due to a mismatch between the dynamic properties of the model containers. In this series of tests the higher frequency components are not of great importance and therefore the variation is insignificant, however, in future tests when earthquake signals with higher frequency components are input it may become an issue requiring further investigation.

U. Cilingir (✉) • S. Haigh • C. Heron • G. Madabhushi
Department of Engineering, University of Cambridge, High Cross, Madingley Road,
Cambridge CB3 0EL, UK
e-mail: u.cilingir@sheffield.ac.uk; skh20@cam.ac.uk; cmh78@cam.ac.uk; mspg1@cam.ac.uk

J.-L. Chazelas • S. Escoffier
Division Reconnaissance et Mecanique des Sols, IFSTTAR, Route de Bouaye, BP4129,
Bouguenais 44341, France
e-mail: jean-louis.chazelas@ifsttar.fr; sandra.escoffier@ifsttar.fr

M.N. Fardis and Z.T. Rakicevic (eds.), *Role of Seismic Testing Facilities
in Performance-Based Earthquake Engineering: SERIES Workshop*, Geotechnical,
Geological and Earthquake Engineering 22, DOI 10.1007/978-94-007-1977-4_5,
© Springer Science+Business Media B.V. 2012

5.1 Introduction

In recent years the geotechnical insight obtained from centrifuge studies has contributed towards shaping many design codes. The question however arises as to how the results would vary if the same test was carried out at different facilities and, consequently, which results would be taken to be reliable and utilised when writing codes of practice. Variations in model preparation techniques, centrifuge type, earthquake actuator, instrumentation and test methodology can affect the observed geotechnical mechanism and hence the quantitative results obtained.

The primary objective of SERIES Task JRA3.2 is to examine this problem specifically for dynamic centrifuge testing. Two centrifuge facilities at IFSTTAR and the University of Cambridge (UCAM) have performed initial tests which were designed to expose to what extent the results can vary between facilities. The specific problem being examined is the non-linear dynamic response of a single degree of freedom structure on a shallow foundation situated on a dry sand bed.

Initial centrifuge tests carried out at both facilities aimed to examine whether a simple sand column would respond similarly in both centrifuges. Models were hence prepared with no structure located on the surface. Following these tests, further models were prepared with single degree of freedom structures located on the surface, as shown in Fig. 5.1. UCAM tested two structures with bearing pressures of 50 and 100 kPa whereas IFSTTAR tested 100 and 300 kPa model structures. This allowed for soil-structure-interaction to be examined over a wide range of bearing pressures but also provided a data overlap with the 100 kPa structure to allow result comparisons to be made between the two facilities. All the model structures were tuned to have a fixed-base natural frequency of 50 Hz (1 Hz at prototype scale).

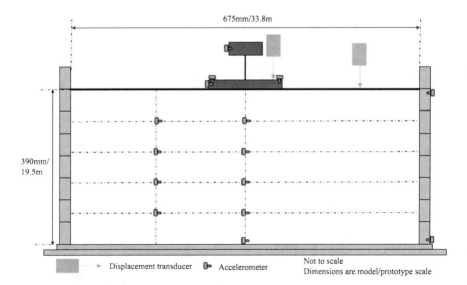

Fig. 5.1 Typical model layout

All the other details relating to the tests were shared between the two facilities, so that the models were prepared to be as close as possible to the agreed parameters.

5.2 Facility Comparison

Both facilities have large beam centrifuges. IFSTTAR has a 5.5 m radius, moveable counterweight Actidyn centrifuge with a 200 g-ton capacity. UCAM has a 10 m diameter balanced centrifuge with a 150 g-ton capacity. The equipment used to induce earthquakes at both facilities is also different; IFSTTAR have an Actidyn QS80 servo-hydraulic shaker (Chazelas et al. 2008) capable of inducing earthquakes of any form; multiple frequency and varying amplitude. UCAM have a stored angular momentum (SAM) actuator (Madabhushi et al. 1998) which is capable of inputting sinusoidal motions with a fixed frequency and fixed amplitude. Alternatively a 'sine-sweep' motion can be created which has a decreasing magnitude and decreasing frequency. Due to the limitations of the SAM actuator, UCAM conducted tests in advance of IFSTTAR which allowed the measured base input acceleration data to be passed from UCAM to IFSTTAR so that an identical trace could be demanded from their earthquake actuator. Example acceleration traces are shown in Fig. 5.2.

Flexible model containers were used in both facilities and varied predominately only in size, with the IFSTTAR box being approximately 25% larger than the UCAM box. Instrumentation at IFSTTAR involved piezoelectric accelerometers and laser displacement transducers. Similar accelerometers were utilised at UCAM, however linear-variable-differential transducers were used to measure displacement. In addition, UCAM also made use of Microelectromechanical system (MEMS) accelerometers which are smaller, lighter and cheaper than their piezoelectric counterparts and are therefore useful for measuring accelerations on the model structures. When locating instruments during the sand pluviation process, it is possible that the sensitive axis of the instruments do not get positioned perfectly

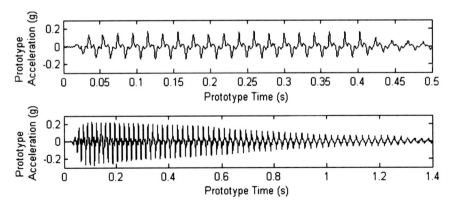

Fig. 5.2 Example input traces. Sinusoidal input and sine-sweep input

aligned with the direction of shaking. Every effort is made to minimise this alignment issue and it is not considered to be a significant source of error.

In respect to sand pluviation, UCAM have an automated sand pourer (Madabhushi et al. 2006) which ensures consistent relative density throughout the depth of the model. LCPC use a flow-controlled raining sand hopper which also ensures a uniform soil density in the model container (Ternet 1999; Garnier 2001).

5.2.1 Test Methodologies

Test procedures at the two facilities do not vary significantly. IFSTTAR take models through stabilisation stress cycles prior to testing by increasing and decreasing g-level in order to ensure that following pluviation of the sand, all local equilibriums between sand grains are stable. UCAM do not perform this cycling procedure as the process of simply loading the model onto the centrifuge will induce enough small amplitude vibrations in the model to allow sand grains in un-stable equilibrium states to rearrange.

In addition, IFSTTAR prepared a new model for every test scenario and only placed one model structure on the sand surface for each test. UCAM aimed to maximise the data collected from each prepared model by having several centrifuge flights per model, i.e. UCAM tested the free-field (no surface structure) scenario, then added two structures to the surface and tested the same model again. An investigation was carried out at UCAM to ensure that no structure-soil-structure interaction would occur as a consequence of placing two structures within one model.

It is also important to note that the two facilities operate their centrifuges at slightly different centrifugal acceleration levels. At a point two thirds down the soil column, UCAM models experience 46 g and IFSTTAR models experience 50 g. This difference will have an influence on the stress levels in the soil and hence the stiffness. Consequently this can be expected to have some impact on the results from the soil-structure interaction as observed through the moment-rotation loops shown later in this paper.

5.3 Bearing Capacity

During discussions relating to the choice of structural bearing pressure for this series of cross-validation tests it was considered useful to examine the bearing capacity of the sand through dedicated centrifuge tests, these were conducted by IFSTTAR. The notion of bearing capacity in dry sands is quite ambiguous. As shown in Fig. 5.3, the bearing pressure can be increased to around 2.5 MPa with approximately linearly increasing settlements before a decrease in stiffness is observed. However, at 2.5 MPa, Fig. 5.3 shows that half a metre of settlement has occurred which is clearly excessive and would be classed as failure. The decision

5 Cross-Facility Validation of Dynamic Centrifuge Testing

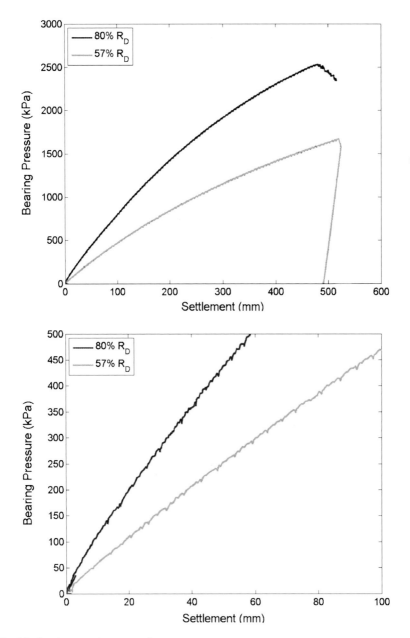

Fig. 5.3 Bearing capacity test results

as to where along this bearing pressure-settlement plot 'failure' would be considered to have occurred is normally dependent on the size of the footing, with around 1% of the footing width being a typical allowable settlement. In the results discussed later in this paper, structures with three different bearing capacities, 50, 100 and

300 kPa are examined. As shown in Fig. 5.3, the static settlements at prototype scale for each structure would be 10, 18 and 60 mm respectively. Sixty millimeters (1.5% footing width) of settlement is quite significant and is on the boundary above which 'failure' would be considered to have occurred. Therefore, testing model structures with bearing pressures from 50 to 300 kPa allows the effect of bearing pressure on the soil-structure interaction to be investigated over the entire range of likely prototype structures.

5.4 Free-Field Site Response

5.4.1 Input Earthquake Replication

As mentioned in the introductory paragraphs to this paper, UCAM performed tests prior to IFSTTAR so that the input acceleration trace UCAM had measured could be replicated. However, unlike 1-g servo-hydraulic shakers, where the input frequencies are lower and hence precise acceleration control is possible, in centrifuge based servo-hydraulic shakers it is very difficult to ensure perfect replication of the demand signals without multiple repetitions and re-tuning of the demand signal. Figure 5.4 shows two comparable input traces in acceleration-time and frequency domains. As can be observed, there are minor variations in the magnitude of the accelerations measured, however, the variation is within the error of the instrument calibration factors. The variation in measured acceleration results in the small variability in the magnitude of the FFT peaks shown in the lower plot of Fig. 5.4. Importantly however the frequency levels match precisely which is vital in these tests due to the proximity of the excitation frequency to the soil-structure systems natural frequency. It is encouraging that such replication can be conducted as it simplifies the process of comparing the soil-structure interaction results between the two facilities.

5.4.2 Free-Field Soil Column Response

Before examining how the free-field soil column response varied between the two facilities, it is informative to compare how the response varied between two separate models prepared at the same facility. IFSTTAR prepared two identical models for each of the initial free-field tests; two at 57% relative density and two at 80% relative density. Taking the magnitude of the FFT spikes at the fundamental earthquake frequency and the second harmonic and normalising these relative to the deepest placed accelerometer produces the plots shown in Fig. 5.5. The FFT peak at the first harmonic (2 Hz) is significantly smaller than the FFT peaks at 1 and 3 Hz, hence why the 2 Hz component is not plotted. Figure 5.5 shows that although not perfect,

5 Cross-Facility Validation of Dynamic Centrifuge Testing

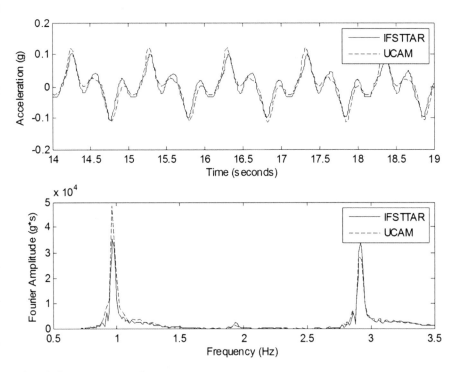

Fig. 5.4 Input trace comparison

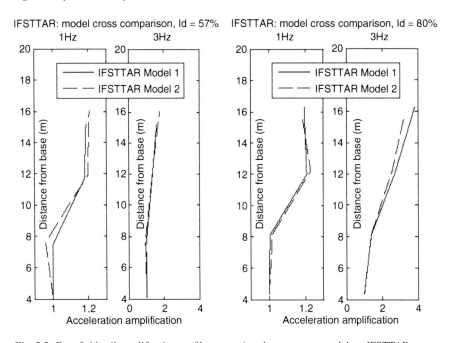

Fig. 5.5 Free-field soil amplification profile comparison between two models at IFSTTAR

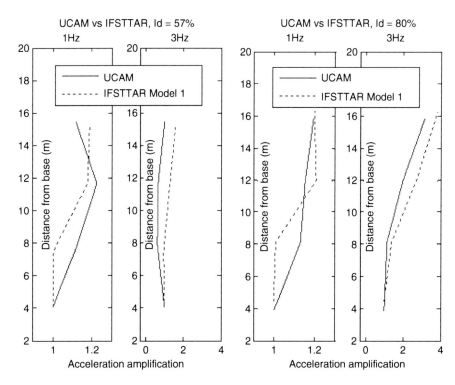

Fig. 5.6 Free-field soil amplification profile comparison between UCAM and IFSTTAR

the amplification profiles match well at both frequency levels and follow similar shapes for both loose and dense sand models.

Observing the repeatability of a single model test allows us to better quantify the differences between models prepared at two different facilities. Figure 5.6 compares the acceleration amplifications, derived as above, as measured in loose and dense models prepared at IFSTTAR and UCAM. The 1 Hz acceleration components in the loose sand models match well. However, the slight differences that exist in the 1 Hz component in the loose model are exaggerated in the dense sand tests. Again, the UCAM trace is relatively linear, as would be expected in a test of a uniform sand column. The largest variation between the two models is at 8 m from the base of the model, at which point the IFSTTAR amplification is approximately 10% greater than the UCAM amplification. However, such a small variation could simply be as a result of measurement errors. Interestingly, towards the surface of the model, the acceleration amplification levels become closer again, which is important when testing soil-structure interaction.

The 3 Hz acceleration amplification through the models shows greater variation between the two facilities than the 1 Hz component. UCAM's data shows initial attenuation of the signal followed by amplification towards the upper sections of the soil column. The data from IFSTTAR shows a slight variation between the two models, with the signal in the loose model being slightly attenuated at the greater

5 Cross-Facility Validation of Dynamic Centrifuge Testing

depths, followed by linear amplification. The signal in the dense model on the other hand shows an increasing degree of amplification moving from deeper to shallower soil. The model containers have been designed in different ways resulting in the IFSTTAR container having a natural frequency of 0.74 Hz and the UCAM container having a natural frequency of 1.34 Hz (Brennan and Madabhushi 2002). This variation will result in differences in the mode shape of the model containers at the 1 Hz input frequency and subsequent harmonics. At higher harmonics the disparity increases between the input harmonic frequencies (2, 3 Hz etc.) and the container harmonic frequencies (IFSTTAR: 2.22, 3.7 Hz, UCAM: 4.02, 6.7 Hz) resulting in a more significant variation in the amplification profiles as observed in the 3 Hz plots.

5.5 Soil-Structure System Response

5.5.1 Structural Response

Prior to examining the comparison between the results of the soil-structure interaction experiments at the two facilities, it is important to confirm the simple dynamic response of the structure is as would be expected. There are a variety of ways in which this can be investigated, though here we will simply look at the frequency response of the model structure. As detailed earlier, the fixed base natural frequency of the model structures is 1 Hz (50 Hz model scale). The frequency at which the structure vibrates can be determined from the acceleration data collected from the superstructure. Placing the model structure on the sand surface will decrease the foundations rotational stiffness and therefore be expected to decrease the natural frequency of the system relative to the fixed base case. Shear modulus degradation during earthquake shaking will cause a further reduction in the foundation's rotational stiffness and hence a further reduction in the model structure's resonant frequency. During the earthquake the structure undergoes forced vibration, and hence except for the transient behavior at the beginning of the earthquake it must respond at the earthquake frequency. During sine-sweep earthquakes owing to the non-stationary nature of the input motion, this transient behavior will persist for the entire earthquake duration. Hence, three different frequencies should be apparent from the data; the forced superstructure vibration frequency during a single frequency input motion, the post shaking natural frequency and a resonant frequency during sine-sweep shaking.

Figure 5.7 displays data from a simple sinusoidal earthquake which had a primary driving frequency of 1 Hz with a peak input acceleration of 0.15 g and a duration of 30 s. The Fourier transforms show the frequency content of the signals over the two periods highlighted in the acceleration-time trace. Initially examining the post earthquake phase, it can be observed that the natural frequencies of the systems are lower than the fixed base frequency (1 Hz) even when no shear modulus degradation is occurring after the earthquake ends. The larger structure's natural frequency (0.9 Hz) deviates further from the fixed based natural frequency than the lower mass system's

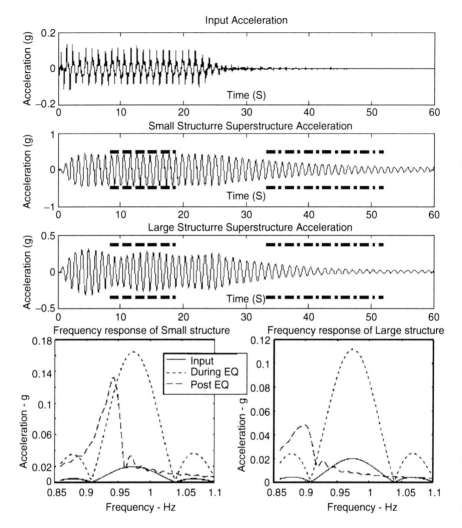

Fig. 5.7 Frequency analysis of superstructure during sinusoidal earthquake

natural frequency (0.93 Hz). This is in line with expectations, given the natural frequency of a system is dependent on the mass, as the mass increases the natural frequency decreases assuming an approximately constant system stiffness.

The natural frequency of the systems during the earthquake can be determined by considering the data collected during the sine-sweep earthquakes, as these motions scan over a range of frequencies. Hence the point at which resonance occurs in the structures can be observed. Figure 5.8 shows the data from one such earthquake. The natural frequencies of the systems are 0.78 and 0.9 Hz for the large and small systems respectively. These values are lower than the post-earthquake frequencies discussed above due to shear modulus degradation in the foundation

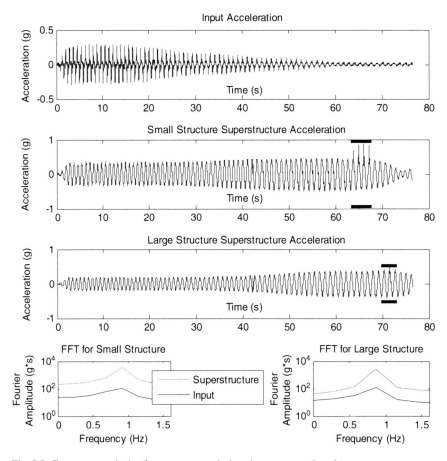

Fig. 5.8 Frequency analysis of superstructure during sine-sweep earthquake

sand. Clearly knowing the frequency at which a system is vulnerable is important in seismic design, and this analysis has highlighted the effect of soil softening in reducing this relative to the fixed base case.

5.5.2 Soil Column Response

Earlier, the free-field soil column response was compared between the two facilities. To further this, a similar analysis can be conducted for the soil column directly beneath the model structures. This analysis can be performed with the results from the tests involving the common 100 kPa bearing pressure structure. Figure 5.9 shows the acceleration amplification beneath the 100 kPa structure at both facilities in loose and dense sand models. The UCAM data shows an increase in soil amplification

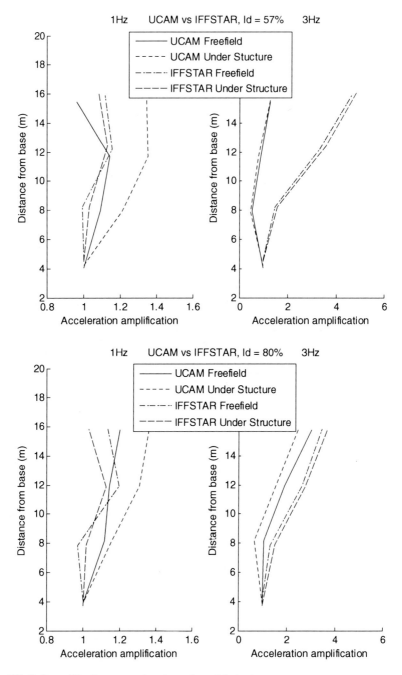

Fig. 5.9 Soil amplification comparison beneath model structure

beneath the structure at 1 Hz in both the loose and dense sand models. However, the IFSTTAR data shows a variable amount of amplification at 1 Hz. In both the IFSTTAR dense and loose models a slight attenuation rather than amplification is observed close to the sand surface.

As discussed earlier, the natural frequencies of the model containers are not perfectly matched. This results in the containers adapting a different mode shape which is likely the cause of the variation in soil amplification beneath the model structure. Given that this work relates to soil-structure interaction, it has to be assessed what impact this variation between the soil amplifications at the two facilities will have on the results obtained. It is also important to determine which, if either, is modeling the true prototype scenario more accurately and hence how well the results will relate back to real design scenarios. We have already observed that the input motions are non-identical in respect to acceleration magnitude. With the amplification factors being relatively small (1.0–1.4 for 1 Hz), the amplification profile may not have such a significant effect on the results obtained. Provided that acceleration levels are measured close to the sand surface, the system response can be normalised to be independent of base input magnitude and the degree of amplification in the soil. This also makes it easier to relate the results back to the prototype scenario as seismic input motions for structural design are often measured at a relatively shallow depth.

This discussion has focused on the 1 Hz fundamental component of the earthquake input motion as it is considered of primary importance in terms of investigating the soil structure interaction. As observed in Fig. 5.9, there is little variation in the 3 Hz acceleration component between the free-field and under the structure. The slight variation that does occur could simply be a function of the accelerometers being located at different distances relative to the boundaries of the model container. Given that the model containers are tuned to match only the fundamental mode of the soil column and not explicitly the higher modes, it is likely that the boundaries of the container influence the response of the sand and hence the applied surcharge has a lesser effect on the measured amplification factors.

5.5.3 Soil-Structure Interaction

Having examined how repeatable the preparation and testing of a soil model are at two different facilities, it is important to investigate how the minor variations discussed previously will affect the soil structure interaction occurring between a shallow foundation and the dry soil on which it is founded.

An area of soil structure interaction currently attracting much attention is the ability to dissipate energy at the soil structure interface by permitting rocking and relative sliding between the footing and the soil beneath. This dissipation reduces the amount of energy transmitted into the superstructure and hence reduces the ductility demand. The mechanism works on a similar basis to base isolation systems used in seismic design, where large rubber bearing are installed between the foundation and the superstructure. In this case, the entire foundation and superstructure are isolated with the soil acting as the 'rubber bearing'. A method to examine the amount of energy being dissipated in the soil structure interface during the earthquake is to plot moment-rotation plots which correlate the moment being applied by the footing to the soil beneath against the rotation of the footing.

The rotations are estimated based on the double integral of acceleration traces from two vertical accelerometers located on either side of the base of the model structures. As the centre of rotation is not known there is a degree of inaccuracy in these results. However, both the IFSTTAR and UCAM data was processed in identical manners and hence comparisons can still be directly made between the two sets of results. The moment is evaluated from the inertial acceleration of the superstructure.

5.5.4 Moment-Rotation Loop Comparison

Moment-rotation plots for comparable tests at UCAM and IFSTTAR are shown in Figs. 5.10 and 5.11. Although a great number of comparisons can be made by comparing results from different tests within each facility, we will focus here on comparing tests between facilities. As discussed previously, these plots are for a model structure with a 100 kPa bearing pressure. The gradient of a moment rotation loops provides an indication of the rotational stiffness experienced between the foundation and the soil.

Figure 5.10, which compares results for a loose model (~50% I_d) would imply that there is a 20% greater rotational stiffness in the IFSTTAR test. Precise measurements of relative density showed that the IFSTTAR model was prepared at a density approximately 5% higher than that of UCAM's, which is the likely cause of the slight slope variation shown in Fig. 5.10. In addition, a certain amount of the increase in stiffness can be attributed to the higher stress levels in the IFSTTAR model due to higher g-level during the IFSTTAR tests as discussed earlier. Due to required improvements in how the rotations are calculated, it is too soon to make significant conclusions from the slight variations in the shape of the moment rotation loops. However, the area contained within each loop is remarkably similar at both facilities which implies similar amounts of energy are being dissipated through the soil-structure interface in the models at both facilities.

With regard to Fig. 5.11, which shows moment-rotation loops for a dense model, similar conclusions as were made from Fig. 5.10 are possible. The slopes are approximately identical and the area contained with the each loop is similar. However, the peak moments calculated for the IFSTTAR model are approximately 20% lower than those calculated for the UCAM model. The way in which the moments are calculated relies on accurate acceleration measurements. The accelerometers used at the two facilities have different manufacturers and different frequency responses; however, both are of excellent quality so this should not affect the acceleration measurement significantly.

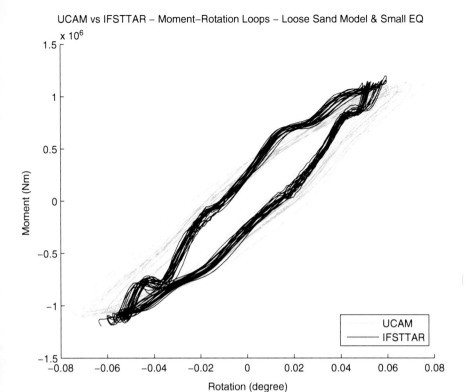

Fig. 5.10 Comparison of moment-rotation loops for loose sand test

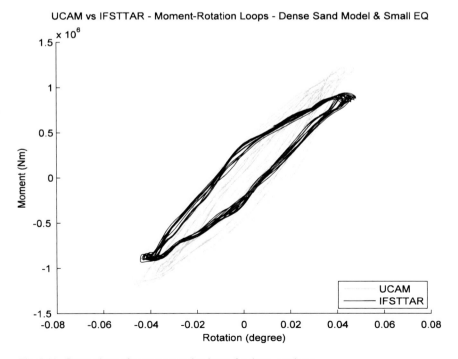

Fig. 5.11 Comparison of moment-rotation loops for dense sand test

5.6 Conclusions

This paper has discussed the reproducibility of geotechnical centrifuge experiments investigating soil structure interaction between two independent facilities. From the results presented and the experience gained in conducting the tests we can conclude that it is possible to obtain largely comparable results. It is important that identical model structures are used and the same instrumentation layouts are used to allow direct comparisons to be made. We have found that the dynamic properties of the model containers are important not only at the fundamental frequency, in this case 1 Hz, but also at higher harmonics. Model containers are often tuned to the primary mode only and higher frequencies are not considered. This results in variations in higher frequency soil acceleration amplification values between different model containers. The moment-rotation cycles agreed well with small variations attributed to the accuracy of the instruments used to make the measurements. It is likely that more accurate ways to measure the rotation and moment would decrease the disagreement found between the two facilities, in fact, more accurate measurement would allow higher frequency components to be included which would smooth the moment-rotation loops, making them more elliptical.

Acknowledgments The research leading to these results has received funding from the European Community's Seventh Framework Programme [FP7/2007-2013] under grant agreement n° 227887 for the SERIES project. The authors would like to acknowledge also the useful discussion with the other JRA 3 partners in SERIES.

References

Brennan AJ, Madabhushi SPG (2002) Design and performance of a new deep model container for dynamic centrifuge testing. In: Proceedings of the international conference on physical modelling in geotechnics, Hong Kong

Chazelas JL, Escoffier S, Garnier J, Thorel L, Rault G (2008) Original technologies for proven performances for the new LCPC earthquake simulator. Bull Earthquake Eng 6(4):723–728

Garnier J (2001) Physical models in geotechnics: state of the art and recent advances. In: First Coulomb lecture, Proceedings of the Caquot Conference, Paris

Madabhushi SPG, Schofield AN, Lesley S (1998) A new stored angular momentum (SAM) based earthquake actuator. In: Proceedings of Centrifuge 98, Tokyo, Japan

Madabhushi SPG, Houghton NE, Haigh SK (2006) A new automatic sand pourer for model preparation at University of Cambridge. In: Proceedings of 6th international conference on physical modelling in geotechnics, Hong Kong

Ternet O (1999) Reconstitution and characterisation of sand sample: application to centrifuge tests and calibration chamber tests. Thesis, Caen University

Chapter 6
Towards a European High Capacity Facility for Advanced Seismic Testing

Francesco Marazzi, Ioannis Politopoulos, and Alberto Pavese

Abstract The activities carried out in the framework of the FP7 project EFAST (design study of a European Facility for Advanced Seismic Testing) are highlighted. The objective is to determine the general characteristics of a new European world-class facility for seismic testing. To this end, the demands for testing necessary to support the modern earthquake engineering research have been investigated and compared to the current capabilities of laboratories in Europe. The performance objectives and the requirements of the facility are therefore established. On the basis of the needs assessment and taking into account the technological advances in experimental techniques and equipment (hardware and software) for seismic testing, a modern facility for experimental seismic research should comprise, mainly, an array of high performance shaking tables and a large reaction structure where both traditional (pseudo-static/dynamic) and innovative testing techniques (e.g. real-time hybrid testing) can be applied and combined. A tentative layout of the facility is proposed and issues related to its optimal utilization are discussed.

F. Marazzi (✉)
European Commission, Joint Research Centre, European Laboratory for Structural Assessment (ELSA), Institute for the Protection and Security of the Citizen (IPSC), via Enrico Fermi 2749, TP 480, 21027 Ispra (VA), Italy
e-mail: francesco.marazzi@jrc.ec.europa.eu

I. Politopoulos
Commissariat à l'Énergie Atomique, DEN/DANS/DM2S/SEMT, Bâtiment 603, CEA Saclay, 91191 Gif-sur-Yvette Cedex, France
e-mail: ioannis.politopoulos@cea.fr

A. Pavese
Eucentre – European Centre for Training and Research in Earthquake Engineering, Via Ferrata 1, 27100 Pavia, Italy

Department of Structural Mechanics, University of Pavia, Via Ferrata 1, 27100, Pavia, Italy
e-mail: alberto.pavese@eucentre.it; a.pavese@unipv.it

M.N. Fardis and Z.T. Rakicevic (eds.), *Role of Seismic Testing Facilities in Performance-Based Earthquake Engineering: SERIES Workshop*, Geotechnical, Geological and Earthquake Engineering 22, DOI 10.1007/978-94-007-1977-4_6,
© Springer Science+Business Media B.V. 2012

6.1 Introduction

Seismic events of the recent past have proved that European and neighbouring countries, especially those in the Mediterranean area, are exposed to a high seismic risk (University of Grenoble 2005, http://sesame-fp5.obs.ujf-grenoble.fr/index. htm). Surprisingly, the number of victims and the overall economic losses are important compared to industrialized countries like Japan and the United States which are often faced with higher levels of shaking. This fact can be explained by considering the higher population density, and the fact that the high number of damaged buildings is due to the large number of monuments or ancient masonry buildings often vulnerable to earthquake loading.

It is readily apparent that, in developed countries, although the numbers of victims of major earthquakes is tending to drop, the costs of the consequences are constantly rising (UNDP 2004). These costs, due to significant damage and widespread disorganization in the area, are constantly rising. Damage to plants may cause loss of data and drops in production, which are extremely costly. A recent example is the social and economic impact of the L'Aquila (Italy) earthquake in April 2009. It is therefore indispensable for Europe to intensify research and development in the field of earthquake engineering (Wenzel 2007).

In the last decades considerable advances have been achieved in the Earthquake Engineering (EE) field. The research results have contributed to the preparation of modern design codes, to the identification of several problems in the existing structures and to innovative solutions for the structural assessment. Despite this huge amount of improvements, there are still several open questions. Some examples are listed below. Predictive models have frequently been calibrated on experimental results obtained from scaled structures. Several innovative technologies for building constructions are entering the market and require careful evaluations to verify the level of safety. The experimental validation of the behavior of large infrastructures (bridges or retaining walls) often requires multi-support excitation. Further insight into soil structure interaction requires testing of heavy models, etc. Available data and future results need to be organized into databases, in order to disseminate them, optimize their use, and provide relevant information for risk oriented approaches. Research to address these issues requires a large amount of analytical and experimental studies.

Moreover, a look at the international EE landscape reveals that, outside Europe, there are several high performance seismic testing facilities either already operating or under construction. There is a trend, mainly in the USA and Asia (China, Japan and Korea), towards facilities with an array of shaking tables which increase operating ease and enable multi-support excitations. The objective is to test structures at the largest possible scale in order to avoid scaling effects. Table 6.1 presents the characteristics of some of the major shaking table laboratories in the world. It is observed that European shaking table laboratories have considerably lower capacities than those of their counterparts outside Europe. This means that, unavoidably, if the situation does not change, Europe will cumulate a considerable lag in experimental earthquake EE and in EE in general, with respect to the USA and Asian countries.

Table 6.1 Major shaking table facilities in the word (Taucer and Franchioni 2005)

Country	Location	No. of DoF	No. of tables (array) (array)	Dimensions [m]	Max payload [ton]	Max overturning moment [ton m]	Max stroke [mm]	Max velocity [mm/s]	Max acceleration [g]	Max frequency range [Hz]
USA	University of Nevada at Reno	2	3	4.3×4.5	45	54 (Yaw add Roll) 135 (Pitch)	±300	±1,270	±1	0–50
	State University of New York at Buffalo	6	2	3.6×3.6 (up to 7×7)	50	46	±150 (vertical: ±75)	±1,250 (vertical: ± 500)	±1.15	0.1–100
	University of California at San Diego	1	1 (outdoor)	7.6×12.2	2,000	5,000 (for a specimen of 400 ton)	±750	±1,800	±1	0–20
Japan	E-Defense in the city of Miki	6	1	20×15	1,200	15,000 (Yaw=4,000)	±1,000 (vertical: ±700)	±2,000 (vertical: ± 700)	±0.9 (vertical: ±1.5)	0–50
China	Tongij University in Shanghai	3	4	4×6	30–70	200–400	±500	±1,000	±1.5	0.1–50
Korea	Network lead by the Korean government	2	3 (1 fixed, 2 movable)	5×5 (fixed) 3×3 (movable)	–	–	–	–	–	–
Europe	France: CEA – Commissariat à l'Énergie Atomique	6	1	6×6	100	500	±125 (vertical: ±100)	±1,000 (vertical: ± 700)	±1 (vertical: ±2.5)	0–100
	Italy : Eucentre – European Centre for Training and Research in Earthquake Engineering	1	1	5.6×7	70	400	±500	±2,200	±6	–
	Portugal : LNEC – Laboratório Nacional de Engenharia Civil	3	1	4.6×5.6	40	–	±290	±700 (vertical: ±420)	±2 (vertical: ±1)	0–50

In addition, it is worth noting that there is an emergence of advanced experimental techniques, such as real-time substructuring and advanced measurement techniques that are being explored in the most innovative laboratories. This is an important point since the new experimental methods, based on the substructuring technique, have the advantage of reducing the specimen size allowing a better use of the available testing machines' capacity.

For all the aforementioned reasons, a new platform for seismic testing in Europe is not just useful but necessary. A new high performance testing facility will enable studying a large variety of structures and systems. In fact, such a facility is an indispensable tool to calibrate and validate new conceptual approaches in modelling and simulations developed for performance-based analysis and design of new structures or retrofitting interventions of safe structures in Europe and even worldwide. It will also contribute to the increase in the competitiveness of European science and industry worldwide. Therefore, the European Commission has granted, within the 7th Framework Program, the EFAST project (design study of a European Facility for Advanced Seismic Testing). The main objective of EFAST is to determine the general characteristics of this new European level facility and to propose a preliminary design of it (http://efast.eknowrisk.eu/EFAST/).

The following presentation begins with a very short overview of testing methods in EE. Then, on the basis of the work done within EFAST, the European needs for seismic testing are discussed and the general characteristics and requirements for an advanced testing facility are presented.

6.2 Testing Methods

In the following, a brief overview of the main experimental techniques and their application field are given, with the aim to analyze the advantages and disadvantages to the prospect of designing a new innovative multipurpose facility. Nevertheless the following apercu is not an exhaustive list of all existing experimental techniques. For instance, some methods which have been applied only rarely or are not sufficiently mature yet are not discussed (e.g. Diming et al. 1999).

6.2.1 Shaking Table

Shaking table testing is one of the oldest and more frequently used experimental methods in earthquake engineering. The method is a direct simulation of the physical phenomenon since it imposes a base induced loading as do real earthquakes. Shaking table tests are real-time dynamic tests and they can deal with phenomena that cannot be accounted for by other testing methods. Their main advantages are that:

- they are more representative of real behaviour in case of rate-dependent constitutive laws

6 Towards a European High Capacity Facility for Advanced Seismic Testing 103

- they can apply horizontal and vertical earthquake excitations. Rotational input is also possible if needed.
- they are very well suited to systems with distributed mass giving rise to distributed inertial forces (e.g. liquid storage tanks, out of plane bending of masonry walls, etc.)
- they are well suited for seismic qualification of sensitive equipment
- they can account for dynamic coupling between horizontal excitation and vertical response (e.g. rocking and overturning of solids, dynamic variation of axial force due to cracking of reinforced concrete walls subjected to horizontal excitation).

As is the case for all the testing methods, their drawbacks are often the other face of their advantages:

- Since the supporting and hydraulic system must have the capacity of carrying vertically the structure and supplying the power corresponding to the earthquake input, relatively big specimens cannot be tested. Therefore, often only a small part of the structure or a scaled model of it can be tested. In both cases questions may arise regarding the match between models and prototype.
- In the case of scaled models, similitude laws have to be used. Nevertheless, in some cases, the applied practical similitude law cannot satisfy the invariance of all the relevant dimensionless parameters associated to different physical phenomena. As an example, let us consider a model of a building on sliding bearings having a velocity dependent friction coefficient. If a gravity similitude is considered, in order to satisfy the ratio between gravity and earthquake induced forces, the model velocity will not fit the scaled velocity. This means that a discrepancy between the actual and the experimental sliding behaviour will occur. In general, scaling also cancels partially the reliability of the method regarding rate-dependent behaviour. Moreover, even when theoretical similitude laws can be respected, the constitutive laws of some materials (mainly brittle materials) do change with the scale due to the so-called size effect.
- In general, for fast testing systems, such as shaking table and real-time hybrid testing, the control is by far less accurate than for quasi-static methods, such as the pseudodynamic (PsD) technique. In particular, spurious base motions, such as rocking, may be induced due to the admittance of the supporting system and/or the actuators because the control cannot fully compensate undesirable motions (Molina et al. 2008). Spurious motions also may occur due to the deformability of the table plate itself (Le Maoult et al. 2009).

6.2.2 The Pseudo-Dynamic Method

The pseudo-dynamic (PsD) testing technique is based on a numerical integration of a discrete-DoF equation of motion containing a theoretical mass matrix and seismic-equivalent external loads, but with a physical model for the restoring forces.

For every time-discrete state, the computed displacements are imposed to the physical model by means of actuators and the corresponding restoring forces are measured by load cells. The method is called PsD because the displacements are applied quasi-statically and, consequently, without the presence of physical inertial forces. The main advantages of the method are the following:

- Specimen dimensions can be very large, so many tests can be carried out without size reduction.
- Since there is no physical base input motion, the specimen can be put on a stiff foundation, thus avoiding spurious base motion.
- Since the experimental time is much longer than real time, the operator can stop the test if an undesirable behaviour is observed during the test. In general, due to the elongation of the time scale, control difficulties can be handled more easily than in real time tests.
- Specimen can be limited to a small substructure (the part that is difficult to model numerically) and the remaining restoring and inertial forces can be simulated numerically in the equation of motion. This technique is called substructuring and has been successfully implemented with the PsD method for many years for important real applications (Dermitzakis and Mahin 1986; Pegon and Pinto 2000; Pegon et al. 2008), while for shaking table and hybrid tests it is still under development.

The main drawbacks are:

- The method is restricted to systems with inertial effects that can be modelled accurately with a small dimension lumped mass matrix (e.g. buildings with masses concentrated to the floors). It is not suited for distributed mass systems.
- Rate-dependent effects cannot be accounted for, unless they are not too strong and some compensation techniques are applicable (Palios et al. 2007).
- Even though vertical dynamic excitation can in theory be applied, it would require installing at least as many additional vertical actuators as vertical degrees of freedom. Therefore, to the authors' knowledge, because of this practical limitation, such tests have never been carried out.
- Since forces are applied by a limited number of actuators, the state of stresses may be different from that of the prototype in the vicinity of the loading points. Therefore, special load-distributing attachments, inserted between the actuators and the specimen, may be needed to minimize local effects.
- The method cannot reproduce the dynamic coupling between horizontal excitation and dynamic vertical response.

It is worth noting that methods for real-time pseudodynamic tests are receiving growing interest (Nakashima and Masaoka 1999). With real-time pseudodynamic tests, rate dependent effects are in principle correctly reproduced (if velocity is tracked satisfactorily). Nevertheless, these methods face problems, such as computation time and compensation of time delay, which are common to all real-time control problems and in particular to real-time hybrid tests, discussed in the following subsection.

6.2.3 Real Time Hybrid Testing

The idea of the PsD test can be extended to a more general category of tests called hybrid. In a hybrid test, by definition, a part of the force terms appearing in the equation of motion are experimentally measured, whereas the other part is numerically modelled. The PsD method described in the previous subsection is an example of hybrid test in which all the inertial forces are numerically simulated. On the other hand, regarding the restoring forces, sometimes some of the structural elements are in the physical model (experimental substructure), while some others are numerically simulated (numerical substructure). Depending on the rate of loading these tests may be either "real-time" or "extended time scale" (fast or quasi-static) tests. The interest of hybrid methods in general is obvious since the "easy" or "known" part of the structure can be modelled numerically and the capacity of the experimental devices can be devoted exclusively to the "difficult" or "unknown" substructure. Pseudodynamic (hybrid extended time scale) substructuring tests have been successfully carried out on real applications as it is mentioned in the previous subsection. Real time substructuring methods appeared recently and for the time being only relatively simple tests are feasible. Actually many difficulties arise during this kind of tests. Among others, one can mention problems related to (Marazzi and Molina 2009a; Ahmadizadeh et al. 2008; Blakeborough et al. 2001; Bonnet et al. 2008; Kim et al. 2008; Tu et al. 2010; Wu et al. 2007):

- the accuracy of the integration algorithms, in particular in the case of rate-dependent behaviour. In fact, in that case the controlled variables of interest are not only the applied actuators' displacement or force increments in a time step, but also their rate.
- Time delay and response time of the actuators.
- Influence of measurement noise on the control action.
- Accurate control of very small relative displacements using large stroke actuators if the physical substructure is on a shaking table and the actuators applying the reactions of the numerical substructure are attached to the reaction wall.
- Realistic boundary conditions that are different from a simple force or displacement as, for instance, when a full distributor (displacements and rotations) or wrench (forces and torques) should be applied.
- Short required computation time implying that numerical substructures with only a few degrees of freedom can be considered since the computation must be done in real time. Alternatively, special computational techniques should be adopted to increase computational efficiency.

Despite the aforementioned problems, as a strategic investment, the design of new testing facilities should anticipate future evolution and progress in that field, making this kind of testing more "common practice" in the future. Moreover, it should be pointed out that progress in this field has a beneficial impact to "conventional" experimental techniques also, as they benefit from advances made in control, measurements etc.

6.2.4 Geographically Distributed Testing

A geographically distributed test is a particular case of hybrid test in which some of the substructures (experimental or numerical) coupled during the experiment are placed in different laboratories separated by a physical distance (Marazzi et al. 2010). The technological challenge for this testing technique is high since it requires, in addition to hybrid testing capabilities in the laboratories, not only performance in the hardware networking and protocols but also capabilities for collaboration in standards for communication and exchange of information. Considerable effort has been put in the research and some successful experiments have been performed, mostly using the PsD quasistatic approach due to the considerable distance among the involved laboratories (http://nees.unr.edu/aboutnees.html, http://www.series. upatras.gr/overview). Nevertheless, as it has been recognized by the experts attending the 1st EFAST Workshop (Marazzi and Molina 2009a), the main interest of this testing technique is the cooperation aspect among the laboratories. The proposed European facility should thus have the internal and networking capabilities for carrying out this type of test. Therefore, it should have capabilities for hybrid testing, as mentioned above, and be endowed with the best available networking technology. Currently, a considerable collaborative work is carried out in this field at a European level, in the framework of the research project SERIES (http://www. series.upatras.gr/overview).

6.3 Testing Needs

This section presents the needs for further experimental EE research in Europe. It is a brief summary of the discussions and conclusions of the 1st International EFAST Workshop that assembled about 30 experts on earthquake engineering. It also reflects the results of an inquiry sent to three target groups of current or potential users of earthquake experimental facilities (Marazzi and Molina 2009b). The target groups include research laboratories, industry (mainly nuclear industry companies and institutions) and construction companies. Unavoidably, this presentation also reflects the personal point of view and professional experience of the authors. The needs inventory is split into different problem classes.

One common point to all classes of problems is that, in order to conduct a meaningful probability risk assessment, the actual available margins of structures have to be estimated. This holds for all structures but it is even more critical for structures of major importance (e.g. power generation facilities, hospitals etc.). Depending on the tested structure of interest (building, equipment or secondary structure), failure can be defined as loss of operational function, collapse or relevant significant damage or collapse. To this end, tests with excitation level up to failure should be possible in future.

It is also worth mentioning that needs, testing technology and methods are tightly interconnected. An improvement of testing methods can suggest new type of tests and thus new needs. In a similar way, new needs result in research on new technologies which, in turn, improve the capability and efficiency of testing methods.

6.3.1 Objectives of Seismic Tests

Generally, according to their main objectives, seismic tests can be classified in one of the following three categories:

1. Qualification tests aiming at qualifying the seismic response of equipment or valuable goods. The scale must be necessarily 1:1, because scaling effects must be completely avoided since the prototype has to be qualified. Tests of this kind are often required by regulations. They are necessary when it is extremely difficult to build an analytical model (e.g. electric cabinets in power generation facilities) and when a yes or no answer is needed or is the only possible realistic answer.
2. Demonstration tests motivated by dissemination purposes or by strategic reasons, such as to convince policy makers. Usually, the tested models are at the biggest possible scale.
3. Research and development oriented tests which are necessary for one or more of the following purposes:

   - to obtain a better physical insight for both local and global structural behaviour
   - to validate or calibrate the available numerical models, the modelling techniques and the level of inherent approximation
   - to propose or validate codes, recommendations and guidelines.

All three categories may need comparable capacity of the experimental facility. For instance, the qualification of heavy equipment at scale 1:1 may need a force capacity comparable to that of a frame building model at a smaller scale. Consequently, high performance is needed in all cases.

Both categories 1 and 3 need a high quality accurate control software and hardware to avoid considerable discrepancies between target motion and achieved motion. In addition, for category 3, accurate, reliable and advanced measurement techniques are necessary to enable a satisfactory exploration of the experimental results.

6.3.2 New Buildings

Despite the considerable progress made in the design and analysis of new buildings as well as the improvement of regulations, further research is necessary in several fields. An overview of research needs for normative purposes is presented in (Pinto et al. 2007). Among others one can mention the following topics:

6.3.2.1 In-Plan Irregular Buildings

Research is necessary to establish how to measure the static or natural eccentricity of a building structure, and how it is affected by the changes in strength distribution. It is also important to set the grounds to move from a torsional failure criterion based on ductility ratio demands to a more rational one based on member absolute deformations, such as drifts.

6.3.2.2 Buildings with Flat Slab Systems

Due to the lack of sufficient available data, current regulations impose, conservatively, a low behaviour factor for flat slab systems which are very common in several European countries. Further experimental research is needed in this field to increase their ductility class.

6.3.2.3 Precast, Prestressed Concrete Elements and Systems

Assemblies of precast elements should be further assessed to properly understand their behaviour under seismic loads. Effective models for both simulation and design of precast structures are also needed, especially regarding joints and foundations. Prestressed technology for post-tensioning of concrete and lamellar wood is now emerging. Therefore a reliable experimental evaluation of their effective behaviour and of the overall safety of the proposed conceptual design and detailing is needed.

6.3.2.4 Masonry Buildings

The seismic response of masonry buildings and buildings with masonry infills needs to be further investigated. In particular, there are very few tests carried out on masonry structures with more that two storeys. The pseudodynamic method can give useful insight into the response of such structures but, because of the distributed mass of walls and infills, shaking table tests would be also relevant. Obviously, dynamic tests on a shaking table are meaningful if the model scale is not too small. A considerable number of tests should be conducted because of the variability of construction techniques and materials. The out of plane behaviour of masonry structures and rules for "simple buildings" should be assessed. A further insight into the role of non-structural masonry infills is also necessary. It is worth noting that Eurocodes (http://eurocodes.jrc. ec.europa.eu/) do not include any provision that accounts, explicitly, for the infills' contribution to the building capacity. In fact, regarding new buildings, EN 1998-1:2004 considers infills qualitatively as a second line of defence, but there is no guidance enabling the assessment of the actual building capacity. The lack of such guidance for existing buildings in EN 1998-3:2005 is even more critical, since the contribution of infills to the capacity of the building may be substantial.

All the aforementioned aspects deal with quite complex phenomena which cannot be investigated reliably with small scale models. Therefore, a high capacity experimental facility is needed in order to carry out tests at the biggest possible scale.

6.3.3 Infrastructures

There is a strong need for further experimental and analytical studies on the seismic response of bridges and viaducts with special attention to the three-dimensional character of their response, the constraint and isolation devices, the flexibility of their foundations, and the actual behaviour of the most critical elements. Soil-structure interaction and non-synchronous ground motion must also be investigated further. The aforementioned topics are of interest not only for new bridges, but also for existing bridges, due to degradation of their properties (ageing) and/or higher required performance level (increased traffic, increased seismicity etc.). Even though all these aspects cannot be completely studied experimentally by means of full scale models, a new facility should have the possibility to investigate some of the aforementioned questions as, for instance, asynchronous base motion of reduced scale models. This implies that a multiple support excitation is necessary.

6.3.4 Retrofitting

Repairing and/or strengthening of existing structures is a major concern of earthquake engineering during the last decades. In fact, there is a plethora of new techniques and materials (e.g. fibber reinforced polymer materials, structural glass, wood-based composites etc.) which are added to the already long list of more traditional techniques. The combination of different materials and the uncertainty on the characteristics of the existing structure makes the retrofitting a complicated problem that cannot be treated only analytically. The aforementioned aspects are of major importance in the case of retrofitting of structures related to cultural heritage. Presently, there are no Eurocodes on cultural heritage interventions, but it is commonly accepted that interventions should be low intrusive and take into account long term consequences. The wide variety of these constructions makes it very difficult to write a dedicated code. On the contrary, each structure is unique and needs a specific study.

Therefore, a further experimental research is necessary in the aforementioned issues. Testing may be required to assess either the original or the retrofitted structure or both. Reliable design codes for new structures have entered into force in Europe only in the last decades and the number of non-seismically-designed structures is larger than in the USA or Japan (Marazzi et al. 2000). To be representative of the actual global and local behaviour of real retrofitted structures, the models should be tested in laboratory at the biggest possible scale (http://www.ectp.org/default.asp).

6.3.5 Seismic Protection Devices

Seismic protection devices, like isolation bearings, have to be systematically tested for qualification purposes. In Europe, currently, the less demanding tests are usually carried out either by the manufacturers themselves or in the existing facilities. Tests on large dimension bearings were not very common in the past but nowadays big isolators are used more frequently in several projects. These qualification tests are done with dedicated experimental set-ups as, for instance, those at the University of San Diego, USA, or at EUCENTRE in Pavia, Italy. However, even the EUCENTRE testing machine, which is the biggest in Europe, cannot test very big bearings. The demand for tests on seismic protection devices will tend to increase with entering into force of the European Standard EN 15129 and the increase of the number of structures incorporating such devices. Therefore, Europe should be also endowed with the capacity for doing large scale tests in the future.

In addition, the influence of these devices on the structure response as well as on sensitive equipment (e.g. equipment of nuclear plants) or secondary structures (e.g. valuable objects like statues in museums etc.) has to be studied further. This implies dynamic shaking table tests. Shaking tables in this case should have a large displacement capacity in order to reproduce correctly earthquake records having significant low frequency content.

6.3.6 Equipment and Components

The proper operation of industrial and power generation plants depends to a considerable extent on the capacity of sensitive equipment to respect the required performance criteria under severe earthquake events. The same holds also for other buildings of major importance as, for instance, hospitals, or even museums where the integrity of statues or other valuable objects is of paramount importance. Therefore, the behaviour of secondary structures has to be studied further. To this end, numerical and experimental research projects and/or qualification tests have to be conducted. This implies that the facility should be able to reproduce the adequate floor signal. Dynamic floor amplification may be considerable and components must be tested under high acceleration (of about 4 g or more) in the frequency range form 0.1 to 50 Hz. For equipment in seismically isolated buildings, as already mentioned, a large displacement capacity shaking table is necessary to reproduce the adequate floor input motion.

6.3.7 Soil-Structure Interaction

Soil-Structure Interaction (SSI) is a common issue to the majority of EE problems. The two major aspects of soil structure interaction are energy radiation and possible nonlinear behaviour of the soil or of the soil-structure interface (nonlinear soil deformations, sliding and/or uplift of the foundation etc.). The effects of SSI on the structure response are a major source of uncertainty. Consideration of the structure, of its foundations and of the soil as a system may lead to alternative cost-effective seismic design concepts that will enable – under certain conditions – concentration of nonlinearity and energy dissipation in the soil or in the foundation. It may also change the ductility demand in the superstructure, leading to a modification of the recommended values of the behaviour factor.

Unfortunately, there are not satisfactory experimental methods to deal with SSI. One method is on site measures of responses to real earthquakes or to explosions or even ambient noise (Tang et al. 1990, 1991; Morisita et al. 1993). Another method is testing of very small scale models in laboratory centrifugal machines. However, although centrifugal tests may be well-suited for studying the soil behaviour, the structure scale is so small that it is no more representative of the prototype.

Tests on shaking tables are not very frequent as it is extremely difficult to carry out tests of models in a meaningful scale. In fact, tests should be as close as possible to real scale in order to avoid scaling effects, but technological problems still prevent from a complete satisfaction of this requirement. Efforts to deal with energy radiation by means of shear boxes, having adequate passive or active lateral boundary conditions, do not give a satisfactory solution because the spurious wave reflexions on the lateral boundaries cannot be completely eliminated. Moreover, wave reflexions also occur at the table plate. However, in the case of local nonlinear soil and/or soil-structure interface behaviour, nonlinearity may dominate the response, putting energy radiation in a secondary role. Soil structure interaction earthquake tests on shaking tables could mainly deal with this class of problems where the distance of boundaries (lateral boundaries and table surface) from the foundation need not be very big. However, even in this case, several problems persist. In fact, full scale models are not feasible because of the huge soil dimensions and weight this would imply. On the other hand, in the case of scale models, similitude law cannot satisfy static soil stresses. The smallest meaningful scale would be of about 1:4 (Marazzi and Molina 2009a). It is worth noting that even this scale results in very heavy models even for simple prototype structures. For instance, let us consider that the prototype structure is a bridge pier 12 m high on a foundation 7 m wide supporting a deck with a weight of 1,000 ton. A reasonable 1:4 scale specimen would consist of a soil container 3 m high having horizontal dimensions of about 7 m × 7 m. The total model weight would be of about 5,000 kN. This means that only very simple, elementary configurations can be tested on shaking tables.

A bigger scale could be envisaged using the PsD method with the test superstructure sited on or embedded in a soil contained in a pit surrounded by the soil of the facility. Two major drawbacks of this method are the difficulty of determining accurately the inertial loading to be considered and the rate-dependent behaviour of many SSI problems which does not comply with the PsD technique (Sect. 6.2.2). However, despite the aforementioned drawbacks and given that there is no ideal SSI testing method, the suggested method could be used on occasion as an alternative method.

For all methods, comparison with numerical simulations is of a paramount importance in order to develop and validate accurate numerical models that could allow us to investigate much more complicated configurations than those which can be physically tested.

6.4 Characteristics of a New European World Class Facility

As already mentioned in the introduction, one of the reasons which motivated the reflexion on a new seismic testing facility of a European level is that there is a risk for Europe to suffer a considerable delay with respect to Asian countries and the USA in EE research based on shaking-table testing. Regarding pseudodynamic testing, the ELSA laboratory (http://elsa.jrc.ec.europa.eu/) of the JRC of the EC in Italy, with its 16 m high and 20 m long reaction wall, is one of the main seismic testing facilities in the world. The situation is different for European shaking table facilities, which have considerably lower performances than those of the major shaking table laboratories in the world. Therefore, it is proposed that the new facility should be, mainly, a shaking table facility. It should allow the scientific community to carry out tests that will meet as closely as possible the needs presented in the previous section. The new facility should also comply as far as possible with the requirements of flexibility, adaptability and operational ease.

6.4.1 Capacity and Lay-Out of the Experimental Facility

Table 6.2 shows some indicative performance parameters for different classes of tests. The given numbers are reasonable rough estimates as a trade-off between needed performance and cost. Obviously it is not feasible, either for technological or financial reasons, to build a facility so big that everything could be tested therein. The objective is to propose a design that will enable to carry out meaningful tests using conventional and/or more recent techniques and technology which has already demonstrated its efficacy and reliability. EFAST is mainly a design study project; it does not aim at investigating and developing new revolutionary experimental techniques which are not consolidated yet.

The acceleration values in Table 6.2 may seem to be unrealistically high. However, it is worth noting that (a) several records of real earthquakes revealed very high acceleration values (e.g. 0.98 g Northridge earthquake, 1994, 0.85 g Kobe, 1995), (b) in the case of scale models, if a velocity similitude is considered, the table acceleration should be multiplied by the inverse of the scale ratio i.e. table acceleration will be higher than ground acceleration of the prototype, and (c) the values in Table 6.2 are conventional acceleration values corresponding to a rigid specimen. Consideration of the dynamic amplification of the specimen results in a higher demand of force capacity which is equivalent to a higher demand of conventional acceleration capacity.

6 Towards a European High Capacity Facility for Advanced Seismic Testing

Table 6.2 Performance demand for possible classes of tests

	Soil-structure interaction	Tests on civil engineering structures	Secondary structures or equipment
Height of specimen:	6 m	15 m	10 m
Mass of specimen:	500 ton	200 ton	1–100 ton
Number of directions:	1	1–3	1–6
Displacement:	±1 m	±1 m	±1 m
Velocity:	±2 m/s	±2 m/s	±2 m/s
Acceleration:	±1.5 g	±2 g	±2 g (100 ton) ±6–7 g (10 ton)
Frequency range:	0.2–50 Hz	0–50 Hz	0–100 Hz

In the case of SSI tests, the major part of the mass on the table is due to the weight of the soil itself and its container which will have, in general, a weak dynamic amplification. Therefore, in that case the required shaking table acceleration could be smaller. Regarding secondary structures and equipment, because of the amplification of the shaking motion at the floor level, floor accelerations to reproduce on the table may be much higher than ground accelerations.

Velocity values are also in agreement with actual recorded velocities (e.g. 1.4 m/s North\-ridge earthquake, 1994, 1.5 m/s Kobe, 1995). High displacement values are also necessary for the shaking table motion to be representative of strong, low frequency ground motions or of floor motions of low frequency buildings (in the case of secondary structures or equipment tests).

As a first step towards the design of a future experimental facility of European interest, the schematic lay-out in Fig. 6.1 is proposed. As already mentioned, the proposed lay-out is the result of iterations between performance needs and cost. It is based on a preliminary cost estimate which is not communicated here since it is still under further investigation. In fact, a reliable cost estimate should include several aspects such as construction, linking to infrastructures, software, handling, operation, maintenance and risk assessment. This task is currently under progress. The laboratory will consist mainly of:

1. two 6 DOF 6 m × 6 m shaking tables. The payload of each table would be of about 100 ton. The tables can be positioned in any place within the trench. They will be able to operate independently or be linked and operate as a single table. Obviously a higher number of tables, possibly in different trenches, would enable testing more complex configurations (see Tongij University in Marazzi and Molina 2009a). However, as already mentioned, the proposed facility is a trade-off between performance-capacity and cost, that is why only two tables are proposed;
2. one 1 DOF (horizontal) shaking table with high payload of about 500 ton. This shaking table will be intended for heavy specimen and in particular for SSI tests like the one described in Sect. 6.3.7;
3. a big strong floor area where dedicated experimental set-ups and testing machines could be flexibly mounted on;

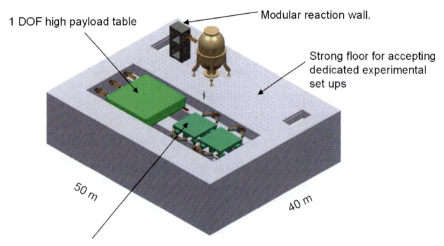

Fig. 6.1 Schematic lay-out of a new European level facility

4. a modular reaction wall allowing for real-time hybrid testing possibly combining shaking tables and actuators attached on the wall;
5. a hydraulic system composed of piping, actuators, pumps and accumulators with a capacity close to the performance criteria given in Table 6.2;
6. a high capacity crane bridge spanning the whole width of the working area and an outdoor area devoted to the construction of specimens.

If necessary, the maximum payload of the two 6 DOF tables could be increased with minor modifications of the supporting system. However, unless the actuators' capacity increases, this will be done at the price of a smaller maximum acceleration at full payload. In addition, adequate control hardware and software has to be specified, purchased or developed to obtain satisfactory control of the tables, dedicated set-ups and additional actuators in the case of hybrid testing.

6.4.2 Numerical Facility

A strong interaction between the aforementioned "purely experimental" facility and a numerical high computational capability facility, on site or remote, is necessary. Actually, high performance and accuracy numerical simulation is necessary not only for advanced experimental methods, like real-time substructuring involving complex numerical substructures, but also for conventional tests. Being able to quickly obtain accurate results of predictive analyses before testing and interpretation analyses after testing is of a paramount importance for successful testing of models with complex behaviour. Predictive analyses are necessary to define the whole testing configuration (model geometry, boundary conditions, properties, input characteristics, measurement technology, sensor locations and calibration etc.). Interpretation analyses may be useful for the detection of problems or unexpected response that occurred during the test (e.g. unsatisfactory behaviour of sensors, actual boundary conditions different from that considered etc.) and can be used as a tool of quality control between two successive tests.

6.4.3 Sensors, Data Processing and Structural Monitoring

Measurement techniques and systems evolve very rapidly. These new devices enable not only better quality measurements, but also new type of measurements, as for example low invasive or no contact measurements etc.

For instance, rapid advances in optical fibers, optical sensors, wireless communication, Micro-Electro Mechanical Systems (MEMS) and information technology – software frameworks, databases, visualization, Internet/Grid computations – have the potential to significantly impact the way complex structural systems are tested in large-scale facilities. Along these lines dense arrays of sensors are envisioned for large civil engineering structures/infrastructures, including foundations and the soil. To cope with the large amount of data generated by a measuring system, on-board processing at the sensor itself allows a part of the computation to take place locally by the sensor's embedded microprocessor. Remarkable improvements are also envisaged from remote measurements when direct contact with the structure should be avoided, e.g. for safety reasons or in emergency situations, or when installation of conventional instruments is impractical for technical or economic reasons. Structural monitoring methods, including system identification techniques, damage assessment and model updating tools, are evolving very rapidly and will play a major role in the near future.

6.4.4 Database and Telepresence

Efficient and better dissemination and use of experimental results should be a requirement for a new facility. An easy and fast access to data coming from experimental testing campaigns will foster the impact of earthquake engineering research on practice, innovation and earthquake risk mitigation, and will provide a broad and solid base for the calibration of numerical models. A web portal should give user-friendly access to the data contained into the database. The database and the user interface should be integrated into the European scientific experimental database network. The new testing facility should provide remote users access to the data during the experiment itself (on-line access), virtual access to the equipment and enable collaborative decision making regarding the testing activities (telepresence). This is also essential for tests involving concurrent use of geographically distributed platforms (distributed testing).

6.5 Conclusions

This paper presents a synthesis of the work done in the framework of the European project EFAST to define the general characteristics of a new European world-class facility for seismic testing of structures. The need for such a facility comes mainly from the increase, in the last decades, of the needs for further experimental research with big scale models and strong input shaking motions. This is also reflected in the increasing number of new high-performance shaking table facilities worldwide that will soon restrict Europe to a secondary role in the field of EE if there is no reaction.

With the ELSA laboratory of the Joint Research Centre of the European Commission, Europe is equipped with a world-class facility for pseudodynamic testing. However, this is not the case for shaking table facilities whose performances did not follow the evolution that took or is taking place in foreign countries. Therefore, it is proposed that the future facility should be a new generation dynamic testing facility. It will be composed, mainly, of high performance shaking tables and a large strong floor where dedicated experimental set-ups could be mounted on. In addition to the capacity for testing relatively big models under severe shaking motions, the new facility should meet, as far as possible, the requirements of flexibility and operational ease. It should also enable hybrid real-time testing, for research purposes in the near future and for purposes of practical interest in the future. Provision should also be made for the coupling of this facility with a high computational capacity numerical facility. For the European community to fully benefit from the future facility, networking between the new laboratory and existing facilities should be developed.

The described proposal is under further investigation to define more precisely the dimensions, the preliminary design of the various components (reaction mass, hydraulic system, tables' plates, modular reaction wall, network connections etc.) and estimate cost. In addition, other complementary aspects, necessary for successful experiments, are under study, such as, for instance, measurement techniques and technology and control issues.

Acknowledgments The work presented in this paper has been carried out in the framework of the European project EFAST (grant agreement no: 212109). This financial support is gratefully acknowledged by the authors. The authors would like to thank also all their partners of the EFAST project and in particular Dr. Javier Molina and Dr. Pierre Pegon for their useful remarks and suggestions.

References

Ahmadizadeh M, Mosqueda G, Reinhorn AM (2008) Compensation of actuator delay and dynamics for real-time hybrid structural simulation. Earth Eng Struct Dyn 37:21–42

Blakeborough A, Williams MS, Darby AP, Williams DM (2001) The development of real-time substructure testing. Philos Trans R Soc Lond A 359:1869–1891

6 Towards a European High Capacity Facility for Advanced Seismic Testing

Bonnet PA, Williams MS, Blakeborough A (2008) Evaluation of numerical time-integration schemes for real-time hybrid testing. Earth Eng Struct Dyn 37:1467–1490

Dermitzakis SN, Mahin SA (1986) Development of substructuring techniques for on-line computer controlled seismic performance testing. UCB/EERC-85/04, Earthquake Engineering Research Center, University of California at Berkeley, Berkeley, CA

Diming J, Shield C, French C, Bailey F, Clark A (1999) Effective force testing: a method of seismic simulation for structural testing. Struct Eng 125(9):1028–1037

Kim CY, Park YS, Kim JK (2008) Hybrid testing facilities in Korea. In: Saouma VE, Sivaselvan MV (eds) Hybrid simulation; theory, implementation and applications. Taylor & Francis/Balkema, London, pp 91–97. ISBN ISBN 978-0-415-46568-7

Le Maoult A, Bairrao R, Queval JC (2009) Dynamic interaction between the shaking table and the specimen during earthquake tests. In: 20 SMIRT, 9–14 Aug, Espoo, Finland

Marazzi F, Molina FJ (2009a) 1st EFAST workshop – challenges, needs and open questions, EUR 23822 EN. Publications Office of the European Union, Luxembourg. JRC5 51632. http://elsa.jrc.ec.europa.eu/list_pub.php?id=195&year=2009

Marazzi F, Molina FJ (2009b) EFAST inquiry. EUR 23998 EN. Publications Office of the European Union, Luxembourg. JRC 54178. http://elsa.jrc.ec.europa.eu/publications/JRC54178.pdf

Marazzi F, Magonette G, Anthoine A, Pinto A, Tirelli D, Pegon P, Renda V, Bono F, Molina FJ (2000) The role of ELSA laboratory in the field of cultural heritage protection. In: "Quarry – Laboratory – Monument" international congress – PAVIA 2000, Pavia, 26–30 Sept 2000

Marazzi F, Molina FJ, Pegon P, Politopoulos I, Casarotti C, Pavese A, Le Maoult A, Atanasiu G, Nguyen V, Dorka U (2010) 1st year EFAST annual report. EUR 24354 EN. Publications Office of the European Union, Luxembourg. JRC 57809. http://elsa.jrc.ec.europa.eu/list_pub.php?id=355&project=19

Molina FJ, Magonette G, Viaccoz B, Geradin M (2008) Apparent damping induced by spurious pitching in shaking-table testing. Earth Eng Struct Dyn 37:103–119

Morisita H, Tanaka H, Nakamura N, Kobayashi T, Kan S, Yamaya H, Tang HT (1993) Forced vibration test of the Hualien large scale SSI model. In: 12th SMiRT, vol K, Stuttgart

Nakashima M, Masaoka N (1999) Real-time on-line test for MDOF systems. Earth Eng Struct Dyn 28:393–420

Palios X, Molina J, Bousias S, Strepelias E, Fardis M (2007) Sub-structured pseudodynamic testing of rate-dependent bridge isolation devices. In: 2nd international conference on advances in experimental structural engineering, vol 1, Shanghai, 4–6 Dec 2007, pp 262–269

Pegon P, Pinto AV (2000) Pseudo-dynamic testing with substructuring at the ELSA Laboratory. Earth Eng Struct Dyn 29:905–925

Pegon P, Molina FJ, Magonette G (2008) Continuous pseudo-dynamic testing at ELSA. In: Saouma VE, Sivaselvan MV (eds) Hybrid simulation; theory, implementation and applications. Taylor & Francis/Balkema, London, pp 79–88

Pinto A, Taucer F, Dimova S (2007) Pre-normative research needs to achieve improved design guidelines for seismic protection in the EU. EUR 22858 EN. Publications Office of the European Union, Luxembourg. JRC 007741. http://elsa.jrc.ec.europa.eu/publications/JRC37741.pdf

Tang HT, Tang YK, Stepp JC (1990) Lotung large-scale seismic experiment and soil-structure method validation. Nucl Eng Des 123:297–412

Tang HT, Stepp JC, Cheng YH, Yeh YS, Nishi K, Iwatate T, Kokusho T, Morishita H, Gantenbein F, Touret JP, Sollogoub P, Graves H, Costello J (1991) The Hualien large scale seismic test for soil-structure interaction research. In: 11th SMiRT, vol K, Tokyo, Japan

Taucer F, Franchioni G (2005) Directory of European facilities for seismic and dynamic tests in support of industry. CASCADE series report no. 6, LNEC, Lisbon, Portugal, ISBN 972-49-1970-6

Tu JY, Lin PY, Stoten DP, Li G (2010) Testing of dynamically substructured, base-isolated systems using adaptive control techniques. Earth Eng Struct Dyn 39:661–681

United Nations Development Programme (2004) Reducing disaster risk. A challenge for development. Bureau for Crisis Prevention and Recovery. New York, N.Y. ISBN 92-1-126160-0 Printed by John S. Swift Co., USA. http://www.undp.org/cpr/whats_new/rdr_english.pdf

University of Grenoble (2005) The SESAME project. Site effects assessment using ambient excitations. http://www.seismo2009.ethz.ch/hazard/risk/SESAME.html

Wenzel H (prepared by) (2007) European strategic research Agenda, "Earthquake engineering – vision – strategic research Agenda – roadmap for implementation". European Association for Earthquake Engineering, July 2007. www.eaee.boun.edu.tr/earthquake-research-agenda-V9-August_Part\%2012.pdf

Wu B, Wang Q, Shing PB, Ouet J (2007) Equivalent force control method for generalized real-time substructure testing with implicit integration. Earth Eng Struct Dyn 36:1127–1149

Chapter 7
Performance Requirements of Actuation Systems for Dynamic Testing in the European Earthquake Engineering Laboratories

Luiza Dihoru, Matt S. Dietz, Adam J. Crewe, and Colin A. Taylor

Abstract A review of the current actuation technology in the European laboratories of earthquake engineering is presented. The existing laboratory infrastructures, the current types of dynamic tests and actuation requirements are investigated. The needs of the earthquake engineering community that are not met by the currently available actuation devices are explored. User opinions are investigated in relation to the desirable performance enhancements and potential optimization solutions for hydraulic, electrical and hybrid actuation devices that would expand the experimental capabilities for dynamic testing. Various avenues for improving the current actuation capabilities are explored and several technical solutions are proposed. The future direction of actuation technology is discussed.

7.1 Current Actuation Technology

7.1.1 Types of Actuators and Selection Criteria

The technical literature on actuator selection and attributes is very rich (see, for example, Nasar and Boldea 1976; Brunell 1979; Huber et al. 1997; Zupan et al. 2002). Actuators are generally defined as controllable work producing devices.

L. Dihoru (✉) • M.S. Dietz • A.J. Crewe • C.A. Taylor
Department of Civil Engineering, University of Bristol, Queen's Building,
University Walk, Bristol BS8 1TR, UK
e-mail: Luiza.Dihoru@bristol.ac.uk; M.Dietz@bristol.ac.uk; a.j.crewe@bristol.ac.uk;
colin.taylor@bristol.ac.uk

M.N. Fardis and Z.T. Rakicevic (eds.), *Role of Seismic Testing Facilities in Performance-Based Earthquake Engineering: SERIES Workshop*, Geotechnical, Geological and Earthquake Engineering 22, DOI 10.1007/978-94-007-1977-4_7,
© Springer Science+Business Media B.V. 2012

In the particular area of seismic testing, actuators are required to provide force and displacement at a certain rate and precision. The basic characteristics of an actuator are the maximum actuation stress (σ_{max} = the maximum applied force per unit cross-sectional area which produces maximum work output in a single stroke) and the maximum actuation strain (ε_{max} = the maximum value of actuation strain in a single stroke which produces maximum work output). As mass and volume of an actuator are also relevant for its performance, the actuator density (ρ = ratio of mass to initial volume of an actuator) is another criterion for selection. Actuation efficiency is assessed via two additional parameters: volumetric power (p = the mechanical power output per unit initial volume in sustainable cyclic operation) and efficiency (η = the ratio of mechanical work output to energy output during a complete cycle of operation). The above mentioned parameters serve in computing various performance indices which measure actuator effectiveness for a certain application (Huber et al. 1997).

Hydraulic and pneumatic actuators provide force and displacement via the flow of a pressurized fluid. In general, the maximum actuation strain (ε_{max}) is limited by the design of the actuator and the maximum actuation stress (σ_{max}) is limited by the pressure of the working fluid. In hydraulics, the difficulty of high-pressure containment begins to outweigh high pressure advantages at about 40–45 MPa (Brunell 1979), though several proprietary actuators may work with stresses up to 70 MPa (Enerpac Ltd). The actuator volumetric power and speed are limited by the maximum sliding speeds that can be tolerated by the seals. The overwhelming majority of the actuation devices employed in seismic testing are hydraulic as they provide suitable output power in cyclic actuation for the typical frequency range ($f < 100$ Hz).

Electromagnetic linear actuators comprise solenoids, moving coil actuators and magnetostrictive devices. Electromechanical solenoids consist of an electromagnetically inductive coil, wound around a movable steel or iron slug (termed the armature). The coil is shaped such that the armature can be moved in and out of the center, altering the coil's inductance and thereby becoming an electromagnet. The armature is used to provide a mechanical force. In cyclic operation, a resetting mechanism (a second solenoid) is needed. Although typically weak over anything but very short distances, solenoids may be controlled directly by a controller circuit, and thus have very low reaction times. Their frequency response is limited by the inertia of the moving armature (operation range up to 100 Hz). The power output of the commercially available solenoids is only about 10^4 Wm^{-3} (about 1,000 times less than a typical hydraulic actuator output), therefore such actuation devices cannot be used alone for providing force in a typical seismic laboratory test. However, their potential in hierarchical actuation layouts is worth investigating. Moving coil actuators on the other hand present the advantage of a low-inertia moving part (coil), therefore they can operate in higher bandwidths (frequencies up to 50 kHz). A wide variety of designs and operation characteristics for moving coil actuators can be found in literature (Borwick 1988; Wavre and Thouvenin 1995). Magnetostrictive actuators are based on materials that can convert magnetic energy into kinetic energy. Certain ferromagnetic materials (ex. iron-lanthanide compounds, Terfenol-D®)

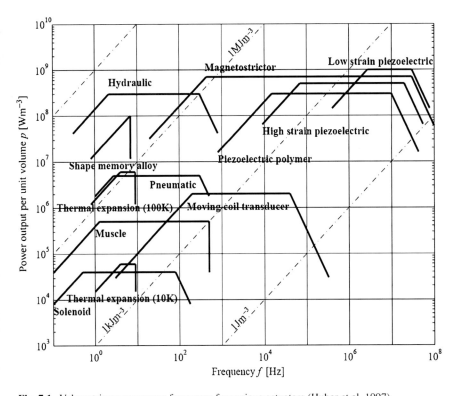

Fig. 7.1 Volumetric power versus frequency for various actuators (Huber et al. 1997)

change dimensions under the influence of an applied magnetic field, caused by the reorientation of magnetic domains. The actuation strain is limited by a magnetic saturation value, therefore such actuators have limited strokes and have found relatively few applications (ex. active aerodynamic surfaces, Bothwell et al. 1995).

Two examples of selection chart for linear actuators according to their power output and frequency range (Fig. 7.1) and specific actuation stress and actuation strain (Fig. 7.2) are given. Heavy lines bound the upper limits of performance.

7.1.2 Actuators Employed in the Earthquake Engineering Facilities

An online survey was carried out between 15/11/2009 and 31/01/2010 by University of Bristol (UNIVBRIS) on the current actuation technology employed by the seismic engineering community in Europe. This work was part of the SERIES Joint Research Activity JRA1 and enjoyed the contribution from ten major dynamic laboratories in Europe. The following institutions took part in the survey: CEA

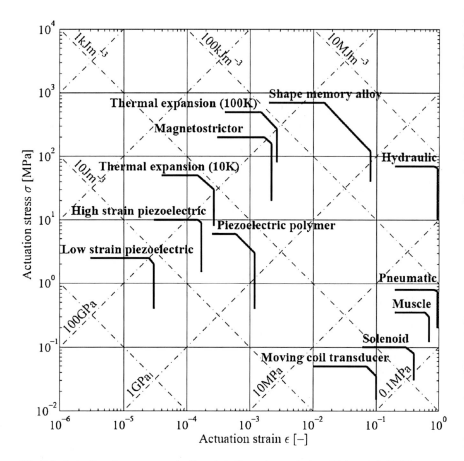

Fig. 7.2 Actuation stress versus actuation strain for various actuators (Huber et al. 1997)

(Commissariat a l'Energie Atomique, France), IZIIS (Institute of Earthquake Engineering and Engineering Seismology, University 'St. Cyril and Methodius', FYROM), JRC (Joint Research Centre Institute for the Protection and Security of the Citizen), METU (Middle East Technical University, Turkey), NTUA (National Technical University of Athens, Laboratory for Earthquake Engineering, Greece), P&P (P&P LMC S.r.l., Italy), UNIVBRIS (University of Bristol, Earthquake Engineering Research Centre, United Kingdom), UNIKA (University of Kassel, Germany), UOXF.DF (University of Oxford, United Kingdom), UPATR (University of Patras, Greece). The current actuation capabilities and performance requirements were compiled using structured questionnaires. Information has been collated about the existing laboratory infrastructures, the current types of tests, experimental configurations and performance requirements. The needs of the earthquake engineering community that are not currently met by available actuation devices have been explored. User opinions have been investigated in relation to performance enhancement and optimisation solutions for hydraulic, electrical and hybrid actuation devices in order to expand the experimental capabilities and meet more sophisticated test needs in the future. The UNIVBRIS survey results have been statistically processed and the emerging conclusions are presented in this paper.

The overwhelming majority of the actuators employed in the European dynamic testing facilities are hydraulic. The hydraulic actuation technology currently in use comes from major worldwide recognized manufacturers such as: MTS Systems Inc, Instron Engineering Inc, Carl Schenck AG, Moog Inc, Bosch Rexroth GmbH, Sandner Messtechnik GmbH, Servotest Ltd and Power Team Ltd.

The current stock of hydraulic actuators is between 10 and 30 years old for 70% of the facilities and less than 10 years old for the rest. The survey results collected for the first five most used types of hydraulic actuators within the laboratory infrastructure show the following technical characteristics:

- Rated Capacity: 50–500 kN (typical), 2,500 kN (maximum)
- Stroke: 50–500 mm (typical), 2,500 mm (maximum)
- Peak Velocity Range: 0.1–4 m/s
- Peak Acceleration (no payload): 2–395 m/s^2
- High Frequency Cut-Off: 5–150 Hz
- Length: 200–3,000 mm
- Mass: 80–2,700 kg
- Type of Controller: controller and actuator made by the same manufacturer.
- Type of Servo-Valves: MTS Systems Inc, Servotest Ltd, Moog Inc
- Flow Rate: 60–640 l/min

The surveyed European earthquake simulators present a wide range of dynamic capabilities:

- Displacement Range: ±50 mm (Thessaloniki) up to ±500 mm (EUCENTRE)
- Peak Velocity: 0.2 m/s (LNEC) up to 2.2 m/s (EUCENTRE)
- Peak Acceleration: 1.8 g (LNEC) up to 6 g (CEA)

The electrical actuators currently in use were supplied by Oswald Elektromotoren GmbH and Linmot Ltd. Their technical specification is summarized below:

- Rated capacity: 0.1–11 kN
- Stroke: 120–160 mm
- Peak velocity range: up to 1.9 m/s
- Peak acceleration (no payload): up to 345 m/s^2
- High frequency cut-off: higher than 100 Hz

7.2 Actuation Requirements for Typical Test Scenarios and Problems of Actuator Performance

Actuators are controllable work-producing devices whose selection is made according to specified test requirements. The following actuation requirements are considered relevant by the seismic testing community.

Peak velocity: Shaking table (ST) testing requires high-capacity servo-hydraulic actuators to drive the large mass of the table and test specimen at the required rate. All shaking table users report that actuator peak velocity is a key parameter in testing. The size of the earthquake that can be reproduced on a shaking table is normally governed by the velocity content, since this is directly related to the oil flow rate that can be provided by the pumping systems and servo-valves (Williams and Blakeborough 2001). High velocity motions cannot always be reproduced by the existing tables due to the flow rate limits imposed by servo-valve design and operation. Problems such as mid-range slack in the servo-valves, backlash in primary swivels and limited flow rate adversely affect the actuator.

Similarly to the ST testing, strong wall/floor real-time dynamic tests with substructuring (RTPD) require low friction devices and high-velocity response. Also, rapid velocity response is often expected to cater for sudden changes in the compliance of the test specimen.

Rated capacity: The participants in the survey reported that high force capacity is the main performance requirement in quasi-static (QS) tests. Cyclic tests for material/component characterization require large forces going in some instances up to 0.5 MN. In ST testing the rated capacity of the actuators was reported to be a limiting factor for the size of the specimen. For example, for the UNIVBRIS earthquake simulator, the maximum overturning moment of a tall and heavy specimen is limited by the capacity of the vertical actuators (4×70 kN).

Maximum stroke: Several ST tests were reported in which the displacement demand was greater than the available actuator displacement range.

Frequency range: The actuator operating frequency range is considered to be very significant by the ST users. ST testing is mostly employed for reduced-scale models, therefore the issue of dynamic input scaling needs to be addressed. Small scale physical modelling on the shaking table implies a reduction of dynamic time step, therefore a requirement for the actuators to perform in higher frequency bands. A soil-structure interaction model (UNIVBRIS) where the scaling factor for length was $SL = 37.5$, required scaling up the frequency of the shaking table input (scaling factor for frequency $Sf = SL = 15.25$). This requirement pushed the actuators towards the high end of their normal operating frequency range (0–60 Hz).

Control speed and linearity: All facilities in the survey consider that actuation should be fast, linear and repeatable (except for QS and pseudo-dynamic (PD) tests where speed is not a requirement). In ST testing, the actuation performance has been limited by the effectiveness of fixed gain control algorithms that assume that both table and specimen behave linearly during a test. In reality hydraulic actuators are not linear devices and the table-specimen system is likely to be non-linear, especially when the model starts to exhibit plastic behaviour. The traditional control strategy (Severn 1997) is based on linear controllers with gains that do not change during a test. An inverse kinematic solver converts the desired table displacements into driving signals for the actuators. At UNIVBRIS eight actuators are employed for a six DOF table, so the solver must carry out a transformation from six DOF to

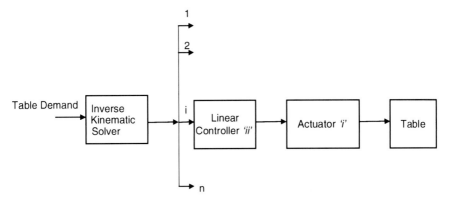

Fig. 7.3 Basic shaking table control strategy (Severn 1997)

eight DOF. Each actuator has in-built transducers that provide displacement and/or acceleration feedback signals to the linear controllers (Fig. 7.3).

In recent years some modifications have been made to this approach. For example, most modern control systems now provide closed-loop control directly on the shaking table degrees of freedom, rather than on the individual actuators. Also, modern controllers employ a three-variable control parameter that is a weighted combination of position, velocity and acceleration (Williams and Blakeborough 2001). However, these recent improvements still cannot alleviate the main problem associated with the linear controllers: the impossibility of catering for the non-linearity of the table-specimen dynamics.

The actuation performance requirements become even more demanding in real-time tests (real time pseudo-dynamic (RTPD) and real time dynamic hybrid (RTDH) tests). The actuator delay (also called 'time lag' or 'phase lag') is reported to be the most significant problem in dynamic testing with hydraulic actuators. The dynamic characteristics of hydraulic actuators inevitably include a response delay, which is equivalent to negative damping (Horiuchi et al. 1999). Time lags of 8–40 ms are reported for the commonly used hydraulic systems in civil engineering applications (Darby et al. 1999, 2000; Horiuchi et al. 1999; Williams and Blakeborough 2001). In order to be successful, RTPD and RTDH testing necessarily require some form of delay compensation. Usually, actuator 'proprietary' controllers implement PID control schemes that provide sufficient level of accuracy and stability for standard dynamic testing. However, in RTPD and RTDH tests, there is a second (interface force) feedback loop through the numerical model in addition to the controller displacement feedback loop. The implication of the second feedback loop is that the control accuracy required for real-time tests is far greater than for standard dynamic testing. It is now well established that a 'proprietary' controller is not sufficient to compensate for the actuator dynamics (Lim et al. 2007), particularly when testing lightly-damped structures where it has been reported that a delay of less than 1 ms between the desired and achieved displacement can be sufficient to cause system instability (Bonnet et al. 2007). To reduce the effect of actuator dynamics a range of control strategies have been proposed (Reinhorn and Shao 2004; Neild et al. 2005), which are implemented as a supplementary control loop around the proprietary controller. One approach is to adopt a modified version of the technique developed for testing a helicopter lag damper (Wallace et al. 2009), in which the proprietary controller is tuned such that the overall actuator dynamics replicate a first order transfer function. An inverse of this transfer function is then included within the numerical model integration routine to compensate for these dynamics. If the actuator is used over a large proportion of its loading range, it is envisaged that the technique will have to be modified. An adaptive delay compensator (Wallace et al. 2005) can be used to compensate for any remaining transfer system dynamics resulting from any deterioration of the first order transfer function fit to the actuator and proprietary controller dynamics due to high loading.

Researchers at the Kassel University in Germany report the use of compensation algorithms based on on-line system identification in the time domain that present advanced features such as adaptive capability for time-varying systems, both linear and non-linear. Their compensation method is developed in the context of sub-step control in substructure tests (Nguyen and Dorka 2008).

Feedback transducer accuracy: The internal and/or external transducers associated with the actuation systems influence significantly the control accuracy and stability of actuation. Ideally, such transducers must be low-noise and highly accurate. In particular, real-time testing methods (RTPD and RTDH) may suffer from noise and/ or insufficient resolution of feedback transducers (internal linear variable displacement transformers (LVDTs) and load cells). For example, a strong floor RTPD test investigating the distributed substructuring test method was reported to be influenced by feedback transducer accuracy (UNIVBRIS). Poor feedback signals from internal LVDTs often make displacement control difficult to achieve and lead to choosing force control in testing (Oxford University). The use of external small stroke LVDTs or linear optical encoders was reported by several facilities as being among the solutions to improve displacement control in actuation (Oxford University, Kassel University).

Flexibility of bearings: Most shaking table actuation systems employ mechanical spherical bearings that present a certain amount of axial flexibility in order to allow rotational movement. By design, the bearing has to be tight enough to prevent axial movements, but flexible enough to allow rotation. When axial flexibility is not properly adjusted, then sudden opening and closing of the joints of the bearings ('backlash') may create high-frequency shock pulses into the table motion. As the backlash is highly non-linear, it is impossible to compensate for it via feedback control algorithms. Minimization of bearing backlash can be achieved by a regular and effective maintenance programme. All the shaking table users that took part in the survey reported that backlash in bearings may affect the actuation control in their facility.

Servo-valve operation may affect the actuation performance through insufficient response rate, mid-range slack and limited flow.

7 Performance Requirements of Actuation Systems for Dynamic Testing...

Specimen-shaking table interaction: ST users often have to address the issue of specimen-shaking table interaction. Larger specimens can pose problems in control especially when motion needs to be prevented in the unwanted degrees of freedom (DOF). The shaking tables in Europe are a wide variety of simulators ranging from one DOF to six DOFs. The modern shaking tables usually provide active control in all degrees of freedom, even though in the passive directions the motion is controlled to be zero. An exception is the table at LNEC Lisbon that is a three DOF simulator which provides active control only for translations on X, Y, and Z, while a physical restraint system consisting of large torque tubes prevents the rotational motions (Bairrao and Vaz 2000; Williams and Blakeborough 2001). Whilst tuning performance during test preparation, shaking table users taking part in this survey employ dead weights to emulate the specimen mass added to the platform (i.e. during the 'matching' process). The ratio between the specimen and the table mass is often reduced by adding static mass to the table in order to improve the shaking table performance. For non-linear specimens, the actuation control can become challenging when the specimen is changing its stiffness during loading (stiffness degradation).

Oil-column resonance: The oil-column resonance was reported to be another potentially-detrimental factor in testing. For example, the UNIVBRIS old shaking table's resonant frequencies in the X, Y and Z directions were measured as being 16.50, 15.75 and 22.75 Hz, respectively (Crewe 1998). These frequencies could overlap onto the dynamic response of the model making the model's true behaviour difficult to interpret.

Usually, an accurate characterization of the shaking table performance involves the measurement of the oil column resonant frequencies in the active degrees of freedom. By knowing the exact values of these frequencies, a more accurate interpretation of test results can be performed: the true model behaviour can be separated from the dynamics of the shaking table.

Axial and lateral stiffness of actuator: The actuators for shaking tables are usually long and thin, because of requirements of long stroke and high velocity. This means that the actuators will have some axial flexibility and will also have the potential for bending laterally like pin-ended struts. The natural frequency of the axial mode of a normally proportioned actuator is generally high (>100 Hz), while the natural frequency of the bending mode is likely to be lower (generally two to three times higher than the natural frequency of the oil column (Crewe 1998). These frequencies are relatively high in relation to the operating frequencies of a table and hence are unlikely to be particularly significant.

The reported factors of influence on the hydraulic actuator performance are summarized below:

- Backlash in swivels (2 reports)
- Slack in servo-valves (2 reports)
- Specimen-shaking table interaction (2 reports)
- Insufficient frequency range for the actuator (2 reports)
- Control algorithms not sophisticated enough (2 reports)

128 L. Dihoru et al.

- Internal LVDTs not precise enough for control (1 report)
- Delays in control system (1 report)
- Oil-column resonance (1 report)

As regards the electric actuation, the main reported limitations were: small work output, electromagnetic interference and high noise at low frequencies.

7.3 User Needs – Current and Future Actuation Requirements

The survey investigated the user needs with respect to the current and future actuation requirements: 70% of the facilities that took part in the survey reported that they experienced demands for tests that could not be met. The test features beyond the current actuation capabilities involve: large displacements (e.g. 0–2,000 mm), large velocities (greater than 1 m/s), high operating frequencies (up to 200 Hz), large forces (up to 4,800 kN), large accelerations (up to 6 g).

In general, there is a need for actuators to be made faster, higher capacity and more efficient in terms of power. The ST users often experience requests for testing specimens with masses beyond the capacity of their shaking table or specimens whose height would lead to larger than allowable overturning moments. There are also requests for shaking table tests to envelope required-response-spectra that contain significant energy at very low frequencies. The requisite stroke can be in excess of that available.

True distributed loading is another area of testing currently unattainable with the present actuation systems. In general single actuators apply loads at points, simulating concentrated loads or the inertial loads of lumped masses. However, walls, dams and bridges are subjected to distributed loads. Within laboratories, pressure bags have been used on slabs with a certain degree of success, but such systems cannot reproduce spatially varying loads and also present the disadvantage of subjecting the walls to unwanted confinement. At present, there is no way to test continuous structures realistically except on a shaking table.

In the field of real-time testing, there is room for improvement in terms of speed, accuracy and stability of actuation. Faster RTPD and RTDH tests present real challenges for most of facilities. The required displacement control accuracy was often not achievable because of the actuator response delay and non-linearity of servo-valves (UNIVBRIS). RTPD tests on devices sensitive to deformation rate are among the tests that are currently unattainable.

7.4 Solutions for Actuator Performance Enhancement

The Laboratories that took part in the survey report that the following actuation features would lead to an increase of performance in earthquake engineering testing:

- Better performing servo-valves (80% of facilities)
- Wider frequency bandwidth (80% of facilities)

7 Performance Requirements of Actuation Systems for Dynamic Testing...

- Less latency in response (80% of facilities)
- Lower cost (80% of facilities)
- Better performing electrical actuators (70% of facilities)
- Lighter actuators (70% of facilities)
- Increased peak velocity (60% of facilities)
- Increased stroke (60% of facilities)
- Increased peak acceleration (60% of facilities)
- Increased force capacity (50% of facilities)
- Smaller geometrical size for actuators (50% of facilities)

In general, the ST users report the need for high velocity, high stroke and high force actuators. The use of additional (pneumatic or electrical actuators) to increase the maximum overturning moment on the shaking table has been suggested. Also, the use of large capacity accumulators in order to meet higher velocity demands (1.2–1.5 m/s) has been proposed.

In parallel with the improvement of mechanical behaviour of actuators, there is a need for more reliable models of servo-hydraulic systems in order to effectively control the non-linear effects that occur on the loading paths. The traditional control strategy for shaking tables involves linear controllers whose fixed gains remain a major limitation especially for specimens that become non-linear during testing. One possible way of overcoming this limitation is to use adaptive control strategies in which the internal parameters (gains) can be tuned on-line during the course of the test. Among the adaptive techniques, the Minimal Control Synthesis (MCS) algorithm (Stoten and Benchoubane 1990) has been successfully tested on shaking tables (UNIVBRIS, National Technical University of Athens (NTUA)). MCS can be used in inner loop, where the algorithm is applied to each individual actuator separately (Fig. 7.4a) or in outer loop where the overall table motion is compared with the overall table demand (Fig. 7.4b).

The latter strategy is similar to the matching procedure used for conventionally controlled tables, with the difference that the control is carried out in real time (Williams and Blakeborough 2001). The architecture and the control scheme of the MCS algorithm are described in detail in several publications (Stoten and Benchoubane 1990; Stoten and Hodgson 1998; Stoten and Gómez 2001). The advantage that MCS presents for shaking tables resides in its robustness and stability, features that are essential in experiments that are often expensive and complex to build. MCS can work wrapped around existing linear controllers and may be used as a retrofit to existing shaking-table-controller systems (UNIVBRIS, NTUA). Kassel University reports on various control techniques employed for actuation control enhancement. Advanced algorithms of digital control (Roik and Dorka 1989) have shown to increase accuracy and stability in continuous hybrid testing with hydraulic actuators. Also, some improvements have been reported in their real time substructure tests following the use of PID error force compensation (Bayer et al. 2005) and phase lag compensation (Van Thuan and Dorka 2008).

Great improvement in performance of real time testing is also possible by reducing the noise in feedback transducers. In principle, electrical filtering induces a phase lag that makes control more difficult. Load cells, for instance, are almost all strain

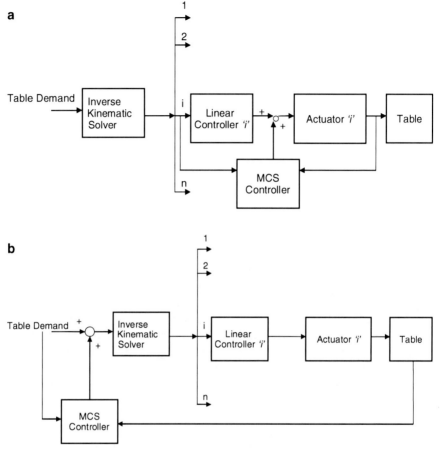

Fig. 7.4 (**a**) Inner and (**b**) outer loop implementation of the MCS adaptive control algorithm (Williams and Blakeborough 2001)

gauge devices which require high amplification to produce measurable signals. This does introduce noise which must then be filtered. More sophisticated devices based on digital techniques such as optical encoders would produce much lower noise levels and make faster testing possible.

Electro-mechanical screw devices currently employed by Oxford University, although currently difficult to control at speed in real time, are considerably lighter for the same level of output force, and extending the stroke comes at a very much reduced cost over that for hydraulic and electro-magnetic devices. Improving the response time of these devices would bring them into the range to be considered for real-time testing.

Lighter weight materials (e.g. aluminium alloy or steel liners/composite containment) are regarded as candidate materials in the manufacturing of hydraulic actuators that may lead to reduction of mass and inertia forces generated on the shaking table.

7.5 Future Direction of Actuation Technology

In order to meet increasingly sophisticated future testing needs, the seismic engineering research community needs to address the issues of optimization, enhancement, adaptation and/or replacement of their current actuation technology. The facilities that took part in the UNIVBRIS survey suggested several actuation enhancements that they consider necessary to meet their future testing needs.

The performance of the existing hydraulic actuation systems can be increased via improved servo-control algorithms, digital noise-free transducers and dual servo-valves. Since 70% of users consider that lighter actuators are needed for better control and minimization of inertial forces, the future use of high-strength composites in actuator manufacturing may provide the solution to this demand.

True distributed loading is currently an area of testing insufficiently covered by the present actuation capabilities. The future use of arrays of small electro-magnetic or electro-mechanical devices is considered a promising solution towards a more realistic simulation of uniformly distributed loading.

Hierarchical layouts that improve actuator performance through increments in geometric and control complexity are potential candidate solutions in actuation. A typical hierarchical configuration may use a servo-hydraulic actuator for coarse positioning and a piezoelectric stage for fine positioning. The principles of hierarchical actuation have been explained in several publications in the last decade (Huston et al. 2002, 2005). The concept of hierarchical design is based on the following notions:

- The actuators are configured in layers of super-actuators based on sub-actuators
- The ground level sub-actuators are configured to optimize the power output
- The level of integration in hierarchy should provide high redundancy
- Sub-actuators with differing capabilities can be used for overall performance enhancement

Four examples of hierarchical architecture consisting of sub-actuators connected in series or in parallel by whiffletree style links are shown in Fig. 7.5. A mathematical theory of hierarchical actuators has not yet been developed, though graph theory is believed to be a possible approach to represent the kinematics of such systems. At present, the hierarchical design is still in its infancy, however its concept may be a promising solution for energy and control optimization.

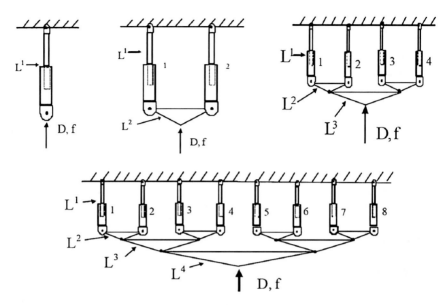

Fig. 7.5 Configurations of hierarchical actuators (Huston et al. 2002, D displacement, f force)

7.6 Conclusions

The present survey has identified a need of high-force electrical actuators providing capacity and precision alike to enter the mainstream research. It is envisaged that future high-performance actuation systems will see electrical actuators as direct replacements for servo-hydraulic actuators or as subcomponents in hybrid actuation systems that employ both servo-hydraulic and electrical actuators. In terms of power output per unit volume, the hydraulic actuators overtake other types of actuators (hydraulic: $p > 108$ Wm^{-3}, pneumatic: $p = 10^6$–10^7 Wm^{-3}, moving coil actuator: $p = 10^6$–10^7 Wm^{-3}, solenoid: $p < 10^4$ Wm^{-3}). In terms of frequency response, the hydraulic actuators perform best in the 10–150 range, while moving coil and piezoelectric actuators can be used in higher bandwidths (10^4–10^7 Hz). The majority of users consider that combinations of hydraulic and electric actuators combining high force capacity with displacement precision in higher frequency bandwidth may lead to an increase of performance and versatility in testing. Research and application driven manufacturers have already presented a new generation of linear electrical actuators that have strokes and velocities compatible with the seismic testing range. The most significant benefit of such actuators is the enhanced high-frequency response characteristics that make them superior to the existing servo-hydraulics systems. However, electrical actuators do not yet have the load capacity range of the latter, although this will continue to improve.

Acknowledgments The research leading to these results has received funding from the European Community's Seventh Framework Programme [FP7/2007-2013] under grant agreement n° 227887.

References

Bairrao R, Vaz CT (2000) Shaking table testing of civil engineering structures – the LNEC 3D simulator experience. In: Proceedings of the 121th world conference on earthquake engineering. NZ Society for Earthquake Engineering, Auckland

Bayer V, Dorka UE, Füllekrug U, Gshcwilm J (2005) On real-time pseudo-dynamic sub-structure testing: algorithm, numerical and experimental results. Aerosp Sci Technol 9:223–232

Bonnet PA, Lim CN, Williams MS, Blakeborough A, Neild SA, Stoten DP, Taylor CA (2007) Real-time hybrid experiments with Newmark integration, MCSmd outer-loop control and multi-tasking strategies. Earthquake Eng Struct Dyn 36(1):119–141

Borwick J (1988) Loudspeaker and headphone handbook. Butterworth, London

Bothwell CM, Chandra R, Chopra I (1995) Torsional actuation with extension torsion composite coupling and a magnetostrictive actuator. AIAA J 33:723–729

Brunell R (1979) Hydraulic and pneumatic cylinders. Trade and Technical, Channel Industries Inc, Santa Barbara

Crewe AJ (1998) The characterisation and optimisation of earthquake shaking table performance. PhD thesis, University of Bristol

Darby AP, Blakeborough A, Williams MS (1999) Real-time substructure tests using hydraulic actuators. J Eng Mech 125:1133–1139

Darby AP, Blakeborough A, Williams MS (2000) Improved control algorithm for real-time substructure testing. Earthquake Eng Struct Dyn 30:431–448

Horiuchi T, Inoue M, Konno T, Namita Y (1999) Real-time hybrid experimental system with actuator delay compensation and its application to a piping system with energy absorber. Earthquake Eng Struct Dyn 28:1121–1141

Huber JE, Fleck NA, Ashby MF (1997) The selection of mechanical actuators based on performance indices. Proc R Soc Lond A 453:2185–2205

Huston D, Esser B, Werner MH (2002) Hierarchical actuators. In: Proceedings of the first world congress on biomimetics and artificial muscles, Albuquerque

Huston D, Esser B, Kahn, Spencer G, Burns D (2005) Hierarchical actuator systems. In: SPIE 5762–42, smart structures and materials: industrial and commercial applications of smart structures technologies, San Diego

Lim CN, Neild SA, Stoten DP, Drury D, Taylor CA (2007) Adaptive control strategy for dynamic substructuring tests. ASCE J Eng Mech 133(8):864–873

Nasar SA, Boldea I (1976) Linear motion electric machines. Wiley, New York

Neild SA, Stoten DP, Drury D, Wagg DJ (2005) Control issues relating to real-time substructuring experiments using a shaking table. Earthquake Eng Struct Dyn 34(9):1171–1192

Nguyen VT, Dorka UE (2008) Phase lag compensation in real-time substructure testing based on online system identification. In: Proceedings of the 14th world conference on earthquake engineering, Beijing, paper 12-01-0148

Reinhorn A, Shao X (2004) Advanced dynamic testing techniques in structural engineering. In: CIE616, Department of Civil, Structural and Environmental Engineering, University of Buffalo

Roik K, Dorka UE (1989) Fast online earthquake simulation of friction damped systems. SDB151 report no. 15, Ruhr-University Bochum

Severn RT (1997) Structural response prediction using experimental data. 6th Mallet-Milne Lecture. A.A. Balkema, Rotterdam

Stoten DP, Benchoubane H (1990) Robustness of a minimal controller synthesis algorithm. Int J Control 51:851–861

Stoten DP, Gómez EG (2001) Adaptive control of shaking tables using the minimal control synthesis algorithm. Philos Trans R Soc Lond 359:1697–1723

Stoten DP, Hodgson SP (1998) Passivity-based analysis of the minimal control synthesis algorithm. Int J Control 63:67–84

Van Thuan N, Dorka UE (2008) Phase lag compensation in substructure testing based on online system identification. In: 14th world conference on earthquake engineering, 12–17 Oct 2008, Beijing

Wallace MI, Wagg DJ, Neild SA (2005) An adaptive polynomial based forward prediction algorithm for multi-actuator real-time dynamic sub structuring. Proc R Soc Part A 461(2064):3807–3826

Wallace MI, Wagg DJ, Neild SA, Bunniss P, Lieven NAJ, Crewe AJ (2009) Testing coupled rotor blade-lag damper vibration dynamics using real-time dynamic substructuring. J Sound Vib 307:737–754

Wavre N, Thouvenin X (1995) Voice-coil actuators in space. In: Proceedings of the 6th European space mechanisms & tribology symposium, Technopark Zurich, 4–6 Oct 1995

Williams MS, Blakeborough A (2001) Laboratory testing of structures under dynamic loads: an introductory review. Philos Trans R Soc Lond A 359:1651–1669

Zupan M, Ashby MF, Fleck NA (2002) Actuation classification and selection – the development of a database. Adv Eng Mater 4(12):933–940

Chapter 8
Model Container Design for Soil-Structure Interaction Studies

Subhamoy Bhattacharya, Domenico Lombardi, Luiza Dihoru, Matt S. Dietz, Adam J. Crewe, and Colin A. Taylor

Abstract Physical modelling of scaled models is an established method for understanding failure mechanisms and verifying design hypothesis in earthquake geotechnical engineering practice. One of the requirements of physical modelling for these classes of problems is the replication of semi-infinite extent of the ground in a finite dimension model soil container. This chapter is aimed at summarizing the requirements for a model container for carrying out seismic soil-structure interactions (SSI) at 1-g (shaking table) and N-g (geotechnical centrifuge at N times earth's gravity). A literature review has identified six types of soil container which are summarised and critically reviewed herein. The specialised modelling techniques entailed by the application of these containers are also discussed.

8.1 Introduction

8.1.1 Physical Modelling in Earthquake Engineering

The structural failures and loss of life in recent earthquakes have shown the shortcomings of current design methodologies and construction practices. Post earthquake reconnaissance investigations have led to improvements in engineering

---

S. Bhattacharya (✉) • D. Lombardi • L. Dihoru • M.S. Dietz • A.J. Crewe • C.A. Taylor
Department of Civil Engineering, University of Bristol, Queen's Building,
University Walk, Bristol BS8 1TR, UK
e-mail: s.bhattacharya@bristol.ac.uk; domenico.lombardi@bristol.ac.uk;
Luiza.Dihoru@bristol.ac.uk; M.Dietz@bristol.ac.uk; a.j.crewe@bristol.ac.uk;
colin.taylor@bristol.ac.uk

M.N. Fardis and Z.T. Rakicevic (eds.), *Role of Seismic Testing Facilities*
in Performance-Based Earthquake Engineering: SERIES Workshop, Geotechnical,
Geological and Earthquake Engineering 22, DOI 10.1007/978-94-007-1977-4_8,
© Springer Science+Business Media B.V. 2012

Table 8.1 Historical development of earthquake engineering practice after Bhattacharya et al. (2011a)

Earthquake	Remarks	Post earthquake developments
28th Dec, 1908 Reggio Messina earthquake (Italy)	120,000 fatalities. A committee of nine practising engineers and five professors were appointed by Italian government to study the failures and to set design guidelines	Base shear equation evolved, i.e. the lateral force exerted on the structure is some percentage of the dead weight of the structure (typically 5–15%)
1923 Kanto earthquake (Japan)	Destruction of bridges, buildings. Foundations settled, tilted and moved	Seismic coefficient method (equivalent static force method using a seismic coefficient of 0.1–0.3) was first incorporated in design of highway bridges in Japan (MI 1927)
10th March, 1933 Long Beach earthquake (USA)	Destruction of buildings, especially school buildings	UBC (1927) revised. This is the first earthquake for which acceleration records were obtained from the recently developed strong motion accelerograph
1964 Niigata earthquake (Japan)	Soil can also be a major contributor of damage	Soil liquefaction studies started
1971 San Fernando earthquake (USA)	Bridges collapsed, dams failed causing flood. Soil effects observed	Liquefaction studies intensified. Bridge retrofit studies started
1994 Northridge earthquake (USA)	Steel connections failed in bridges	Importance of ductility in construction realised
1995 Kobe earthquake (Japan)	Massive foundation failure. Soil effects were the main cause of failure	Downward movement of a slope (lateral spreading) is said to be one of the main causes. JRA (1996) code modified (based on lateral spreading mechanism) for design of bridges

analysis, design and construction practices. A brief historical development of earthquake engineering practice illustrating how earthquake engineers have learned from failures in the past is outlined in Table 8.1. Therefore, before applying any earthquake resistant design method to practical problems, or establishing a proposed failure mechanism, it is better to seek verification from all possible angles. Obtaining data from physical systems (i.e. physical modelling) on the performance of the design method is essential to that verification. Such data ideally could be acquired from detailed case histories during earthquakes but only by putting society at risk in the meantime. The fact that large earthquakes are infrequent and that most structures are not instrumented makes this method of verification difficult. Alternatively, a record of observations of the response of a physical system can be acquired by modelling.

8 Model Container Design for Soil-Structure Interaction Studies

Fig. 8.1 Damage predominantly due to structural inadequacies observed during the 2001 Bhuj earthquake

Fig. 8.2 Collapse of Bio-Bio Bridge during the 2010 Chile earthquake

8.1.2 Seismic Soil-Structure Interaction Problems

Structural failure during earthquakes can result from inadequacies of either the structure itself, its foundation, or a combination of both. Figure 8.1 shows the failure of a residential building predominantly due to structural inadequacies such as poor ductility and improper beam-column detailing. On the other hand, the failures shown in Fig. 8.2 are probably due to foundation failure or soil-structure interaction. In such failures the soil supporting the foundation plays an important role. The behaviour of foundations during earthquakes is often dictated by the response of the supporting soil. In general, there are two problematic types of ground response: (a) liquefaction, such as seen in the 1995 Kobe earthquake; (b) amplification of the ground motion, such as seen during the 1989 Loma Prieta earthquake in California or the 1985 Mexico earthquake.

Fig. 8.3 Kandla port building

8.1.3 Physical Modelling of Seismic-Soil Structure Interaction Problems

Figure 8.3 shows a piled building in Kandla, which tilted towards the Arabian Sea during the 2001 Bhuj (India) earthquake. The operative mechanism that enabled the tilting is hidden and since earthquakes are very rapid events and as much of the damage to piles occurs beneath the ground, it is hard to ascertain the failure mechanism without undertaking a costly excavation. Alternatively, physical modelling can be used to gain an understanding of the operative mechanisms.

8.1.4 Structure of the Chapter

The chapter is structured in the following way. Section 8.2 presents the two types of geotechnical seismic testing carried out: shaking table test at 1-g and geotechnical centrifuge testing. Some modelling issues are also discussed. Section 8.3 develops the requirements of a model soil container from first principles which are valid at 1-g (shaking table) or N-g testing (centrifuge). Section 8.4 of the chapter critically reviews the six different types of model container used for testing.

8.2 Physical Modelling

8.2.1 1-g Model Tests Using Shaking Table

The behaviour of soil is non-linear and stress dependent. Soils that show contractive behaviour (loose to medium dense sand) under high normal stress may exhibit a dilative behaviour at low stress level. Therefore, physical modelling of geotechnical problems should ensure that stress dependent behaviour is correctly accounted for. High gravitational stresses cannot be produced in a shaking table test. The typical height of soil container varies between 0.5 and 6.3 m and therefore the vertical effective stress is limited to a maximum of 5 kPa (saturated sand) to about 120 kPa (dry sand). The effect of stress dependency on soil strength dominated problems (for example slope failure) in 1-g testing can be addressed through the change in soil density in model ground. It must be mentioned that in some dynamic soil-structure-interaction problems stiffness rather than strength governs and this issue is dealt in the subsequent paragraphs.

In shallow soil container, the isotropic stress level controlling the mechanisms under investigation is low, leading to higher friction angle but at the same time low shear modulus when compared to its equivalent prototype. Many researchers, such as Kelly et al. (2006), Leblanc et al. (2010), address the issue by pouring the sand at lower relative density. Figure 8.4 shows the variation of the peak friction angle (φ')

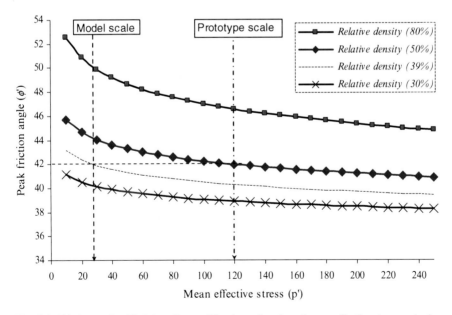

Fig. 8.4 Friction angle of Leighton Buzzard Sand as a function of mean effective stress and relative density

with mean effective isotropic stress (p') for silica sand having various relative density based on Eq. 8.1 following the stress-dilatancy work of Bolton (1986).

$$\phi' = \phi_{cv} + 3\left[R_D\left(9.9 - \ln p'\right) - 1\right] \qquad (8.1)$$

where:

R_D = Relative density of the sand
ϕ_{cv} = Critical State Angle of Friction of the sand; it is 34.3° for Leighton Buzzard sand
p' = Isotropic stress in kPa

Based on this approach, if 120 kPa of mean effective prototype stress at 50% relative density is to be modelled in a small scale laboratory model at 25 kPa stress, the sand is to be poured at about 39% relative density ensuring that the peak friction angle is the same. There is limitation in this approach in the sense that there is a minimum density beyond which sand cannot be poured and other secondary effects become important.

While the effect of peak friction angle in 1-g testing can be resolved using the method presented in Fig. 8.4, further thought is required to address the issue of shear modulus, G. The shear modulus of a soil is function of effective stress shown by Eq. 8.2.

$$G \propto p'^n \qquad (8.2)$$

where the value of n depends on the type of soil. The value of n varies between 0.435 and 0.765 for sandy soil (Wroth et al. 1979) but a value of 0.5 is commonly used. For clayey soil, the value of n is generally taken as 1.

While deriving the scaling laws or non-dimensional groups for a physical model the above concepts should be taken into consideration. These are necessary to interpret the model test results in order to scale up the results for prediction of prototype consequences. Every physical process can be expressed in terms of non-dimensional groups and the fundamental aspects of physics must be preserved in the design of model tests. The necessary steps associated with designing such a model, to be implemented either in 1-g or a multi-g testing (centrifuge) environment, can be stated as follows:

1. To deduce the relevant non-dimensional groups by thinking of the mechanisms that govern the particular behaviour of interest both at model and prototype scale.
2. To ensure that a set of scaling laws are simultaneously conserved between model and prototype through pertinent similitude relationships.
3. To identify scaling laws which are approximately satisfied and those which are violated and which therefore require special consideration.

While deriving the non-dimensional groups, appropriate soil stiffness should be taken into consideration. Examples of derivation of scaling laws for general dynamic problems can be found in the literature, see for example Iai (1989), Muir Wood et al. (2002), Bhattacharya et al. (2011b).

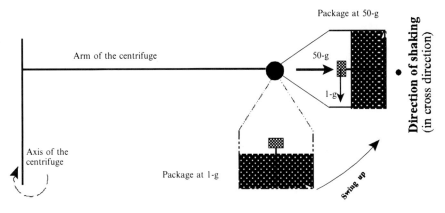

Fig. 8.5 Schematic diagram showing the working principle of a geotechnical centrifuge

8.2.2 Dynamic Geotechnical Centrifuge Tests

An alternative way of modelling is through the use of a geotechnical centrifuge which enables the recreation of the same stress and strain level within the scaled model by testing a 1:N scale model at N times earth's gravity, created by centrifugal force (see Fig. 8.5). In the centrifuge, the linear dimensions are modelled by a factor 1/N and the stress is modelled by a factor of unity. Scaling laws for many parameters in the model can be obtained by simple dimensional analysis, and are discussed by Schofield (1980 and 1981).

8.3 Requirements of a Model Container

Soil strata within the ground or underneath a prototype structure have infinite lateral extent while a model test will have a finite size. The challenge is therefore to replicate the boundary conditions of a ground in a container with finite dimensions. Figure 8.6 shows the pattern of soil deformation along the soil layers. The theoretical pattern of deformation is dependent on the assumption one makes regarding the variation of shear modulus with depth. It may be noted that the displacement is constant at a particular horizontal plane and the amplitude of displacement varies with depth. The soil column can therefore be idealised as a shear beam. Figure 8.7 shows the comparison of a shear beam with Euler-Bernoulli beam. The design of the soil container should be carried out in such a way to replicate as close as practicable the stress-strain condition of an infinite lateral extent and finite depth soil profile, when subjected to a 1-dimensional horizontal shaking, see Fig. 8.6.

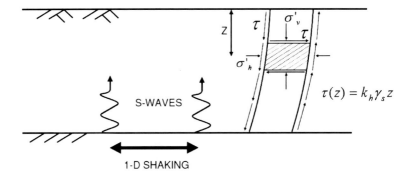

Fig. 8.6 Soil layer of infinite lateral extent and finite depth subjected to a base shaking at its bedrock

Fig. 8.7 Shear beam and Euler-Bernoulli beam

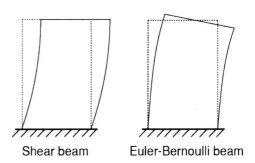

8.3.1 Before Shaking (Geostatic Condition)

The stress field at any point in a given plane in a mass of the soil can be represented by normal and shear stresses. The effective vertical and horizontal stresses are given by:

$$\sigma'_v = \gamma z \quad (8.3)$$

$$\sigma'_h = K_0 \sigma'_v \quad (8.4)$$

Where γ and K_0 are the unit weight of the soil and the coefficient of lateral earth pressure at rest respectively.

In the geostatic condition the vertical and horizontal planes are principal stress planes and the normal stresses acting on them are principal stresses. The vertical and horizontal stresses are also principal stresses as can also be deduced from the pole (*P1*) of the Mohr circle illustrated in Fig. 8.8.

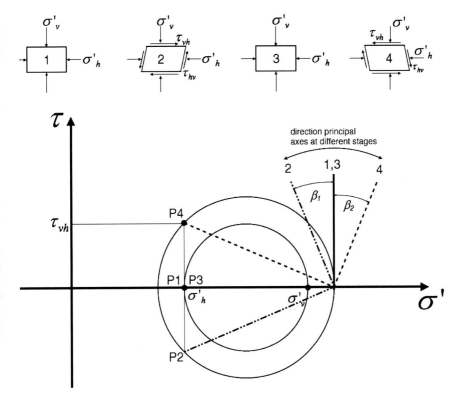

Fig. 8.8 Stress path of a soil element when subjected to horizontal shaking (Adapted from Kramer (1996))

8.3.2 During Shaking

As the horizontal shaking starts, shear waves (indicated by S-waves in Fig. 8.6) propagate vertically upward within the soil. Normal stresses remain constant while shear stresses in both vertical and horizontal planes increases. The shear stress induced by the horizontal shaking may be estimated by Eq. 8.5.

$$\tau(z) = k_h \gamma z \qquad (8.5)$$

where k_h represents the coefficient of horizontal acceleration. The stress field induced by the shaking causes a rotation of the vertical planes while the horizontal planes remain horizontal. This state of stress is also shown in Mohr circle (Fig. 8.8). Since the horizontal and vertical stresses are constant, the Mohr circle increases in size while its centre remains unchanged. Therefore the principal axes will rotate (anticlockwise in this case) and the new position of the pole is now indicated by *P2* (Fig. 8.8). Taking into account the cyclic nature of the shear stresses the principle axes will continuously change denoted by angle β.

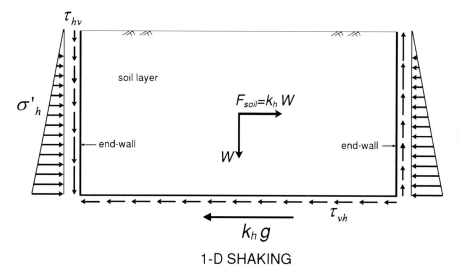

Fig. 8.9 Schematic representations of shear and normal stresses generated within the soil layer due to the inertia of the soil (Adapted from Zeng and Schofield (1996))

8.3.3 Stress Similarity

The horizontal shaking generates shear stresses in both vertical and horizontal planes. If the container end walls are frictionless, vertical stresses cannot develop and the stress field near the boundaries will be different from that of the prototype. However, if the model is tested in the central part of the box at a considerably large distance from the end-walls, such effects can be assumed to be minimal. The evaluation of the distance beyond which the stress field is not affected by the boundaries is complex and requires detailed experimental or numerical analysis. During shaking, the mass of the soil generates an inertia force that can be considered as a horizontal load given by Eq. 8.6 and acts at the centre of inertia of the soil layer (see Fig. 8.9)

$$F_{soil} = k_h W \qquad (8.6)$$

Figure 8.9 shows that (for the particular case considered here) the inertia force of the soil generates a clockwise overturning moment. For the stability of the system this overturning moment must be balanced by the shear stresses acting on the vertical plane. Therefore in absence of adequate friction between the end-walls and the adjacent soil, the shear stresses may not be capable of balancing the overturning moment.

8.3.4 Propagation of the Shaking to the Soil Layer

An important feature of the soil container is represented by the fact that the shaking applied to the base of the container should be transferred to the upper layer of the soil.

8 Model Container Design for Soil-Structure Interaction Studies 145

This condition can be accomplished by the use of a rough base, which enables the generation of shear stresses in the horizontal plane (i.e. at the interface between the soil and the base of the container).

8.3.5 Strain Similarity

The displacement profile of the soil induced by the shaking has to satisfy the condition that at a particular depth the displacement is constant. In other words, the horizontal cross section must remain horizontal. In the model container, the finite dimension of its width (the dimension orthogonal to the shaking) may cause an alteration of the plane-strain conditions. This may be avoided by making the side walls smooth for 1-dimensional shaking tests.

8.3.6 P-Waves Generation and Reflection Problems

S-waves will propagate through the soil layer (see Fig. 8.6). However, during shaking, the soil next to the boundaries may undergo under compression and extension and P-waves may be generated. Therefore the response of the model will be affected by the interaction between P and S waves. Another consideration that must be taken into account is therefore the reflection of the waves from the artificial boundaries. In an infinitely extended soil layer this phenomenon is absent since there are no boundaries and the energy of the waves diminishes with distance. The attenuation of the energy may be explained considering two different mechanisms: The first mechanism is the friction generated by the sliding of the grain particles which converts part of the elastic energy to heat. This dissipation may be considered as function of the hysteretic damping of the soil. The second mechanism is due to the radiation damping which is related to the geometry of the propagation of the waves. As waves propagate, their energy will spread to a greater volume of soils and this is also known as geometric attenuation which can occur even in absence of damping. Further details can be found in Kramer (1996).

8.3.7 Water Tightness

If the tests are carried out in saturated soils, the soil container must be watertight.
In summary, a model container has to satisfy the following criteria:

1. Maintain stress and strain similarity in the model as in the prototype
2. Propagate the base shaking to the upper soil layers
3. Reduce the wave reflections (energy) from the sidewalls which would otherwise radiate away in the prototype problem. Also to ensure negligible P-waves generation due to the presence of artificial boundary.

4. For saturated soil tests (i.e. liquefaction tests), the soil container should be watertight.

Some additional conditions need to be satisfied:

1. The model container must have adequate lateral stiffness so that a zero lateral strain (K_0 condition) condition can be maintained. This is particularly pronounced in centrifuge testing during the centrifuge spin-up.
2. The frictional end walls must have the same vertical settlement as the soil layer so that no additional stresses are induced in the soil. This effect is particularly important in centrifuge testing during the swing up. However, in 1-g testing any unwanted component of vertical motion should not induce additional stress in the soil at the boundaries.

Further details on the conditions related to the centrifuge testing can be found in Brennan (2003), Zeng and Schofield (1996).

8.4 Different Types of Container

The design requirements illustrated in the previous section can be achieved through various types of container design. It must be mentioned however that each of the designs has its own advantages and disadvantages, and they satisfy the different design requirements to a different degree. The different types of soil container available can be summarised as follow:

1. Rigid container
2. Rigid container with flexible boundaries (e.g. duxseal or sponge)
3. Rigid container with hinged end-walls
4. Equivalent Shear Beam (ESB) container
5. Laminar container
6. Active boundary container

8.4.1 Rigid Container

Figure 8.10 shows a schematic diagram of a rigid box. In this design, the shear stiffness of the end walls is much higher than the stiffness of the layers of soil contained by it. The end walls and the base of the container are also designed to be rough which is achieved by gluing a layer of sand onto it. This ensures the development of shear stresses in vertical plane at the interface between the container and soil. As discussed in the previous section, the rough base enables the base shaking to propagate through the soil layer. In order to maintain the plane strain condition, the surfaces of the side-walls must be very smooth which is often accomplished by treating the

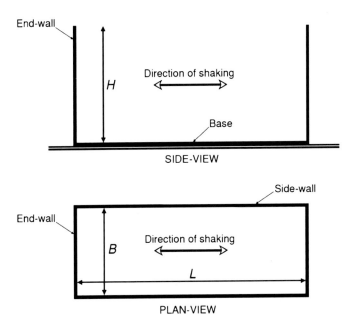

Fig. 8.10 Schematic diagram of a rigid container

inner surfaces of the container with grease or oil or by a plastic material (e.g. teflon or latex). An alternative design that achieves a similar effect uses smooth glass as the side-walls.

An important issue that has to be taken into account during the design is the ratio between the length and the height of the container. Numerical studied conducted by Fishman et al. (1995) and Whitman and Lambe (1986) quantified the zones close to the end walls that are affected by the artificial boundary. Their study suggests that this zone extends from the end walls into the container a distance of up to 1.5–2.0 times the height of the container. Based on these considerations, it may be suggested that the ratio of the length, L, of the container to the height, H, (see Fig. 8.10 for the notation of the symbols) should be more than 4. Table 8.2 lists several examples of rigid soil container founded in the literature.

8.4.2 Rigid Container with Flexible Boundary

In order to limit the reflection of the waves from the rigid boundary (one of the limitations of a rigid container) some adjustments can be made. Essentially the end-wall conditions are modified by the use of soft materials such as *sponge* that are glued along the end-walls of the container (see Fig. 8.12). The benefits are the following: (a) a partial reduction of the reflection of the waves, (b) reduction of the lateral

Table 8.2 Example of rigid containers presented in the literature

Shape	Shaking direction	L-B-H [mm]	L/H	Side-walls	Base & end-walls	Testing	Reference
Rectangular	1-D	597-270-150	4.0	Teflon	Rough sand paper	Centrifuge	Adalier and Elgamal (2002)
Rectangular	1-D	500-565-190	2.6	No-details	No-details	Centrifuge	Whitman and Lambe (1986)
Rectangular (Fig. 8.11a)	2-D	712-432-440	1.6	Smooth plastic membrane	Base covered by sand-glue mixture.	Centrifuge	Ng et al. (2004)
Rectangular (Fig. 8.11c)	1-D	1,500-400-1,000	1.5	Perspex and wood plates	Terram geotextile membrane	Shaking table	Norton (2008)
Rectangular (Fig. 8.11b)	1-D	450-240-400	1.1	Perspex	PTFE (poly-tetra-fluoro-ethylene) sheets	Shaking table	Dash (2010)

8 Model Container Design for Soil-Structure Interaction Studies

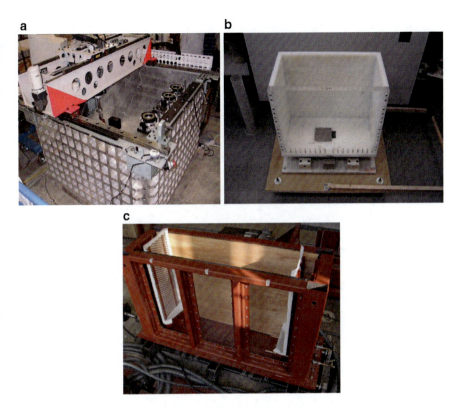

Fig. 8.11 Examples of rigid container: (**a**) Rigid container used in centrifuge at the Hong Kong University of Science and Technology (HKUST) (Courtesy of Prof Charles W.W. Ng). (**b**) Rigid box used in the small shaking table at the University of Bristol. (**c**) Rigid box used in the shaking table at the University of Oxford

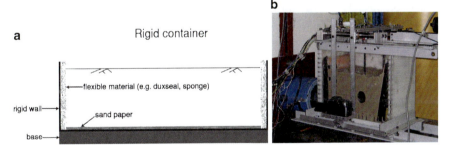

Fig. 8.12 Rigid container with flexible boundaries: (**a**) Schematic diagram (**b**) Example used in the Bristol Laboratory for Advanced Dynamics Engineering (BLADE)

stiffness of the end wall. The effect of sponge on K_0 condition is not yet established and needs further research. A pipe sealant material such as Duxseal has been used in the centrifuge modelling. Cheney et al. (1988) and Steedman and Madabhushi (1991) suggested that a third of the incident waves are reflected by the duxseal boundaries. It must be mentioned that duxseal material being much stiffer than normal sponge is recommended for centrifuge applications. However, the application of softer materials at the end-walls of the soil container raises some uncertainties regarding the actual boundary conditions. Moreover, the reflection of the waves is not completely eliminated. Table 8.3 lists some examples of rigid container with flexible boundaries.

8.4.2.1 Advantages

The main advantage of this type of design when compared with a rigid box is the reduction of the wave reflection and the P-wave generation.

8.4.2.2 Limitations

There are two main limitations. Firstly the actual boundary conditions introduced by either the sponge or duxseal are unknown. This uncertainty may become critical if the tests are to be modelled analytically or numerically. Secondly, the sponge and the duxseal material only reduce the wave reflection from the artificial boundaries, therefore such phenomenon cannot be considered completely absent.

8.4.3 Rigid Container with Hinged End-Walls

Figure 8.13 shows a schematic diagram of a rigid container with hinged end-walls. In such a design, the end-walls are permitted to rotate about the base due to the hinged connection. In order to have unison movement, the two end-walls may be connected together by a tie-rod. An example of this type of container can be found in Fishman et al. (1995) where the walls were hinged but also had some degree of flexibility. They reported strains in the walls which permitted evaluation of lateral earth pressure.

8.4.4 Equivalent Shear Beam (ESB) Container

In this design, the stiffness of the end walls of the container is designed to match the shear stiffness of the soil contained in it (Fig. 8.14). In other words the fundamental frequency of the container and soil deposit are matched by design. It is assumed that the soil behaves as assemblage of equivalent shear beams and so are the end-walls. If this condition is satisfied, the interaction between the container and the soil may be considered negligible and the stress and strain similarities can be considered accomplished. However, the shear stiffness of the soil varies during shaking depending

Table 8.3 Example of rigid containers with flexible end-walls

Shape	Shaking direction	L-B-H [mm]	L/H	Side-walls	End-walls	Testing	Reference
Rectangular	1-D	4,270-910-1,220	3.5	Tempered plate-glass	Hinged to the box	Shaking table	Fishman et al. (1995)
Rectangular (Fig. 8.12b)	1-D	450-240-400	1.1	Perspex	Sponge (30 and 60 mm thick at top and bottom respectively)	Shaking table	Dash (2010)
Rectangular	1-D	1,920-440-600	3.2	Acrylic plate	Sponge (50 mm thick)	Shaking table	Ha et al. (2011)

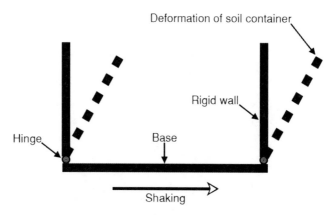

Fig. 8.13 Schematic diagram of rigid container with hinged end-walls

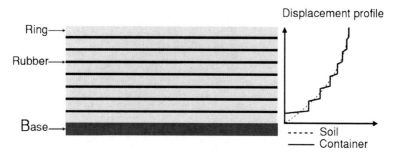

Fig. 8.14 Schematic diagram showing the Equivalent Shear Beam container

on the strain level. Therefore the matching of the two stiffnesses (end-wall and the soil) is possible only at a particular strain level. Schofield and Zeng (1992) explain the boundary conditions to be met by the ESB container:

1. The boundary must have the same dynamic stiffness as the adjacent soil to minimise energy reflection in the form of pressure waves.
2. The boundary must have the same friction as the adjacent soil to sustain complementary shear stresses.
3. The sidewalls should be frictionless to have plane strain condition.

In the literature different design methods are available. The ESB box at the University of Cambridge (Zeng and Schofield 1996) was designed to match the stiffness between the container and soil over a limited range of strains in the zone of interest. On the other hand, the shear stack at the University of Bristol (Dar 1993), shown in Fig. 8.15b, is designed considering a value of strains in the soil close to the failure (0.01–1%). Therefore the Bristol container will be much more flexible than the soil deposit at lower strain amplitudes and, as a consequence, the soil will always

8 Model Container Design for Soil-Structure Interaction Studies

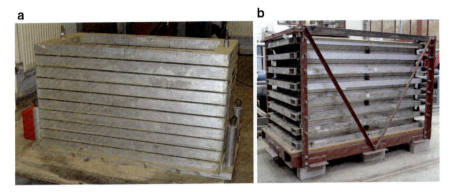

Fig. 8.15 Examples of equivalent shear beam container: (**a**) ESB used in centrifuge testing, University of Cambridge (**b**) Shear stack used in 1-g testing, University of Bristol

Table 8.4 Example of Equivalent Shear Beam container

Shape	Shaking	L-B-H [mm]	L/H	1-g/N-g	Reference
Rectangular	1-D	2,000-750-1,750	1.1	1-g	Carvalho et al. (2010)
Rectangular	1-D	1,200-550-800	1.5	1-g	Dar (1993)
Rectangular	1-D	4,270-910-1,220	3.5	1-g	Fishman et al. (1995)
Rectangular	1-D	4,800-1,000-1,200	4	1-g	Crewe et al. (1995)
Rectangular	1-D	560-250-226	2.5	N-g	Zeng and Schofield (1996)
Rectangular	1-D	800-350-600	1.3	N-g	Madabhushi et al. (1998)

dictate the overall behaviour of the container. In a recent study using the Bristol shear stack, see Bhattacharya et al. (2010), it was observed that the ratio between the container stiffness and initial stiffness of the granular deposit determined the success in reproducing the true shearing behaviour of the granular material under seismic shaking. In particular, it was observed that when a very stiff deposit was used (for example rubber granule was used to create a large stiffness contrast in a multi layered deposit), the container behaved almost like a rigid container. Under such condition, the container drives the soil shearing.

In terms of manufacturing, ESB container consists of a rectangular box made from aluminium rings separated by rubber layers. The rings provide lateral confinement of the soil in order to reproduce the K_0 conditions (zero lateral strains), while the rubber layers allowed the container to deform in a shear beam manner. Table 8.4 lists some examples of ESB container described in the literature.

8.4.4.1 Limitations

Rubber has a linear elastic stress-strain relationship up to large shear strain level. On the other hand the behaviour of the soil under cyclic loading is highly non-linear and hysteretic particularly at large strains.

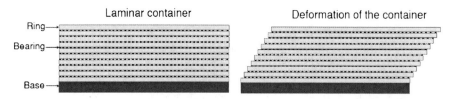

Fig. 8.16 Schematic diagram of laminar container

Table 8.5 Examples of laminar containers

Shape	Shaking	L-B-H [mm]	L/H	1-g/N-g	Design	Reference
Rectangular	1-D	900-350-470	1.9	1-g	Stack of laminae separated by bearing	Gibson (1997)
Rectangular	1-D	1,000-500-1,000	1	1-g	Stack of laminae separated by bearing	Prasad et al. (2004)
Circular	2-D	2,280-2,130 (D-H)	1.1	1-g	Container hanging on the top lamina supported by frame	Meymand (1998)
Rectangular	2-D	1,888-1,888-1,520	1.2	1-g	Laminae supported by a frame and move independently	Ueng and Chen (2010)
Rectangular	1-D	457-254-254	1.8	N-g	Stack of laminae separated by bearing	Van Laak et al. (1994)
Rectangular	1-D	710-355-355	2	N-g	Stack of laminae separated by bearing	Pamuk et al. (2007)
12-sided polygon	2-D	584-500 (D-H)	1.2	N-g	Stack of laminae separated by bearing	Shen et al. (1998)
Rectangular	1-D	900-450-807	1.1	1-g	Laminae are supported individually by bearings and steel guide connected to an external frame	Turan et al. (2009)
Square	2-D	1,000-1,000-1,000	1	1-g	Laminae are supported individually by bearings	Jafarzadeh (2004)

8.4.5 Laminar Container

Figure 8.16 shows a schematic diagram of a laminar container. The most common design consists of a stack of laminae supported individually by bearings and a steel

8 Model Container Design for Soil-Structure Interaction Studies   155

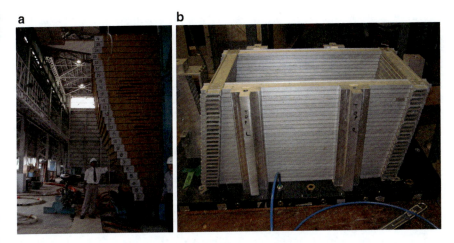

Fig. 8.17 Laminar container: (**a**) Large laminar container used in 1-g testing in Tsukuba, Japan (**b**) Laminar container used in centrifuge, University of Cambridge

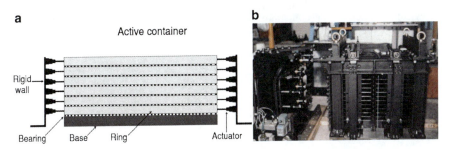

Fig. 8.18 Active boundaries container: (**a**) Schematic diagram of the container (**b**) Example of active boundaries container (Courtesy of Professor Akihiro Takahashi)

guide connected to an external frame. The design principle of a laminar box is to minimize the lateral stiffness of the container in order to ensure that the soil governs the response of the soil-box system.

Table 8.5 lists nine different types of laminar containers presented in the literature. It may be observed that most common shape is rectangular. However for two dimensional shaking tests the shape of the box can be square or circular or 12–sided polygon. Figure 8.17 shows the photograph of two laminar boxes.

8.4.6 Active Boundaries Container

The design principle of an active boundaries container is very similar to that of a laminar box with the only difference that external actuators are connected to each lamina. Different pressure may be applied from the actuator in order to achieve the desired prototype condition. This type of container may be used in a situation where the stiffness of the soil is varying sharply during the shaking (e.g. for liquefaction application). Figure 8.18a shows the schematic diagram of an active boundary box and Fig. 8.18b shows the photograph of the active box from Tokyo Institute of Technology.

8.5 Discussion and Conclusions

There are differences between physical testing and physical modelling. Physical testing is the actual test carried out using a model (either a laboratory scale or full scale) but physical modelling refers to the modelling of a particular aspect (e.g. collapse mechanism or physical process) of a prototype problem under consideration. Physical modelling therefore requires the recognition of the physical mechanisms or processes that control the behaviour of interest and therefore the derivation of the relevant non-dimensional groups. Each physical test can be analysed as a separate prototype on its own merits. As the physical tests are real physical events, the data generated by these tests can be instructive and useful if all the variables and detailed physical parameters that contributed to the test observations are recognised. In this context, it must be mentioned that a number of distortions can be induced by the testing methods and/or artificial boundaries required to model an infinite soil medium. In this chapter, a thorough investigation on the model soil container for seismic testing of geotechnical models is considered.

Soil profile on site can never be replicated in a model and none of the physical modelling techniques are perfect. For example in 1-g testing, it is difficult to model the stress dependency behaviour of granular material. On the other hand, centrifuge testing of geotechnical models are challenging (carrying out all the actuations while the model is spinning at a high angular velocity) and the technique itself introduces errors, such as radial distortion, gravitational distortion, angular distortion, Coriolis distortion. Also, the scaled model in a centrifuge is not free of external vibrations. In this chapter, the issue of model container has been studied. The requirement of a perfect model soil container is first established from first principles. Six different types of model container designs are reviewed and discussed. Based on the review, it is realised that none of the six types of containers are suitable for all types of modelling applications. It is important to choose particular types of container depending on the problem in hand.

References

Adalier K, Elgamal AW (2002) Seismic response of adjacent saturated dense and loose sand columns. Soil Dyn Earthquake Eng 22(2):115–127

Bhattacharya S, Dihoru L, Taylor CA, Muir Wood D, Mylonakis G, Moccia F, Simonelli AL (2010) Kinematic bending moments in piles: an experimental study. In: Proceedings of the

8 Model Container Design for Soil-Structure Interaction Studies

7th international conference on urban earthquake engineering and 5th international conference on earthquake engineering, Tokyo

Bhattacharya S, Hyodo M, Goda K, Tazoh T, Taylor CA (2011a) Liquefaction of soil in the Tokyo bay area from the 2011 Tohoku (Japan) earthquake, Soil Dyn Earthquake Eng, doi:10.1016/j.soildyn.2011.06.006

Bhattacharya S, Lombardi D, Muir Wood D (2011b) Similitude relationships for physical modelling of offshore wind turbines. Int J Phys Model Geotech 11(2):58–68

Bolton MD (1986) The strength and dilatancy of sands, Géotechnique 36(1):65–78

Brennan AJ (2003) Vertical drains as a countermeasure to earthquake-induced soil liquefaction. PhD thesis, University of Cambridge

Carvalho AT, Bile Serra J, Oliveira F, Morais P, Ribeiro AR, Santos Pereira C (2010) Design of experimental setup for 1 g seismic load tests on anchored retaining walls. In: Springman S, Laue J, Seward L (eds) Physical modelling in geotechnics. Taylor and Francis Group, London

Cheney JA, Hor OYZ, Brown RK, Dhat NR (1988) Foundation vibration in centrifuge models. In: Proceedings of the Centrifuge 88, international conference on centrifuge modelling, Paris

Crewe AJ, Lings ML, Taylor CA, Yeung AK, Andrighetto R (1995) Development of a large flexible shear stack for testing dry sand and simple direct foundations on a shaking table. In: Elnashai (ed) European seismic design practice. Balkema, Rotterdam

Dar AR (1993) Development of flexible shear-stack for shaking table of geotechnical problems. PhD thesis, University of Bristol, Bristol

Dash SR (2010) Lateral pile-liquefied soil interaction during earthquakes. DPhil thesis, University of Oxford, Oxford

Fishman KL, Mander JB, Richards R Jr (1995) Laboratory study of seismic free-field response of sand. Soil Dyn Earthquake Eng 14:33–43

Gibson AD (1997) Physical scale modelling of geotechnical structures at one-g. PhD thesis, California Institute of Technology, Pasadena

Ha I, Olson SM, Seo M, Kim M (2011) Evaluation of liquefaction resistance using shaking table tests. Soil Dyn Earthquake Eng 31:682–691

Iai S (1989) Similitude for shaking table test on soil-structure-fluid model in 1 g gravitational field. Soil Found 29(1):105–118

Jafarzadeh B (2004) Design and evaluation concepts of laminar shear box for 1G shaking table tests. In: Proceedings of the 13th world conference on earthquake engineering, Vancouver, paper no. 1391

Kelly RB, Houlsby GT, Byrne BW (2006) A comparison of field and lab tests of caisson foundation in sand and clay. Géotechnique 56(9):617–626

Kramer SL (1996) Geotechnical earthquake engineering. Prentice Hall, Upper Saddle River

Leblanc C, Byrne BW, Houlsby GT (2010) Response of stiff piles to random two-way lateral loading. Géotechnique 60(9):715–721

Madabhushi SPG, Butler G, Schofield AN (1998) Design of an Equivalent Shear Beam (ESB) container for use on the US Army. In: Kimura, Takemura (eds) Centrifuge. Balkema, Rotterdam

Meymand PJ (1998) Shaking table scale model tests of nonlinear soil-pile-superstructure interaction in soft clay. PhD thesis, University of Berkley, Berkley

Muir Wood D, Crewe AJ, Taylor CA (2002) Shaking table testing of geotechnical models. Int J Phys Model Geotech 2:1–13

Ng CWW, Li XS, Van Laak PA, Hou DYJ (2004) Centrifuge modelling of loose fill embankment subjected to uni-axial and bi-axial earthquakes. Soil Dyn Earthquake Eng 24:305–318

Norton H (2008) Investigation into pile foundations in seismically liquefiable soils. MSc thesis, University of Oxford

Pamuk A, Gallagher PM, Zimmie TF (2007) Remediation of piled foundations against lateral spreading by passive site stabilization technique. Soil Dyn Earthq Eng 27:864–874

Prasad SK, Towhata I, Chandradhara GP, Nanjundaswamy P (2004) Shaking table tests in earthquake geotechnical engineering. Curr Sci 87(10):1398–1404

Schofield AN (1980) Cambridge centrifuge operations, twentieth rankine lecture. Géotechnique 30: 227–268

Schofield AN (1981) Dynamic and earthquake geotechnical centrifuge modelling. Proceedings international conference recent advances in geotechnical earthquake engineering and soil dynamics, vol 3, 1081–1100

Shen CK, Li XS, Ng CWW, Van Laak PA, Kutter BL, Cappel K (1998) Development of a geotechnical centrifuge in Hong Kong. In: Centrifuge 98(1), Tokyo

Steedman RS, Madabhushi SPG (1991) Wave propagation in sands. In: Proceedings of the international conference on seismic zonation, Stanford University, California

Turan A, Hinchberger SD, El Naggar H (2009) Design and commissioning of a laminar soil container for use on small shaking table. Soil Dyn Earthquake Eng 29:404–414

Ueng TS, Chen CH (2010) Liquefaction of sand under multidirectional shaking table test. In: Proceedings of the international conference on physical modelling in geotechnics, ICPMG, Hong Kong, pp 481–486

Van Laak P, Taboada V, Dobry R, Elgamal AW (1994) Earthquake centrifuge modelling using a laminar box. Dynamic geotechnical testing, vol 2, ASTM STP 1213. American Society for Testing and Materials, Philadelphia, pp 370–384

Whitman RV, Lambe PC (1986) Effect of boundary conditions upon centrifuge experiments using ground motion simulation. Geotech Test J 9(2):61–71

Wroth CP, Randolph MF, Houlsby GT, Fahey M (1979) A review of the engineering properties of soils with particular reference to the shear modulus. Report CUED/D-SOILS TR75, University of Cambridge

Zeng X, Schofield AN (1996) Design and performance of an equivalent-shear-beam container for earthquake centrifuge modelling. Geotechnique 46(1):83–102

Chapter 9
Computer Vision System for Monitoring in Dynamic Structural Testing

Francesco Lunghi, Alberto Pavese, Simone Peloso, Igor Lanese, and Davide Silvestri

Abstract In combination with standard transducers and data acquisition systems, computer vision can be adopted in order to perform the analysis of the behaviour of structures during dynamic tests such as earthquake simulations on shake tables. The paper describes the design and implementation of a machine vision system aimed at providing bi-dimensional position measurement of reflective markers directly placed on test specimens. The developed solution is composed of a scalable set of acquisition units, each consisting of a high definition digital camera and a personal computer. A sequence of images is acquired by the cameras and the position of the markers in the scene is estimated by means of a software application running on the computers. Each unit can perform measurements in a single plane which is defined in a previous calibration phase. The method has many advantages over the most commonly used acquisition devices such as accelerometers and potentiometers: first, the absence of contact between the acquisition device and the tested structure, which allows the non-invasive deployment of an arbitrary number of measurement targets, which is even more important in destructive tests, for preventing the loss of expensive transducers; second, the direct calculation of the position of an object in length units, without the need of post processing like integration and conversion, as required when using accelerometers in shake table tests. Besides, in the selected plane, thanks

---

F. Lunghi (✉) • S. Peloso • I. Lanese • D. Silvestri
Eucentre – European Centre for Training and Research in Earthquake Engineering,
Via Ferrata 1, 27100 Pavia, Italy
e-mail: francesco.lunghi@eucentre.it; simone.peloso@eucentre.it;
igor.lanese@eucentre.it; davide.silvestri@eucentre.it

A. Pavese
Eucentre – European Centre for Training and Research in Earthquake Engineering,
Via Ferrata 1, 27100 Pavia, Italy

Department of Structural Mechanics, University of Pavia, Via Ferrata 1, 27100 Pavia, Italy
e-mail: alberto.pavese@eucentre.it; a.pavese@unipv.it

M.N. Fardis and Z.T. Rakicevic (eds.), *Role of Seismic Testing Facilities in Performance-Based Earthquake Engineering: SERIES Workshop*, Geotechnical, Geological and Earthquake Engineering 22, DOI 10.1007/978-94-007-1977-4_9,
© Springer Science+Business Media B.V. 2012

to the adoption of infrared illumination and filters to reduce environmental lighting interferences, each unit can follow the movements of a large number of markers (up to 50 for each camera in the performed tests) with a precision of around 0.05 mm. On the other hand, the method is by itself unable to overcome problems deriving from the three-dimensional movement of the selected markers. The paper also explains the different approaches and the corresponding results obtained while solving this issue.

9.1 Introduction

In experimental testing laboratories, especially in those performing dynamic and static trials on building elements or structures, one of the main tasks is to collect information and acquire data form the experiments. Different kinds of sensors are generally employed in order to convert physical variables like deformation, acceleration, force or pressure into analogical or digital electrical signals. Every methodology usually has some advantages as well as some limitations concerning, for example, precision, dynamic range, sampling frequency and other characteristic features of the acquired variable. In seismic engineering and in particular in experiments performed by simulating earthquake events, the positions and movements of elements on the tested specimens are certainly among those carrying most of the information needed for studying the buildings response. It is a common solution to operate the acquisition by means of potentiometers and by integrating the output of accelerometers directly deployed on the structure. Although very precise and capable of performing at high frequencies, both these sensors suffer from a series of issues. Potentiometers are fast and accurate but have very limited extensions, moreover they can measure relative distances between two elements and cannot acquire global positions of elements in the specimen. Accelerometers, on the other way, are designed to acquire the acceleration signal and are therefore subject to high noise and biases when applied to the displacement case. Another common issue is the "invasivity" of these sensors, since they need to be directly connected to the test specimen. This implies, among other things, the risk of losing expensive instruments in events of collapses or building damages.

In light of these considerations, a machine vision system has been developed in order to measure the position of elements on the test specimens and try to overcome the above limitations. A vision system, indeed, can perform on a wide area of the building and doesn't need any operative hardware to be positioned on the tested construction.

This is the description of the work done aiming at implementing hardware and software architecture for bi-dimensional high precision measurements. In more detail, the following is a proposed solution to the task of following markers moving in a scene by acquiring and analysing a sequence of images captured using high definition digital cameras.

Many books and papers in computer science literature focus on the three dimensional case (Pollefeys 2002), and provide a spatial identification of the position of

9 Computer Vision System for Monitoring in Dynamic Structural Testing

markers. With the third dimension comes nevertheless the need for multiple points of view and, having to compensate multiple errors, a diminished accuracy. This paper offers an alternative approach, starting from the very restrictive hypothesis of planar measurements and tries to reach the maximum level of achievable precision, while still requiring a single camera for each target.

The work done will be summarized in the sequence of technical and scientific challenges faced during the development of a scalable system for bi-dimensional measurements. Throughout this process, the complete lack of information coming from the observation of the third dimension proved to be an overwhelming limitation and thus, in the final part of this treatment a couple of methods for the solution of this issue are presented.

9.2 Analysis of the Problem

As already stated, the implemented architecture is intended as a practical solution to the problem of identifying the absolute position of markers in a scene taken with high definition cameras and following the evolution of the displacements of these targets during the acquisition of a sequence of images. Some of the words used in this description deserve a little definition.

The scene can be defined as the visual percept of a spatial region. Its digitalization is technically the result of the conversion of the amount of reflected light which reaches the camera sensor into a matrix of numbers. Each number is the discrete numerical representation of a colour or, as in this case, of a quantity of light intensity and its position in the matrix is the location of that quantity in the image. This unit is also referred to as a pixel (picture element) as it represents the atomic element composing a digital image. A marker is an object, applied or however visible in a scene whose pixel representation in the digitalized picture is identifiable from the rest of the scene. Different kinds and typologies of markers are available, depending on the different identification and analysis methods used to process the image. The computer vision workflow considered in this work is then the acquisition of a single frame or a sequence of frames and the application to these images of an identification algorithm in order to estimate the position of the markers in the scene. As a result of the digitalization of a three dimensional real world view into a bi-dimensional digital image, a degree of freedom and its corresponding amount of information gets inevitably lost.

In order to complete the proposed task, the system must be able then to convert the coordinate in the image of a marker into its expected position in the scene. This operation consists in mapping logical coordinates (pixels) in physical coordinates (millimetres or in general length units). For this reason, the computer must acquire and use a certain amount of knowledge concerning the scene, target, camera and their mutual position. The process of computing this information is usually called "calibration".

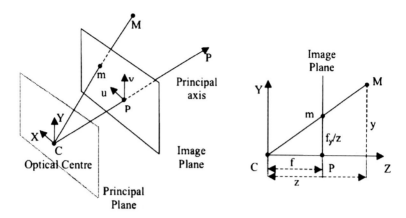

Fig. 9.1 The pin-hole model ignoring the lens non-linearity

9.2.1 Calibration of the Machine Vision

This operation is a very common practice in every field of machine vision because it addresses a series of issues generally deriving from the human interaction in the placement of cameras, markers as well as lighting and distortion problems. For this reason there are many literary works on this topic and many different models to geometrically and mathematically define the parameters of this problem. After one of these models has been chosen, there are several methods for optimizing its free variables and obtain those called the calibration parameters. Some of these parameters will be related to the camera itself, like lens distortion and focuses, others will depend on the scene and the mutual position of target and viewpoint instead.

The one adopted in this work is the "pin-hole" camera model, considering the projections of the visual ray on a nonlinear surface (lens) placed at a distance f (focus) from the optical centre. This camera model is based on the principle of co-linearity, in which each point in the object space is projected by a straight line through the projection centre into the image plane (Fig. 9.1) (Bradski and Kaehler 2008).

The mapping between three dimensional world coordinates (M) and the corresponding coordinates in image pixel (m) is widely treated in literature and is based on the application of four different transformations. Neglecting the effect of the lens non-linearity, the algorithm to convert a physical coordinate into a logical coordinate corresponding to the digitalized position on the camera sensor is described in the following. First of all, the co-linearity principle states that each point over the same visual ray is projected into the same pixel in the image plane. For this reason, the reduction of a degree of freedom is performed by using the normalization factor s in the equation below:

$$s \cdot m' = A[R|t]M' \tag{9.1}$$

9 Computer Vision System for Monitoring in Dynamic Structural Testing

which can be expanded into:

$$s \cdot \begin{bmatrix} u \\ v \\ 1 \end{bmatrix} = \begin{bmatrix} f_x & 0 & c_x \\ 0 & f_y & c_y \\ 0 & 0 & 1 \end{bmatrix} \begin{bmatrix} r_{11} & r_{12} & r_{13} & t_1 \\ r_{21} & r_{22} & r_{23} & t_2 \\ r_{31} & r_{32} & r_{33} & t_3 \end{bmatrix} \begin{bmatrix} X \\ Y \\ Z \\ 1 \end{bmatrix} \tag{9.2}$$

where (u, v) is the pixel representation of the homogeneous world coordinate $(X, Y, Z, 1)$, f_x and f_y are the focal length, expressed in pixels/physical units, (c_x, c_y) is the principal point, which is commonly the centre of the image and $R|t$ is the mutual roto-translation between the target reference system and the camera's one.

The A matrix is called the "camera matrix" and its parameters are considered intrinsic of the camera because they are independent on the actual camera position as well as on the shot. The 3×4 roto-translation matrix is instead called the "extrinsic matrix" since it is the representation of the mutual position (and rotation) between the point of view and the target. The scalar value s is the normalization factor, which is a mathematical tool to reduce the dimension of the coordinate vectors from 3 to 2, which, geometrically, can be described as the visual distance of the object from the optical centre (Zhang 1998).

Thus, after a point M is roto-translated by means of the $R|t$ matrix, before being scaled and centred into the image using the A matrix, it is in the form (x, y, z) and can be converted in pixel dividing by z ($z \neq 0$ because objects placed in the optical centre are not projectable). So, decomposing Eq. 9.2, one can write:

$$\begin{bmatrix} x \\ y \\ z \end{bmatrix} = R \begin{bmatrix} X \\ Y \\ Z \end{bmatrix} + t \tag{9.3a}$$

$$x' = x / z$$
$$y' = y / z$$
$$(s = 1 / z) \tag{9.3b}$$

In order to complete the decomposition of Eq. 9.2, the camera matrix contribution must be taken into account:

$$u = f_x x' + c_x$$
$$v = f_y y' + c_y \tag{9.4}$$

This set of equations is valid for ideal devices, but in fact real cameras are subject to different kinds of distortions in the captured image, mainly because of the shape and positioning of their lenses. This commonly results in a larger radial distortion component and a minor tangential one. In order to consider this non-linear behaviour, the model must be enhanced by introducing, before the application of the intrinsic matrix, a non-linear function.

$$x'' = x'(1+k_1r^2 + k_2r^4) + 2p_1x'y' + p_2(r^2 + 2x'^2)$$
$$y'' = y'(1+k_1r^2 + k_2r^4) + 2p_2x'y' + p_1(r^2 + 2y'^2)$$
$$with \quad r^2 = x'^2 + y'^2 \tag{9.5}$$

where k_1, k_2 are the radial distortion parameters; p_1, p_2 correspond instead to the tangential one.

Of course this implies rewriting Eq. 9.4 as:

$$u = f_x x'' + c_x$$
$$v = f_y y'' + c_y \tag{9.4a}$$

Calibrating a camera consists in identifying the free parameters in the above described model. This is usually done by a procedure which is logically subdivided in two steps, corresponding respectively to the evaluation of the intrinsic and extrinsic parts of the transformation. The first step is executed by presenting a series of known example patterns, usually a chessboard or more generally an array of points lying in a grid of known coordinates, in order to perform a linear optimization of the A, k and p parameters. Once the intrinsic values are estimated using all the provided images, the extrinsic matrix is calculated and, since it depends on the scene, it is referred to a single reference image (Heikkila and Silvén 1997).

9.2.2 Inverse Transformation and Markers Detection

The calibration procedure explained in the previous section allows the projection of the coordinates of an object in the scene into the corresponding pixels in the captured image. This is performed by applying the pin-hole model and by means of visual rays. As mentioned, converting a three dimensional space point into its two dimensional counterpart implies a loss of information, in fact any point lying on a visual ray will be projected into the same pixel on the image for this reason the process is irreversible. This is true if no additional hypotheses hold, but for the development of this work the assumption of knowing the plane where all the markers reside and on which all the movements will be performed is valid. Therefore, adding this constrain means allowing the existence of a solution for the problem. Geometrically, it is like setting a plane on which the visual rays, originated in the optical centre and passing through a pixel, are projected.

At this point, the algorithm which performs the backward identification of the real-world position of any pixel can be written; this is done by reproducing an inverse function of each steps of the abovementioned calibration method.

For convenience, since the chessboard is placed and recorded in different position for the intrinsic identification but only one image is needed for the $R|t$ parameters, the plane identified by the chessboard position in that image will be the one on which the

9 Computer Vision System for Monitoring in Dynamic Structural Testing

markers' positions and movements will be projected. Supposing one wants to know the pixel position (u,v of Eq. 9.4a) of an object in the scene, the algorithm will evaluate the position of that object in physical units, on the measurement plane. The first step to invert in this backward computation is of course the last one, which is the application of the A matrix, which is invertible, by construction. After the application of A^{-1}, the algorithm has to face the problem of reverting the distortion components. This is of course impossible in a closed form because of the involved non-linear Eq. 9.5, but the process is approximated by an iterative algorithm, which is well known in literature (Devernay and Faugeras 2001) and which is used to rectify distorted images after the evaluation of a camera intrinsic parameters set. This first phase is independent of the extrinsic parameters and does not need any planar constraint because the dimension of both the domain and co-domain is the same. But at this point the projection in the real world coordinates must be performed and a three dimensional coordinate must be computed.

Indeed, the only remaining part of the onward process that still needs to be reverted is Eq. 9.3 but, as widely anticipated, substituting Eq. 9.3a for 9.3b results in a set of three equations with four unknown values. In order to evaluate the R and t parameters, on the other side, an image is computed where the chessboard pattern is recognized and its corners coordinates are identified. These coordinates are then given to the algorithm in order to optimize the free parameters. Without any loss of information the plane where the calibration chessboard lies is arbitrarily chosen to have $Z=0$ and thus the same approach can be used for the backward process.

Of course this implies all the limitations imposed by the strong hypothesis of having all the markers lying on the plane of the chessboard and to constrain their movements to exist over that plane. It conversely gives the possibility to solve the system of equations without introducing multiple points of view or additional information coming from outside the calibration process.

The last row of Eq. 9.2 is therefore:

$$R_{31}^{-1}(x-t_1)+R_{32}^{-1}(y-t_2)+R_{33}^{-1}(z-t_3)=Z=0 \qquad (9.6)$$

In the above equation, the only unknown value is z, and, substituting the values of x and y from Eq. 9.3b, z is calculated:

$$z=\frac{R_{31}^{-1}t_1+R_{32}^{-1}t_2+R_{33}^{-1}t_3}{R_{31}^{-1}x'+R_{32}^{-1}y'+R_{33}^{-1}} \qquad (9.7)$$

Trying to give a practical meaning to this result, it could be described like evaluating, for each pixel coordinate (x',y') the corresponding visual distance (the "s" of the pinhole model (9.1)), that is an idea of the position over the visual ray of the object in the real world which is projected in that pixel. An infinite number of points in the scene will project along this visual ray into the starting pixel but only the one lying on the calibration plane will have this z value.

Now, the final step, the computation of the X,Y coordinates (the Z has been imposed to be null) of the object in the target reference system is trivial because now the x and y values can be calculated (Eq. 9.3b) by simple substitution. Once the backward algorithm has determined the undistorted and normalized version of the pixel coordinates and all the exceeding unknown values are set, Eq. 9.2 can be applied to calculate the position of that pixel on the measurement plane.

At this point, the method used to determine the position of an object in the scene starting from its coordinates in the image has been described. In order to complete the task the algorithm needs to be able to identify the marker in the image and determine its position in pixel with the highest achievable accuracy.

This is done with a custom implementation of a blob analysis method. A blob in a digital image can be roughly defined as a set of contiguous pixels having the same colour or at least having colour tones inside a chosen range. The blob can also have minimum area or shape constraints. A blob analysis consists in the process of detecting blobs in an image, that is choosing groups of pixels satisfying the above conditions. When a blob is selected, all the pixels satisfying the background test colour threshold participate in the centroid calculation. Its position is determined by a weighted average of the colour intensity of its pixels. The system can reach an accuracy ranging from 1/10th (scored in the worst performed test) up to a theoretical 1/50th of sub-pixel precision. A simple rough estimate of the time needed for the elaboration of a single 4 M pixels frame is between 100 and 200 ms, depending on the machine overhead, limiting therefore the application to the entire frame to acquisition frequencies of about 5–10 Hz. For this reason, an alternative to the elaboration of the entire frame has been developed, involving the process of blob tracking, which means that the search for a marker is computed only in a restricted area or region-of-interest (ROI) around its previous position.

So far the software methods used to calibrate the camera and to detect the position of a marker in the image returning its estimate position in the measurement plane have been described. In the following section the hardware setup as well as the technical solutions adopted in order to realize the overall system will be explained.

9.3 Implementation

9.3.1 Hardware

One of the biggest advantages coming from the assumption of positions and displacements lying on a single plane is the need for only one point of view per scene. For this reason, it has been thought to develop a scalable architecture, based on multiple independent and self sufficient acquisition elements. Each one of them is composed by a personal computer, bundled in a ruggedized and wheeled case with a UPS, supplying power for the whole unit even if disconnected from the power line. The computer contains X64 Xcelera CL-PX4 frame grabber, plugged into a PCI express 4x socket, which can manage up to two high definition cameras. The selected cameras are PT–40–04 M60 high definition digital cameras (Dalsa). They have grey scale CCD sensor cameras capable of grabbing $2,352 \times 1,728$ (8bpp) images at 60 Hz, scalable up to $2,352 \times 864$ at 120 Hz. Each camera is supplied

9 Computer Vision System for Monitoring in Dynamic Structural Testing

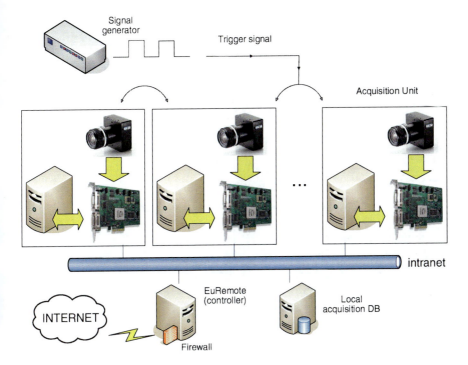

Fig. 9.2 Overall scheme of the vision system

with its own lighting device. In order to perform in different lighting conditions, the system is developed to exploit the camera sensibility to light components near the infra-red spectrum. For this reason, the lighting environment has been chosen to be in that range and NIR illuminators have been adopted. Correspondently, the markers are crafted of a reflective material and, on the camera objective, a NIR filter is plugged. The developed measurement system is employed in different kinds of acquisitions but it is generally used as a multiple point of view monitor.

Another challenging issue to deal with concerns the synchronization among cameras. This problem has been treated combining a trigger signal and a remote control application to manage the different acquisition units as described in more detail in the following section (Fig. 9.2).

9.3.2 Software

As previously mentioned, two are the main software modules acting in the system: the image acquisition and processing tool, deployed on each unit, and the remote controller to manage the whole architecture. The first is a Microsoft Windows application, which can be subdivided into different software layers corresponding to the

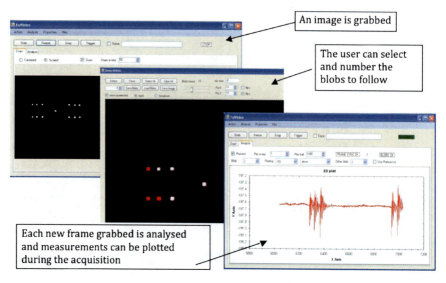

Fig. 9.3 Blob tracking sequence

different levels of hardware abstraction. Therefore, starting from the lower tier, which is the closest to the machine and moving to the highest ones, which are responsible for the interaction with the users, they are:

- Dalsa Drivers (Sapera LT SDK) – interface to frame grabber
- C++/Assembly analysis module – image compression/markers detection
- OpenCV (Intel Corporation 1999–2001) C++ higher level frame and signals filtering management
- C++/CLI .NET front-end
- WCF/Remoting service – listening for remote automation

As mentioned, each acquisition unit acts as a server listening for incoming connections and letting remote users to operate on the camera. On the other side, a thin remote controller application has been developed, which is responsible for issuing commands for starting/stopping the acquisition and setting filenames for saving on multiple remote stations at the same time. During the acquisition, it also polls the state of each unit and shows the events of blobs lost, frame trashed (a condition that happens when the unit is unable to perform the elaboration on schedule) or generic malfunctioning (Fig. 9.3).

9.4 Application to a First Case Study

One of the first tests done using the developed system was on a building in 1:2 scale and consisted in the simulation of a seismic event on the shake table, performed at the Eucentre foundation laboratory in Pavia. At that time the setup was composed

9 Computer Vision System for Monitoring in Dynamic Structural Testing

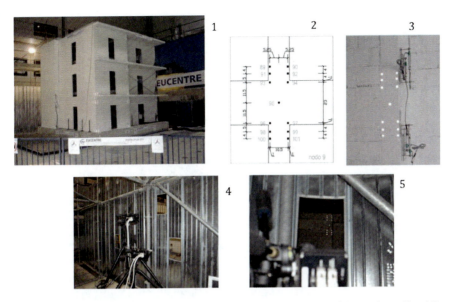

Fig. 9.4 (1) RC building tested; (2 and 3) example scheme and placing of the markers; (4 and 5) position and viewpoint of the cameras

of eight acquisition units, targeting a total of about 250 markers directly applied to the structure. The sampling frequency, due to the high dynamic of the applied accelerogram, was set to 120 Hz and, considering an average distance of 6 m between the cameras and the target, this resulted in 1.8 × 1.0 m wide frames (Fig. 9.4).

The results of this test are of little interest in this chapter but a few considerations must be taken into account concerning the accuracy and efficiency of the vision system. For instance, by examining the difference between the positions and displacements measured by the new methodology in events of "rigid movements", like the ones related to markers placed on the foundations for example, an estimate of the coherence and precision achieved can be retrieved. The acquired signals show a perfect synchronization, with a sample maximum phase error and the average error between the three cameras aiming at the foundations was less than 0.01 mm with a standard deviation of 0.05 mm. Although these values seemed encouraging, it appeared clear that the system error was related, if not directly, to the involved displacements. This resulted from the observation of the error vs. displacement graph, which was almost linear. This issue was at first related to a mis-calibration error but the experience gained focusing on the higher floors signals helped in identifying an intrinsic problem of the model. The point is that the method relies on a strong hypothesis of planar positions and movements which is not always true in both its assumptions. It is easy to see that neither the starting positions nor the displacements lie on a single plane and, additionally, that even the slightest error or imprecision in the chessboard placing will result in incoherent signals between the different cameras. These two issues share a common denominator, which is the contravention

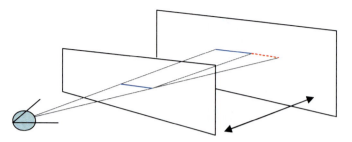

Fig. 9.5 The error (*dashed segment*) due to the parallax effect

of the primary planar hypothesis, but deserve two different approaches in their solution because of the different situations in which they occur. The first one, that is the fact that markers can be stuck on a non-planar surface and that the chessboard can select a wrong or inaccurate measurement plane, can be considered an intrinsic error, which is somewhat related and similar to a calibration issue. The second one is instead highly unpredictable and completely depending on the displacements of the markers towards or away from the camera during the tests. Just to clarify what so far has been only theoretically described, what happened is that the building, due to its asymmetry and subject to the actuator force, twisted rotating around a vertical axis. That resulted in markers getting closer to or farther from the camera which means moving on the visual ray direction. But, by design, this is the third missing dimension and thus the corresponding components of similar displacements cannot be measured by this vision system.

As previously said, both situations depend on the same optical problem called parallax. The parallax is a phenomenon for which an object is perceived in a different position with respect to the background in the event of movements of the observation point. Think of an old analogical clock: since its hands are never completely leaning on the dial, that little distance will cause the parallax effect and the time read will depend on the observation angle. For the same reason, moving the clock will change the apparent position of the hands compared to the background quadrant. This is because the human brain receives the projection of the hand shape on the dial and then evaluates the time. The same problem affects the vision system when only one camera is used and the planar hypothesis is made (Lepetit and Fua 2005) (Fig. 9.5).

9.5 Parallax Error Estimation

The system so far described is affected by errors deriving from its bi-dimensional nature. As previously stated, when the target of the measurements lies or moves outside the calibration plane, the parallax effect leads to erroneous interpretation of its appearing position on the camera sensor. All the different sources of imprecision derive form the fact that the system relies on the projection of an object on the calibration plane and not on its actual position. For shortness, these issues have

9 Computer Vision System for Monitoring in Dynamic Structural Testing 171

been grouped and classified as the out-of-plane error, since their entity is strictly related to the deviation from the original calibration plane.

Additional experiments have been executed in order to evaluate the effect of the parallax issue to the system. Preliminarily, a geometrical model has been developed to determine the relationship between the position of the target with respect to the camera and the error in its evaluation caused by out-of-plane movements. After that, tests were carried out in order to validate the model. Analysing in deeper detail the out of plane error, it can be easily related to the distance of the target from the focal axis which is the visual ray from the camera to the centre of the image. For what concerns a summarized quantization of the effect of the out-of-plane in the above described experiments, by analysing the data, the different impact of the markers' perceived movement in relation to the focal axis distance appears to behave almost linearly. For example, at a focal length of 600 mm, moving the object by 250 mm in a direction perpendicular to the calibration plane, the system measured an erroneous displacement of 33.57 mm. Conversely, executing the same kind of movements at a focal distance of 130 mm, the resulting recorded movement was only 4.55 mm. Considering this error model and the results of the tests done in order to validate it, the out-of-plane error can represent a serious issue in event of measurements performed far from the centre of the image. Unfortunately this is a common situation, because, in order to have both a good precision and a wide dynamic in the markers' detection, the camera sensor is generally used in its complete extension.

9.6 Facing the Out-of-Plane Error

Two solutions have been proposed for the two main aspects of the parallax problem (non planarity of the markers as a starting condition and during the tests), both acting in a post-processing phase. The first one can be described as a calibration fine tuning and is executed only once per camera, the second is a signal processing method that must be executed on each different acquisition in order to compensate the dynamic parallax error resulting from the target movements. The latter can be named "on line recalibration", since it involves the application of the calibration algorithm during the acquisition.

For what concerns the first method, it must be first described in relation to its application field. The error to reduce, indeed, is mainly due to two components: (i) an error in the placing angle of the chessboard, which is particularly relevant when considering multiple cameras acquiring the same movement and (ii) the very likely condition of having markers that do not lie on the same plane.

9.6.1 Calibration Fine Tuning

The first problem is twofold because, besides being the result of human imperfections, it could be also intrinsic in the test structure. This is because even the most accurate placing of the chessboard and therefore even the best definition of the measurement plane can differ from the real movement plane. It is also quite common to have the target moving in a plane that is unpredictable before the actual start of the testing operations. This must be addressed with information coming from the observation of movements executed by the markers.

In order to solve these issues, therefore, an additional amount of information must be provided to the system. The algorithm must know a set of coordinates belonging to the correct location and displacements of a marker in the scene. Amongst the possible choices, those adopted in the experiments carried out in the evaluation of this system have been the selection of a central marker and its arbitrary assumption as a reference marker and the comparison between the recordings of a slow movement with both the machine vision tools and other common sensors like a potentiometer or the same actuator's readout.

Choosing a central marker means reducing the parallax effect which has been demonstrated to be bigger as closer the marker gets to the image border and is remarkably easy because it is a method that needs no external devices interaction.

This doesn't solve the problem of mutual irregularity between the cameras but offers a simple way to fix the intrinsic camera error when markers are not placed on the same plane. Using an external device in a single slow acquisition, on the other hand, can be seen as performing a further calibration step. In this light, the additional sensor is like a different kind of chessboard which gives calibration information concerning the effective movements.

The wider the movement used in this post-calibration is, the better the accuracy of the correction. Moreover, this calibration must be done only once for each test setup since it does not depend on the real displacements performed during the different trials. All markers are considered separately and some additional data concerning them are stored in this new calibration model. For this reason, this extended calibration model is refined to a marker's granularity and does not only refer to the camera. The output of the developed algorithm is therefore divided into two sets of parameters. The first one concerns the angular error of the camera, which is stored for each marker since it results from the execution of the correction of its data (usually very similar for all the markers belonging to a camera). This is also a proof of concept of the algorithm itself, since it appears reasonable that different markers recorded by the same camera showed to evaluate the same angular error, which is physically due to chessboard positioning imperfection. The second group of parameters, instead, define the position of the marker in the optical ray coordinate. A more detailed description is due by this second algorithm because in this case the parameter to estimate is not as clearly evident in the formulas as the R and t matrices, and its variation circumvents the planar hypothesis which is the foundation of the entire mechanism.

Indeed, the starting condition on which the base algorithm relies is the planarity of the movements. The reader will remember how this condition allowed the suppression of the Z component of the marker position in the scene. In this new method, the hypothesis of planar movements still holds but markers are no more imposed to lie on the calibration plane.

Table 9.1 Mean, max and standard deviation (over all the markers with respect to the selected one) of the error before and after the correction

	Before	After
Mean error [mm]	2.01	0.06
Maximum error [mm]	5.988	0.19
Standard deviation of error [mm]	3.578	0.066

The main idea is then to estimate each marker's starting Z by calculating the effect of the parallax on the reference movement. In more detail, a direct calculation of the z (note the lower case) component would be more useful, since it is starting from the estimation of the visual distance that the previous algorithm computes the X and Y. This technique has been successfully applied in the test performed on the shaking table and has fixed most of the measurement noise experimented in the previous tests (Table 9.1).

9.6.2 On-Line Recalibration

The algorithms so far proposed appeared to reduce the error deriving from the out-of-plane effect at a reasonable level but cannot be considered a complete solution to the parallax issue. First of all they still rely on a strong hypothesis: having a marker to trust and having a recording of a stiff movement of the complete markers set. Besides that, the biggest limitation in their appliance is that they cannot overcome additional out-of-plane errors made during the test execution. Indeed, they offer a countermeasure to problems occurring in the camera setup and calibration phase, in other words in an off-line phase. But nothing can guarantee that markers will lie either on the calibration plane or even in a single plane during the whole experiment. In order to overcome this further issue, another methodology has been proposed, consisting in the application of recalibration patterns on the tested structure to be tracked during the acquisition.

This has been done by crafting a set of non reflective stiff plates covered by a grid of markers, called "super-markers". Each super-marker can be considered as a local replacement for the calibration chessboard. Indeed, the solution consists of a series of re-calibrations performed at a local or global level in order to adapt the calibration parameters to the plane resulting from the super-markers' identification. This recalibration can occur both during the acquisition and in a post-processing phase since the markers in the super-marker are built in reflective material and can be detected with infra red illumination. Super-markers can be bound to all or just a set of common markers and multiple super-markers can be employed in the same scene in order to allow tracking markers on different planes. This approach cannot manage a high nonlinear behaviour in the trajectories of the markers and does not take into account local deformations, since it still involves the validity of a looser hypothesis of local planarity.

Table 9.2 Reconstructed x,y ranges (taken from one of the freely moving tests)

Coordinate [mm]	Minimum	Maximum
X (probe)	−10.6	24.27
Y (probe)	−0.4	0.81
X (grid)	−0.04	0.05
Y (grid)	−0.04	0.05

Before its adoption in a real test application, this solution has been first assessed in a simple laboratory experiment. A single super-marker has been built and plugged on a box on top of which a rail with a sliding marker was laid, resulting in a horizontally moving probe over the super-marker.

The experiment was subdivided into two different tests, namely the evaluation of the capability of the system to track the three dimensional position of the super-marker, by reverse evaluation of the recorder $R|t$ matrices and the assessment of the effectiveness of the achieved correction.

The first one was simply qualitatively evaluated by executing simple movements and retrieving the computed 3d path of the super-marker. In order to have a better view of the capabilities of the system, three different recordings have been executed: a motionless super-marker (nailed to the wall), a swinging super-marker, still nailed to the wall pulled by one side and then released (pendulum), and a super-marker freely moved around in the room. During the experiments, the sliding probe was moved and the algorithm reconstructed its relative displacements with respect to the fixed grid.

A quantitative confirmation of the positive effect of the algorithm can be the range of variation of the x and y coordinates of the different markers in the scene after this reconstruction (Table 9.2).

At this point, after validating the methodology, the research aimed at determining a relationship between the number of markers in the super-marker and the efficiency of the reconstruction. Keeping the size of the markers fixed, which was chosen to be similar to that used for standard markers, and the maximum size of the super-marker grid selected as an A4 format in order to leave enough space in the scene and in light of the local planarity hypothesis, the amount of markers needed in order to have a good reconstruction has been searched. Different tests have been performed, each using a different number of markers in the grid (Fig. 9.6).

Unfortunately, the number of markers seemed to be highly related to the max error and therefore a minimum of 28–30 markers has been selected in order to benefit from the algorithm. This method has been tested in a real experiment on another building. Different tests have been performed on this specimen, in a sequence of ascending earthquake magnitude. The super-markers have been applied on the building face in order to have up to two super-markers per camera.

The experiments at lower intensities were performed without any problem at the system's maximum speed (120 Hz). For what concerns, in contrary, the accelerograms with higher PGAs (peak ground acceleration), they needed to be repeated at lower sampling frequencies, because the resulting speed of the specimen caused the super-marker to be lost by the tracking algorithm. This is because in order to keep

9 Computer Vision System for Monitoring in Dynamic Structural Testing

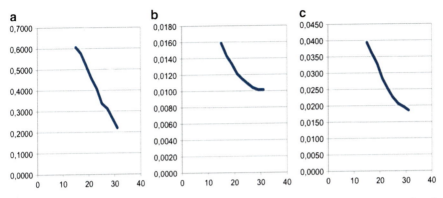

Fig. 9.6 Reduction of the (**a**) maximum, (**b**) mean and (**c**) standard deviation of the error [mm] with the number of the markers

the super-markers small, the corresponding ROIs had to be selected from an insufficient area and neighbour blobs did collapse when the movements reached their maximum speed. By analysing the resulting data, super-markers have therefore been estimated to limit the acquisition sampling frequency to a maximum value of 10–20 Hz for dynamic testing.

9.7 Conclusions

The positive features as well as the limitations of the developed machine visual system have been exposed, in order to provide the audience with a complete overview of the potential of the proposed solution. Summarizing, the opinion of the developers on the actual implementation is that it can be extremely useful when applied to measurements on planar surfaces, because, in that case, the detection of positions on the selected plane has demonstrated to be extremely fast and precise.

For what concerns the applicability of the developed system to dynamic testing of buildings and in general to large specimens, the advantage of being able to perform measurements on a large area without needing to be inside or even near the tested element makes it extremely competitive with respect to the traditional sensors usually adopted in seismic engineering. A couple of other appealing features are the scalability of the system and the large amount of detectable positions for a single point of view (up to 100). Moreover, the performances of the identification are only related to the computational power of the adopted machine vision hardware, making the system easily scalable. Another useful feature is that, once calibrated, the system is able to output directly in physical units the acquired signals, without needing further conversions or processing, promoting its adoption for real time monitoring or tele-presence implementations.

Lastly, for what concerns the possible further developments, an obvious direction to follow could be toward solving the parallax problem in a wider range of conditions. That could be done by implementing out-of-plane cameras giving a sort of three-dimensional information on the evolving calibration plane or by employing different devices both on the cameras and the specimen in order to make the acquisition units aware of the depth component.

Acknowledgements The research leading to these results has received funding from the European Community's Seventh Framework Programme [FP7/2007-2013] under grant agreement n° 227887 for the SERIES project.

References

Bradski G, Kaehler A (2008) Learning openCV computer vision with the openCV library. O'Reilly Media, Sebastopol, CA

Devernay F, Faugeras O (2001) Straight lines have to be straight – automatic calibration and removal of distortion from scenes of structured environment. Mach Vis appl 13:14–24

Heikkila J (2000) Geometric camera calibration using circular control points. IEEE Trans Pattern Anal Machine Intell 22(10):1066–1077

Heikkila J, Silvén O (1996) Calibration procedure for short focal length off-the-shelf CCD cameras. Machine Vision Group, Department of Electrical Engineering, University of Oulu

Heikkila J, Silvén O (1997) A four-step camera calibration procedure with implicit image correction. Machine Vision Group, Department of Electrical Engineering, University of Oulu

Intel Corporation (1999–2001) Open source computer vision

Lepetit V, Fua P (2005) Monocular model-based 3D tracking of rigid objects: a survey. Found Trends Comput Graph Vis 31(9):1552–1566

Li M (2001) Correspondence analysis between. The image formation pipelines of graphics and vision. In: Proceedings of the IX Spanish symposium on pattern recognition and image analysis, Castelló

Pollefeys M (2002) Visual 3D modeling from images. University of North Carolina, Chapel Hill

Pollefeys M, Koch R, VanGool L (1998) Self-calibration and metric reconstruction in spite of varying and unknown internal camera parameters. In: International conference on computer vision, Bombay

Simon G, Berger MO (2002) Pose estimation from planar structures. Comput Graph Appl 22:46–53

Skrypnyk I, Lowe DG (2001) Scene modeling, recognition and tracking with invariant image features. In: International symposium on mixed and augmented reality, Arlington, pp 110–119

Sturm P, Maybank S (1999) On plane-based camera calibration: a general algorithm, singularities, applications. In: Conference on computer vision and pattern recognition, Ft. Collins, pp 432–437

Swaminathan R, Nayar SK (2003) A perspective on distortions. In: Conference on computer vision and pattern recognition, Madison

Tardif JP, Sturm P, Trudeau M, Roy S (2009) Calibration of cameras with radially symmetric distortion. Pattern analysis and machine intelligence, IEEE Trans 31(9):1552–1566

Zhang Z (1998) A flexible new technique for camera calibration. Pattern analysis and machine intelligence, IEEE Trans 22(11):1330–1334

Chapter 10
Quality Needs of IT Infrastructure in Modern Earthquake Engineering Laboratories

Mihai H. Zaharia and Gabriela M. Atanasiu

Abstract The paper presents some ideas about the quality needs for Information Technology (IT) infrastructure as part of the equipment required in modern Earthquake Engineering Laboratories. Considering the size and the services offered today by advanced testing laboratories in earthquake engineering and seismic risk management, many types of IT solutions are available. There is also a large variety of tests, in terms of specimen weight and/or geometry, which requires specific IT infrastructure and corresponding equipment to be deployed. From the point of view of the quality requirements of the specific equipment of IT infrastructure, three different application areas need to be considered: the control of experimental platforms, the data acquisition networks gathering experimental data and the computing power necessary for the simulation of the tests and the post processing of experimental data. After presenting the comparative analyses of IT infrastructure needs in Earthquake Engineering laboratories world wide, some distributed patterns related to the emerging IT infrastructure are analysed. Some quality needs for certain specific IT components are also identified. The questions arising from the variation in the conditions and the needs between the different countries where future laboratories may be located are also addressed; for this, the highlights various quality requirements of standards from the US and Europe, complementary to the usual quality needs of advanced Earthquake Engineering laboratories.

M.H. Zaharia (✉) • G.M. Atanasiu
Multidisciplinary Center of Structural Engineering & Risk Management,
"Gheorghe Asachi" Technical University of Iasi, Bdul. Dimitrie Mangeron, 43,
Iasi 700050, Romania
e-mail: mike@cs.tuiasi.ro: gabriela.atanasiu@gmail.com

M.N. Fardis and Z.T. Rakicevic (eds.), *Role of Seismic Testing Facilities
in Performance-Based Earthquake Engineering: SERIES Workshop*, Geotechnical,
Geological and Earthquake Engineering 22, DOI 10.1007/978-94-007-1977-4_10
© Springer Science+Business Media B.V. 2012

10.1 Introduction

Since the dawn of humanity, there has been a link between natural disasters in earthquakes and socioeconomic systems. People have always struggled to obtain and, after an earthquake, to preserve and protect the most valuable resources of all kinds. Nowadays, research in the field of Earthquake Engineering, including aspects of Risk Management of natural disasters, especially earthquakes, floods, tsunamis, etc., also takes into account the research developments in Information Technology (IT), along with the estimation of the socioeconomic impact of potential disasters on the economy and society in general.

Thanks to advanced IT, the relevant information on earthquake engineering research developed all over the word can be usually stored in a database, together with the analysis of results, the funding offered by the international community within the framework of proactive disaster risk reduction projects, etc.. New technology now offers enhanced capabilities of participation, on and offline, to different categories of users, from professional users, experts, practitioners, students and academia representatives to public categories and other stakeholders interested in preventing, minimizing and learning about earthquakes effects.

In the case of earthquakes, among others, one has to admit that due to the lack of reliable data and of consistent definitions of the means or ways to estimate the real costs of natural disasters, this is a very difficult task. More quantifiable damage and loss calculations are necessary, even nowadays, in order to introduce important changes in the adopted decision making or marketing practices at various levels, for the benefit of the countries or regions affected by major earthquakes, as well as for the financial aid funds from the international community.

One of the ways of developing new knowledge in order to prevent or minimize seismic risk is to develop modern Earthquake Engineering laboratories equipped with new IT infrastructure (Atanasiu and Leon 2007; Atanasiu et al. 2008a, b).

Taking into account the emergency of seismic risk in Europe and worldwide and the concerted efforts made in the USA, Japan, and recently in China and Taiwan, to develop modern laboratories for research in the field of earthquake engineering along with new networking and communications technologies, it is worth mentioning here the efforts of the European Earthquake Engineering Community supported by the European Commission within FP7 program 2007–2013, with relevant projects concerning the design of a new advanced testing platform (EFAST Project, efast.eknowrisk.eu) and the synergy of research which has been developed in this important field, currently in their execution phase, which intend to contribute, in different ways, to the reduction of the effects of earthquakes and to the dynamics of knowledge in the European Research Area of Earthquake Engineering.

One of the first applications of IT to disasters was in the form of voice communications. Since then, many advances have led to the adoption of additional ways of information processing in disaster management (e.g. to include text, geospatial data, video, sensor data, and to collect these and other types of data in databases or other electronic forms).

10 Quality Needs of IT Infrastructure in Modern Earthquake... 179

Other technical disciplines directly or indirectly involved in the advances of disaster management can be transformed by the progress made in the IT field. For example, there is a new range of monitoring and control capabilities available to civil and mechanical engineers thanks to the construction of small sensors, microprocessors and wireless communication devices. Many applications require the employment of sensors on a wide scale. These capabilities begin to be available thanks to the results of research in distributed sensor networks.

The benefits of comprehensive monitoring and management of engineered systems added to an advanced platform, such as EFAST, can extend beyond their own boundaries; for example managing interactions between systems, such as the power grid and the communication networks that rely on them. This underscores the importance not only of collecting system-specific data, but also of normalizing and exchanging real-time assessment data between systems.

Important economic benefits both for the platform itself and also for the users may be obtained by continuous monitoring and analysis of a critical infrastructure in the future operational stage, achieved by extending new instrumentation capabilities. This would enable the routing of sensor information from buildings, bridges, and infrastructure systems, e.g. roads, communications, and power systems, gas or sewer lines, to monitoring locations, providing information about the robustness and safety of the involved infrastructure. Another critical problem that must be dealt with, in order to extend sensor capabilities, is to provide power supplies independently from the general electric grid.

The networking capabilities using IT solutions which should be offered by EFAST advanced platform will add economic benefits to the initial investment made for building EFAST, as well as to the whole range of users, in respect of the global need for reducing the effects of earthquake events in Europe and also worldwide.

Thus, the networking opportunities associated with IT can lead to an important economic impact by creating a critical mass of users that provides a potential point of interoperability and cooperation across national laboratories located in different European countries or elsewhere. For example, even when the communications infrastructure is damaged, the ad hoc use of 802.1 lx wireless capabilities in laptops, peer-to-peer use of Land Mobile Radio System radios, and use of Family Radio Service/General Mobile Radio Service "walkie-talkies" can help to provide communications. Even if these technology options may already be in the hands of the users, they may not be exploited in disaster situations, because policies and procedures for their use are not operable yet.

Other examples of valuable information technologies include information (Rao et al. 2007) such as:

- use of sensors, wikis, blogs, and data-mining tools to capture, analyze, and share lessons learned from operational experiences;
- use of databases, web portal, and call centre technologies to establish a service to provide information about available equipment, material, volunteers, and volunteer organizations;

- use of planning, scheduling, task allocation, and resource management tools to help in formulating disaster management plans and tracking execution of the plans, and to ensure timely recognition of problems and associated follow-up decision making;
- use of deployable cell phone technology to rapidly establish stand-alone communications capabilities for use in disasters where local infrastructure is damaged.

IT can also enable more robust, interoperable and priority-sensitive communications, because disaster management requires such communications systems capable of supporting interoperation with other systems. Thus communication networks must be more flexible, in case of a disruption, than many regular commercial networks. They must work for longer periods of time without power, they must expand capacity to meet emergency needs, and should be able to be self-reconfigurable, and handle the range of communication needs and environmental conditions that occur during disaster situations. Also, they must have some established points of interoperability, and must be able to distinguish between different classes of communications and properly prioritize them.

Simulation systems offer an interesting way to train disaster management professionals to better predict some issues and also to become more adaptable to working across different organizational restrictions, becoming accustomed to different technologies and integrating operations. Simulation systems can simultaneously interface with and drive both planning systems and training systems (National Research Council 2005). This would enable preparatory work to ensure the robustness of plans against multiple scenarios, and support training according to the plans. It could also ensure a smooth transition toward systems for response execution. Using detailed analysis of the response to past disasters, these simulations could have a solid foundation, and their accuracy should be expected to increase after each incident.

The prospect of adaptive planning, scheduling, and resource allocation processes that continually fine-tune logistical support plans to the evolving situation are based on the advances in computing power and the continuous development of new algorithms, which are able to incorporate the risk management of a plan in the presence of different classes of uncertainty. One such uncertainty class refers to the quality of information, for example, the level of certainty that a road is not damaged and a fire department vehicle can pass through. Another class of uncertainty concerns the likelihood of success for potential courses of action, for example since it is unknown whether an aid convoy will be able to get through using a certain route, it could be helpful to divert some of the supplies to another route.

Due to the advances in high-performance computing, it is presently possible to execute and analyze discrete event simulations with millions of parallel computational processes. These have been integrated into war games extending over weeks, involving tens and hundreds of thousands of human personnel. This means that it is becoming technically feasible to run situation analysis systems for disasters that continuously operate on the "best available understanding", i.e. filling in missing information with simulations, models, and forecasts when necessary, and afterwards replacing them with sensor data, situation reports, and incoming supply requests when this information becomes available.

Many changes to organizational structures and processes (e.g. more distributed decision making) have been enabled or even driven by the use of IT solutions. Especially flexibility and agility are very important in disaster situations because a great variety of problems arise in these cases, for which no type of organization or group of organizations is always adequately prepared in advance. Additional problems that can benefit from IT assistance include integrating the operations of multiple organizations more quickly and even increasing the cohesion of people who had not worked together before.

Therefore, it is critical to establish mechanisms that ensure that researchers deal with real-life problems and that practitioners are exposed to new technology opportunities, especially in light of the significant non-technical factors affecting the adoption of IT for disaster management.

The empirical evidence shows, for example, that IT is not simply a tool for automating existing processes, but rather an enabler of organizational changes. It is the complementary investment in decentralized decision making systems, training, and business processes along with technology that allows organizational efficiency improvements (Dedrick et al. 2003).

10.2 Needs for Quality Assurance in Earthquake Engineering Laboratories

Earthquake Engineering (EE) laboratories traditionally presented their function, facilities, performances, research and services offered, location and personnel information in the form of booklets, leaflets or lately audio-video presentations. Until recently, such a presentation could be qualified as being in agreement with the needs of a local regional or even national community at the considered level of technology of the whole society.

The latest developments in the field of Information and Communications Technology, along with the new knowledge in the field of Structural Dynamics and Earthquake Engineering in an emerging global economy recently affected by the global economic crisis, have created a new international context for the operationally active laboratories in Earthquake Engineering. New demands for change in higher education regarding the competence based learning along with the development of the required skills of students involved in Master's or Doctoral Study Programs have also opened a new window to services, training, and videoconferences, new types of users and knowledge developed in these advanced research centers in Earthquake Engineering. Finally, the picture of the new features added to the traditional ones has to be completed by the needs and requirements of state or private companies for materials, structural elements or sub-structures, qualifications for constructions, existing or in the design phase, located in high seismic hazardous areas of Europe or worldwide.

In this context, different classes of users of Earthquake Engineering laboratories and the variety of services offered are practically requiring some guaranties for the activities conducted here, ascertained by the Quality Assurance procedures, made transparent to the whole community of users and stakeholders.

Advanced seismic testing facilities were from the beginning complex electrome-chanical and hydraulic infrastructures. Until the last decades, the IT approaches had been used only at the level of automated experiment control and for some minimal data acquisition interfaces. The data provided by the experiments were separately ana-lyzed and processed using a medium performance computing environment/system.

Nowadays the role of IT has become dominant in each stage, beginning with the experiment prerequisites, followed by the experiment itself, and finishing with the final analysis of the gathered data. The common IT-based infrastructure used in earthquake engineering laboratories consists of: a telepresence room that is usually accessible only to the earthquake engineering community, a mixed platform control equipment, the needed computing systems, the control equipment of audio video infrastructure used in the experiment, and the seismic platform and multiple sensor networks used for data acquisition. Based on this infrastructure, there are two major concerns related to multiple required quality standards that have to comply with the facility itself at each level and to the procedures of systems calibration specifically required for each experiment. These aspects may even increase in complexity; more complex IT, developed to support complex experimental programs, may become available in the future (e.g. real time interaction between laboratories geographi-cally distributed during the experiment in terms of audio video and experiment data or off line experimental data availability using a distributed database.)

One of the first functions of IT infrastructures of Earthquake Engineering is to raise awareness all over the world among future users/beneficiaries (e.g. academic staff and students, researchers, practitioners, various groups of interested people from construction owners to the general public, along with different stakeholders at regional, national and global level.)

In order to take advantage of the IT opportunities in the twenty-first century, a modern and complex Earthquake Engineering Laboratory should have a formal pre-sentation on the WEB page of the Institution/University/Department with a short description of the physical facilities which are the main components of the experi-mental testing. This first function is now utilized worldwide and in Europe, if we were to mention laboratories such as the Joint Research Center of the European Commission, Tamaris Laboratory of Commisariat de l'Energie Atomique in France, EUCENTRE Laboratory in Italy, LNEC Lisbon in Portugal, Structures Laboratory of the University of Patras, and other known European testing facilities involved in the SERIES FP7 Project (www.series.upatras.gr). Usually a short presentation describing the main per-formances of the physical facilities used for advanced seismic testing are visible on the Web page, along with a short presentation of the faculty experts who are develop-ing design lab methodology procedures and conducting experiments together with permanent staff for research and development activities. The physical facilities are described in detail in the Laboratory Manual together with the services offered to the Earthquake Engineering community and other interested stakeholders.

Regarding other laboratories typical of industrial engineering, i.e. aeronautical engineering, the EE Laboratories do not have specific standards for seismic experiments or seismic qualifications. However, in order to ensure a certain stamp of quality for the experimental results and conclusions, a needed set of Internal Procedures of Quality, developed within the Quality Laboratory Manual, would be a necessary asset for an advanced seismic facility.

10.3 European and International Trends

In order to discuss the Quality Assurance of a Testing Laboratory from different fields of engineering or other domains, it is important to clarify first the meaning and content of these two words. Quality Assurance is a complex process composed of internal and external Evaluation aiming to obtain an Accreditation Certificate issued by an accreditation body.

For the accreditation needs of a testing laboratory, necessary mostly for industry and business, the European Association for Accreditation (EA) was set up in 1997 in the European Union and was registered in 2000 as a non-profit association.

This association is practically a European network of nationally recognized accreditation bodies located in the European geographical area, which aims to build confidence based on the independence, competence and impartiality of its members.

The EA mission consists in defining, harmonizing and building consistency in accreditation as a service within Europe, by ensuring common interpretation of the standards used by its members, and transparency in each operational stage. The final goal is to maintain a multilateral agreement on mutual recognition between accreditation schemes and reciprocal acceptance of accredited conformity assessment services and results. The peer evaluation system based on common values, in accordance with international practices should be also part of the mission, since EA is enjoying a regional membership of International Laboratory Accreditation Cooperation (ILAC) and International Accreditation Forum (IAF).

This organization is acting also as a technical resource on main issues related to the implementation and operation of the European policies on accreditation, since EA covers accreditation of Testing Laboratories, testing procedures, and calibration with all processes included. The European Association (EA) is also an umbrella of inspection and certification bodies, going further to cover aspects related to Quality management systems (QMS) and Environmental management systems (EMS) for products and services, according to the European Regulation EMAS.

The EA signatories' national accreditation bodies and organizations dealing with quality assurance and quality management along with their profile are listed on the web site at URL address: http://www.european-accreditation.org/content/mla/what.htm.

The EA members have agreed in the EA Multilateral Agreement (EA-MLA) to recognize each other's granted accreditations and the reports/certificates issued by the entities accredited by it, with periodic mutual evaluation (http://www.european-accreditation.org/content/EAGuide/).

At international level there is also an organization, namely ILAC, offering confidence in the affiliated accreditation bodies and in their ability to determine the competence of the certification bodies, through a similar document of Multilateral Recognition Arrangement IAF-MLA, respectively ILAC-MRA, enjoying in this way a worldwide recognition.

The Earthquake Engineering laboratories in Europe have had, since their establishment, the tradition, maintained during the years of operation in all major centers of Europe, to conduct research and related experiments, whose variety and complexity increased in the last decade due to the enhanced capability of electronic and IT infrastructure. Actually, this classic role of EE laboratories was specific for all classes of laboratories of earthquake engineering, all over the word, in their preliminary years of functioning. Lately, as we witnessed tremendous changes in Information and Communications Technology, the EE laboratories, as well as other laboratories from other fields such as industrial or/and business, have become an open window to the world; and this change, along with the globalization of economy, has managed to induce new opportunities not only in research but also in the field of services to communities of researchers, users from the academia, stakeholders at different levels, business opportunities and industrial products specific for constructions, as well as for other fields of structural dynamics in their need for qualification or even certification.

Actually, after the 1990s, the National Science Foundation initiated and developed a nationwide US network, i.e. the George E. Brown Jr. Network for Earthquake Engineering Simulation, which linked important EE laboratories affiliated to centers of research and universities in United NEES, in the United States of America, using among other things a new concept for the IT infrastructure. The JRC International Workshop, held within the FP7-EFAST project, identified new items for the portfolio of activities along with the classical one of research projects and seismic qualification for structural/non structural or prototype/model for different classes of constructions and equipment, carried out now in EE laboratories in Europe and worldwide.

The EE laboratories have nowadays developed important services, opportunities and capabilities based on the use of enhanced and modern IT infrastructure capabilities. An important role, representing at the same time a message for different classes of users and stakeholders including possible decision makers or finance organizations in use of IT infrastructure, is to build a portfolio of awareness of performances and activities and to enhance the attractiveness of the laboratory.

10.4 IT Infrastructure in Earthquake Engineering Laboratories

Usually, at European level, the EE related IT infrastructures were developed from the initial stages based on incremental approaches, once new or modern IT devices were in place. This was naturally justified in most laboratories by the access to funding which in many countries was strongly depending on local or governmental politics. Yet, there are laboratories which have modern and uniform design infrastructures that entirely replace old infrastructures, the cost of IT infrastructure being usually lower in comparison with the rest of the costs involved. If one analyzes the latest inquiries regarding the state of the art of European Laboratories working in the seismic field along with their facility performance and research/services offered (Marazzi and Molina 2009), one can notice that telepresence and inter-facility distributing testing capability presently exist in a reduced number of European top laboratories.

Considering the Internet broadband options of these laboratories, one can also conclude that only 20% of facilities can rapidly deploy a telepresence room. Also, a more reduced percentage of around 10% from the total number of respondent laboratories have a distributed database available for experiments.

If distributed testing represents a feasible future direction of research for the EE community, in this case one can state that an important number of laboratories in Europe need to increase considerably investments in state-of-the-art IT infrastructure. Fortunately, the scale of these investments is affordable if compared to the cost of seismic experiment itself.

Besides investments, EE laboratories need clear rules of internal procedures, which are useful for internal evaluation as a component of Quality Assurance and build trust among themselves and in front of various classes of present and future users.

In what follows we have focused on a generic IT infrastructure description for an advanced seismic testing facility, corresponding to the momentum level of development in this field. For this generic deployment of EE-IT infrastructure the needed quality requirements both for equipment and for main IT infrastructure components are presented.

10.4.1 IT Infrastructure Description

The IT infrastructure, nowadays an important component of modern Earthquake Engineering and Structural Dynamic laboratories, implies mainly:

- Complex Data Acquisition Systems (DAQ) based on various sensors and required digital acquisition systems;
- A network of Pan-Tilt-Zoom (PTZ) high performance cameras, for the real time video inspection of the experiment site;
- The needed computer systems for experimental storing, processing, and experiment control.

A scheme for the generic IT infrastructure of a modern EE laboratory is presented in Fig. 10.1.

Following the analysis of the layout in Fig. 10.1, some data streams, specific to this type of experiments, could be identified. The bidirectional command and control flow for the seismic platform actuators is usually isolated and supports mostly manual control. There are cases when the experiment conditions and the simulation

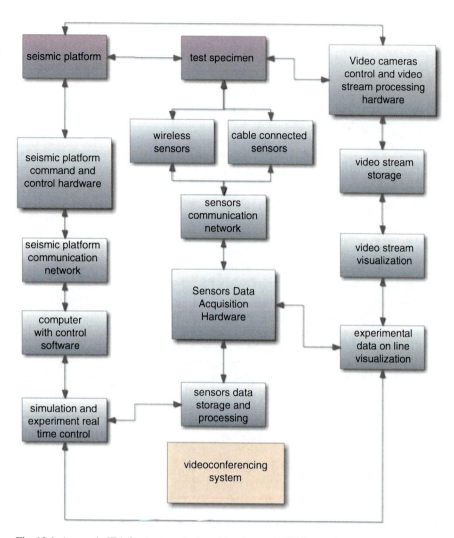

Fig. 10.1 A generic IT infrastructure deployed in advanced EE laboratories

parameters need to be continuously corrected due to the interaction with the simulations, when the computing system, placed locally or remotely (in case of real time distributed testing), is connected to the data flow. Usually the shaking testing platform can be viewed at the black box level because they are usually manufactured by dedicated companies that give to the EE engineering the required amount of maneuverability and performances. In addition, these companies should be able to give explicit or implicit indications along with the procedures needed for the calibration operation, at the required levels of precision.

Another data stream is given by various types of sensors placed on the platform in order to test different classes of specimens. Usually the cable-based communication network is used due to high resistance at the interference with powerful magnetic fields that are generated on the experimental site. There are problems related to communication interference when the wireless sensors are used. As a consequence, those types have not been granted by the EE community. Anyway, if one considers the wide scale adoption of this technology in the last decade by civil engineers (Taylor et al. 2010) this approach should not be neglected in the future. The sensors are connected to the professional data acquisition system that can be deployed in industrial racks, or sometimes installed in personal computers. The approach depends on the laboratory complexity. After the acquisition phase the data are stored and displayed on and off line. This data flow can be used for controlling the simulation in distributed hybrid testing, for manual real time control of the experiment conditions or for further investigations. In the last decade the experiment data are made available outside the laboratory using the Internet. In this case, distributed testing is experimented sometimes; however, it is not widely used because in most situations some glitches in the Internet communication can destroy highly expensive experiments in progress. As a result, the solution is considered yet unfeasible from an economic point of view. There are some successful experiments regarding distributed testing which prove the feasibility of the concept (Pearlman et al. 2004; Yang et al. 2007), but the economic risks are still too high.

The final data flow is provided by video cameras. Hereby, the system used depends on cost and complexity, in accordance with researchers' needs, and may vary within a large interval. Anyhow, the main idea is to have PTZ facility. The surveillance systems development at the market level can offer good solutions for fulfilling the EE requirements. Also, the web control of the camera system is provided without problems. Concerning this aspect of IT Infrastructure in EE laboratories, it has to be mentioned that the George E. Brown Jr. Network for Earthquake Engineering Simulation – NEES and Bristol University have developed a correct integration of web access into their specific systems (Pegon et al. 2010; Zaluzec 2001). The resulted data flow is compressed using specific standards by the use of hardware dedicated boards. Usually the company is responsible for assuring proper technical performance up to this level of data flow processing. The compressed video streams are then displayed and stored on dedicated spaces. If the online access is needed during and after the experiment it is easy to be implemented with a supplementary server.

Usually, an operational Telepresence Room implies only human interaction, done in a virtual space. Within the Earthquake Engineering community, the Telepresence Room is designed based on the concept of experimental virtualization, in terms of full access to live experiments and experimental data, allowing sometimes the control of the experiment too, using simultaneously a virtual space of live interaction for experts from various geographically distributed laboratories. In the future a next step seems to be needed which will integrate the videoconference room in the Telepresence room, now being separately deployed.

10.5 Quality Requirements for Basic IT Infrastructure Components

In this respect, even if the common practice nowadays in this specific field of testing is represented by the advanced platforms equipped with shaking tables, reaction wall and other expensive technological equipment, usual for seismic and dynamic experiments, there is a need for quality assurance concerns, especially at the internal level of each laboratory, in terms of evaluation system procedures, set up internally for each specific laboratory. The minimal level of initiation and further developing of quality is to set up a Quality Manual with specific procedures not only related to the design, functioning or operation during the specific seismic testing, but also considering the evidence of quality measures for accompanied services, among them the IT infrastructure deployment, functioning and operation during the laboratory lifecycle.

According to the SEELS laboratory in Buffalo (SEESL 2011) or the Seismic testing laboratory at Bristol University (BEELAB 2011), the Quality Manual (QM) is documented if mainly the following chapters are included:

(a) Laboratory physical and technical details

  - Laboratory affiliation and Facility description (location, topology, design details, testing area details);
  - Infrastructure details presentation of testing space composed usually of: testing equipment, platforms, reactions frames, and other devices;
  - Infrastructure Instrumentation including Testing set-up, Control room, gantry crane;
  - IT infrastructure description and capabilities including server room location and other IT needed ITC deployed devices;
  - Electronic packaging area;
  - Support area and related support area presentation containing description of all types of needed machinery and tools to support operational experiments, phases of facility welding machinery, gantry crane, etc., including also the delivery area for samples, models or prototypes for testing purposes;
  - Visitors' gallery description.

(b) Laboratory Equipment specific for seismic testing (shaking tables and reaction walls mainly, and also other testing systems, for example soil-interaction testing if available) with performance details data, along with electronic, optic, analogical devices of system instrumentation needed for input and output analysis;

(c) ICT infrastructure description in details needed for different internal operations and external virtual services, communications, real time or/and offline user services, and other functions;

(d) Internal quality procedures for each type of operations described at subchapters (a) to (c) and services online and offline conducted in laboratory.

Regarding these specific subchapters dedicated to ITC infrastructure we considered that due to the new era of economic globalization, the tremendous changes in Information Technology and market-oriented services in a globalised economy and business society, it is relevant to open the discussion on main issues of quality components for the ITC infrastructure of EE laboratories. These quality procedures presented in the chapter of Quality Management Manual related to IT infrastructure could be in view for an internal evaluation or possible external evaluation, in case of a specific EE laboratory whose top management decided to enhance its confidence opening the door for services to industries or business in the field of seismic qualification or other type of required services for industrial needs.

Moreover, the European/International recognition agreements could be considered as possible models to facilitate the access and benchmarking among laboratories in Europe at bilateral level, with extension, if needed, to all members of EA conferring confidence based on a periodic mutual evaluation.

10.5.1 Quality Needs for Sensor Communication Networks

10.5.1.1 Performance Needs

Up to now there is not enough information with a clear description in order to draw up a list of commonly agreed technical specifications for the types of sensors used. Only some general performance indicators like the type, the noise rejection and the precision are taken into account. In fact, some internal quality procedures may need to be generalized after negotiation at EE community level.

Here, a problem related to the noises generated in various forms (like electromagnetic (EM) radiation or parasites injected into the power network) is identified. Professional types of sensors that have a good shielding and internal amplifying, and sometimes calibration, do not create too many problems. Unfortunately, there are many situations in EE laboratories, when the sensors are hand-made at laboratory level (the prototyping approach). In this case the noise influences must be avoided. The approaches are classic: shielding, at the sensor and also at the communication channel level, the ground isolation system, and if it is possible, the signal amplification must be done at the sensor level. Otherwise, there is a good chance that noise could make unfeasible some parts of the measured signal by masking.

10.5.1.2 Quality Needs of Sensor Communication Networks

A search through the available procedures for wireless communication in European EE laboratories revealed a low use or a lack of operational use in place. The high magnetic fields generated by the platform equipment are considered to be the main cause. However, the experts of University of California at Davis (NEES 2004b) have already implemented a customized solution which consists of a distributed high-speed wireless data acquisition system (WIDAQ). A host radio is used to control the modules using a combination of Time-division multiplexing (TDM) and Frequency Division Multiplexing (FDM) to gain a broadband for collecting 56 Mb of data at any individual event (NEES 2004b).

Advanced sensors are a solution to be used in modern laboratory nowadays, comprising of: wireless Micro-electromechanical systems, MEMS-based accelerometers, and piezoelectric transducers for strain and acceleration measurements (NEES 2004c). Related to this solution, it is perhaps suitable to consider an update in the future, using technologies based on Wimax products (www.wimax.com).

Regarding the quality assurance needed for wired communications, there are some sets of standards and related protocols specific to this technology, which should be used by laboratories for compliance purposes. The most used standards are: Serial Interface RS232 (Horowitz and Hill 1989), Universal Serial Bus USB (Wikipedia 2010), and protocols like TCP/UDP (Stevens 1994) or Inter-Integrated Circuit – I2C (Nxp 2007).

From all the classes of standards only RS232 standard needs some supplementary precautions when it is applied. The main aspects of quality in using it are related to possible power surges that may appear even into an environment with a good design regarding ground infrastructure. These power surges on RS232 device might be experienced, unfortunately, more often that one would expect, due mainly to various unexpected high voltage noises that sometimes exist at the level of the ground connector. In order to avoid these technical problems the following standards: DIN VDE 100, DIN VDE800, DIN43629 and DIN EN 50289 presented in DIN VDE 800 2010; DIN VDE 100 2010; DIN cable (2010) should be considered when drafting the Laboratory Quality Manual.

10.5.2 Data Acquisition System DAS Requirements

10.5.2.1 System Performance Requirements

Data acquisition system DAS, including calibration procedure, shall conform to NIST traceable primary standards (NIST 2010) and other applicable standards as IEC 60068-3-3 standard (IEC 1991). The data acquisition system shall be recalibrated at least once every year. These calibrations shall be verified before each test. The sampling rate shall be sufficient to capture all aspects of the load-deformation response relevant to the analytical modeling of the component.

The University of Texas at Austin (NEES 2004d) has, for example, three state-of-the-art data acquisition systems that are operational based on:

- Sercel 408XL System, which is capable of collecting data up to 2000 channels;
- VXI Technology system which is a 48-channel dynamic signal analyzer system;
- Agilent 4-channel Dynamic Signal Analyzer, this equipment being connected to the NEES grid via a satellite modem (NEES 2004d).

In another developed EE laboratory operating in the United States of America, at University of Lehigh (www4.lehigh.edu), there is a digital 8-channel control system with real-time hybrid control packages, with each channel of the controller designed to follow an independent random load or displacement history. Also, a high speed 256-channel data acquisition system, capable of acquiring data at 1,000 Hz (1,000 samples per second) per channel and expansion to 512 channels is installed (NEES 2004c).

As reported in NEES (2004b), the EE laboratory of University of California at Davis is equipped with a large inventory of transducers, and signal conditioning equipment can be used to monitor model behavior during an experiment. This data acquisition system currently records up to 160 transducers and can be expanded, should the need arise, while at University of California at Berkeley, conventional deformation and force measuring instruments are available to support data collection using a new 128-channel data acquisition system that features a hardware-based data ring buffer. Also a hybrid simulation controller with 8 independent control channels enabling quasi-static, near-real-time and real-time hybrid simulation, or by a 4-channel MTS flex-test system (www.mts.com) enabling conventional static and quasi-static testing is in use (NEES 2004a).

In the European Earthquake Engineering Community, there is also some important development of IT-based Infrastructure to be found in Earthquake Engineering laboratories. So, at the Commissariat à l'Energie Atomique CEA, in Tamaris Laboratory (www.cea.fr), there are around 250 channels at 200 Hz within an Acquisition system of Pacific Instrument PI 660, which has enhanced acquisition, filtering and storage abilities (Atanasiu et al. 2010).

At the European Centre for Training and Research in Earthquake Engineering, EUCENTRE (elsa.jrc.ec.europa.eu), the Data Acquisition System includes a 250-channel system based on 18 bit hardware and an advanced wireless system based on 8 high definition, characteristic of digital cameras.

The Laboratório National de Engenharia Civil, LNEC (www.lnec.pt), uses a technology based on National Instruments (NI) boards. This acquisition system allows up to 154 channels for measuring pressures, forces, accelerations, displacements (linear variable differential transform – LVDT and optical) and strains.

At the ELSA Laboratory of the Joint Research Centre, JRC (http://elsa.jrc.ec.europa.eu/), a well-known institution of the European Commission, the IT infrastructure is based on an Acquisition System equipped with a number of 250 channels at 500 Hz including a dedicated acquisition procedure involving 1–4 Advantech PCL816 analog acquisition cards composed of 32 channels and several National Instruments NI acquisition racks, Peripheral Component Interconnect – PCI eXtensions for Instrumentation (PXI), respectively, with corresponding USB devices to host needed connection.

The analysis of the state-of-the-art of DAQ systems, used in modern IT-based laboratory, revealed some minimal technical specifications of the system, which are: 1,024 full available channels; the maximum frequency will be 1 KHz, bearing in mind that each channel should have 16 bits, with possible variations of 14 or 18 bits.

Regarding the qualification of recorded test data, there is a need that these data should be acquired at a sampling rate of at least 10 times the highest fundamental frequency identified in any main direction of testing specimen, with a minimum rate of 200 samples per second. All the acquired data should be low-pass filtered by a block-wall type filter having a corner frequency of at least two times the highest natural frequency of interest in any main direction of interest for the testing specimen, but not exceeding 30 Hz (FEMA 461 2007).

However, for most of the EE laboratories it seems that an externalization of data acquisition systems during the calibration procedures is more effective due to its economic efficiency for the whole process of quality assurance.

10.5.2.2 Quality Needs

Regarding software, the quality assurance is hard to achieve. One situation is when the software used is provided partially or totally by a hardware producer, e.g. National Instruments, NI (www.ni.com). In this case it usually meets the quality assurance standards because in the worst case the provider will give detailed methodologies about how the calibration process or other fine tuning related operations can be done.

In the other case, when the needed software is developed on site by an EE Expert, it is preferable to acquire some certification form. The simplest solution for the EE community is to establish internal sets of quality procedures based on standards included in a Quality Manual with special reference to IT Infrastructure used in the laboratory.

If this approach is insufficient and a supplementary certification is required, then it is recommended to achieve a minimal required quality by complying with the requirements of the following standards: ISO 9001:2008 (ISO 2008) and IEEE/ANSI: 730:2002 (ANSI 2002), and also with Standard of Software Unit Testing 1008 (IEEE 1987) and Standard for Software Test Documentation 829 (IEEE 1998).

10.5.3 Telepresence Room Requirements

10.5.3.1 Telepresence Performance Needs

In what follows, the system based on real time support for live interaction between the researches using telepresence equipment is analyzed. The concept of telepresence must not be understood as videoconference because it is different and more complex.

There is already a perception in the EE community that, in the future, a virtual collaborative environment for live interaction between researchers all around the globe during the distributed testing experiments is required. Some minimal technical requirements regarding the communication network implies: a 10Gbps capability of Internet service provider at the laboratory level and an interval between 1 and 512 Mbps for passive users, such as other researchers who might be interested in the subject. These technical requirements for the equipment must comply, at a minimal level at least, with the following quality requirements: availability of a network broadband, good quality Internet service and multimedia signal.

The fully integrated telepresence solution, as the one of CISCO (CISCO 2010), may be considered as a good approach. The main argument is the reduced know-how required for technicians to use the system after its deployment. Such solutions have already been experienced in the procedure of quick control of PTZ in case of high performance cameras. The only disadvantage of this approach is higher initial costs. It is hard to estimate, if we take into account the life cycle associated costs for this type of infrastructure (if custom-made), if the use of a technological mix from various hardware and software providers will give significant cost decreases. If the initial costs are affordable, a fully integrated solution is recommended.

The telepresence room in the EE laboratories has more complex functions than a classical system of telepresence used for common videoconferences, since this room is based on simultaneous visual information, such as:

- impersonation of the other remotely connected colleagues that attend the experiment;
- digital information given by the platform control systems;
- digital information received from the sensors placed at the experiment site;
- video information about the experiment site;
- information about the results from dynamic simulations from computing cluster during the distributed experiment;
- information on the capability of direct control in order to obtain a fine tuning during the experiment.

In the case of distributed experiments most information has to be shared between all geographically distributed participants in real time. Also, the experiment controls can be provided by the use of touch screen and keyboard combination placed at each working place in the telepresence room. The rest of the information can be displayed by using different screens, one for each type of required visual information.

For the moment there is a need for a supplementary inquiry at EE laboratories level in order to finally define the expected visual performances for each type of video information to be transmitted using the telepresence room facility, since any present solution regarding telepresence is challenged by the tremendous advances in Information Technologies, being at the moment at a new frontier of development.

10.5.3.2 Quality Needs

When discussing the quality of telepresence equipment, we should bear in mind that there are some minimal requirements for the components of telepresence, given by standards with which the equipment should comply. For example, for Television systems, the MPTE 370 M-2006 (SMPTE 370 M 2006) should be considered and the standard IEEE 802.1Q VLAN; for tagging, encapsulation (IEEE 2005); for CoS and trucking encapsulation, 802.1w RSTP (CISCO 2006), 802.3ad LACP and 802.3af PoE for video stream (IEEE 2008) and also British standards such as BS.1284, BS.1387, BS.1423, BS.1548 and BS.1873 for audio streams (ITU-R 2010).

10.5.4 Datacenter Requirements

10.5.4.1 Datacenter Performance Needs

A dedicated Datacenter for the EE laboratory is usually recommended only for large infrastructures where complex tests and over 100° of freedom (DOF) simulations for large test specimens have to be run along with complex distributed testing. A typical Datacenter operating in an advanced EE laboratory consists of several server racks with uninterruptible power supply (UPS) and sometimes a network attached storage (NAS), deployed at the control room level.

Depending on the laboratory complexity, from a construction layout point of view, the computing space is more developed or based on few personal computers only.

In case an EE laboratory upgrades its IT infrastructure an incremental approach should be useful, while in case a new EE laboratory is designed a systematic approach for IT infrastructure design is recommended. Figure 10.1 shows a possible solution for designing the IT infrastructure. Yet if the magnitude of the expected experiments is higher (e.g. nuclear industry or army demands) then a more complex IT infrastructure is needed. A possible solution was proposed in EFAST- FP7 project (http://efast.eknowrisk.eu/EFAST/). In Fig. 10.2 the design proposed for the datacentre of IT infrastructure of EFAST is presented. The existence of a dedicated datacentre for the seismic facility was justified by the high performance computing demands and also because of the required high performing telepresence system. What we see here is an approach derived from the datacentre standard TIA 942 (DiMinico 2006), with some modifications in order to better serve the EE community needs.

An advanced experimental site is based on some dedicated communication networks which require: an Ethernet network composed of routers, hubs or wireless access points, located in different locations of the infrastructure. For these types of equipment, the power supply must be different from the one used by the platform itself. The deployed network shall provide connectivity for any equipment that supports the protocols to transfer information given by the sensors and PTZ cameras. This network should be placed, from an architectural point of view, in the upper side of the platform walls, in order to provide an easy access to video camera and minimize the interference from access point and other platform equipment. In order to increase the flexibility a similar network can be deployed at the ground level also.

The other communication network can be a mix from various cables in accordance with all needed wire communication protocol used for sensor communication (like USB, Serial Interface RS232, UTP or optic fiber or even wireless). The position of this network and the required special entry points is strongly dependent on the solution elected for the platform foundation.

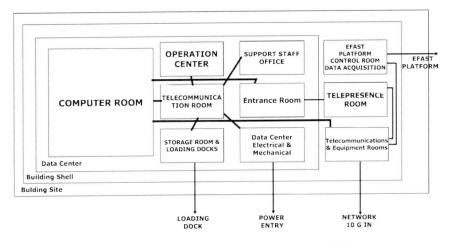

Fig. 10.2 The EFAST datacenter designed architecture (Atanasiu et al. 2010)

10.5.4.2 Quality Datacenter Requirements

Regarding computing power and communication needs for an EE facility, an approach can be to use the organization datacenter and to deploy near the seismic platform only the required hardware and software (telepresence room and some high speed routers). This is currently the most common approach and it is justified by the cash flow of the laboratory that drives to an incremental approach in the investment plan. This is very good for small and medium size laboratories. The communication center is not dedicated especially to an EE laboratory.

For the new research facilities of large dimensions this approach is unfeasible. There are new concepts that the EE community slowly begins to take into consideration like distributed testing and virtual spaces for real time interaction between laboratory experts worldwide.

The main specifications and their minimal requirements are shown in Table 10.1.

In order to check the quality, one has to analyze all entry specifications; a good approach would be to use a datacenter as base core and to complete it with the required supplementary infrastructure. The minimal requirements of datacenter equipment to comply with the standards are:

- Standard TIA942 (DiMinico 2006) in order to assure a proper design for a dedicated datacenter;
- American Society of Heating, Refrigerating and Air-Conditioning Engineers – ASHRAE data books (ASHRA 2010) and the ones of Uptime Institute – TUI (Uptime 2010) that may be used to properly design the cooling system of the datacenter;
- Telcordia standards like GR-63-CORE (Telcordia 2010, 2011) and European standard EN50173-5 (Engelhardt 2010) dealing with proper deployment of all communications cabling infrastructure.

If the requirements of the above standards are met, then it could be considered that the quality needs for the datacenter deployed equipment are fulfilled.

Table 10.1 Input specification for EE laboratory IT infrastructure

Feature	Requirements
High performance telepresence room	At least 4 GB/s better 10 Gb/s direct connection to internet, as for European Union Geant (www.geant.net)
	High performance dedicated routing devices (e.g. Cisco TelePresence Multipoint Switch)
	High performance dedicated hardware for media streams acquisition and processing
	Professional large dimension screens
	Custom or professional software solutions
	Associated redundant data persistence system
The experiment command and control section needed to control the local or distributed experiment	At least 1 GB/s
	High performance routers
	Custom made software like NEES DataTurbine
Large models (e.g. 200° of Freedom per structural model) real time simulation and experiment control	High performance computing facility based on supercomputer or powerful computing cluster
	Unix/Linux based software
On site sensor networks	Wired and wireless complex matrix of custom and/or professional sensors and the on site required communication network
Data Acquisition and platform control hardware	Seismic platform control system
	Professional DAQ solution
High performance 2D and 3D on site vision system	Dedicated hardware systems
	Custom and/or professional software
Access on and off line to all sensors data and video stream regarding all experiments stored data of projects	Powerful Web Server
	At least 1 GB/s better 10 GB/s direct connection to internet
	High performance dedicated routers

10.6   Final Remarks

During the past decades, following the tremendous qualitative leap in Information and Communications Technology, as well as the economic globalization, Earthquake Engineering laboratories all over the world have witnessed a new dimension which exceeds the classic research activities in the field of structural dynamics and EE. This cutting edge approach can be identified both in the NEES network operating in the USA, and in the high performance EE laboratories in Europe, Japan and China.

With a view to increasing the worldwide competitiveness of advanced EE research conducted in the EU countries, as well as aiming to develop the international dimension of these laboratories' capabilities, based on the services provided by the information and communications technology infrastructure, this paper is presenting both modern generic schemes for these IT based infrastructures, and a series of requirements apt to certify the quality assurance of research results, as well as of various services which can be provided nowadays to industry and business alike.

References

ANSI (2002) American National Standards Institute. http://webstore.ansi.org/RecordDetail.aspx?sku=ANSI%2FIEEE+730-2002

ASHRA (2010) American Society of Heating, Refrigerating and Air-Conditioning Engineers. http://www.ashrae.org/aboutus/index

Atanasiu GM, Leon F (2007) A digital methodology for urban risk management. Bul Polytech Inst Jassy, LIII (LVII), Section Civil Engineering and Architecture, Fasc. 1–2, pp 15–19

Atanasiu GM et al (2008a) Decision based risk assessment model for existing damaged infrastructure, application to Iasi city. In: Rostum J, November V, Vatn J (eds) COST Action C19, proactive crisis management of urban infrastructure. SINTEF Byggforsk, COST Office, European Science Foundation, Brussels, pp 211–218

Atanasiu GM, Leon F, Zaharia MH (2008b) Intelligent agents for life cycle management of structures and infrastructures in seismic areas. In: Biondini F, Frangopol DM (eds) Life-cycle civil engineering. CRC Press, Taylor& Francis Group, Boca Raton, pp 921–928

Atanasiu GM, Zaharia MH, Boronea SA (2010) Prototype on required hard- and software for web portal to enable efficient access and networking. Research report, Deliverable D5.1, EFAST project. http://efast.eknowrisk.eu/EFAST/

BEELAB, Bristol Earthquake and Engineering Laboratory Ltd. (2011) Commercial testing at BEELAB. http://www.bris.ac.uk/civilengineering/research/dynamics/eerc/commercial.html

CISCO (2006) Understanding rapid spanning tree protocol (802.1w). www.cisco.com/en/US/tech/tk389/tk621/technologies_white_paper09186a0080094cfas.html

CISCO (2010) Cisco telePresence essential operate service. http://www.ict-partner.net/en/US/services/ps2961/ps7072/services_data_sheet0900aecd80557f72.pdf

Dedrick J, Gurbaxani V, Kraemer KL (2003) Information technology and economic performance: a critical review of the empirical evidence. ACM Comput Surv 35(1):1–28

DiMinico C (2006) Telecommunications infrastructure standard for data centers ANSI/TIA-942. www.ieee802.org/3/hssg/public/nov06/diminico_01_1106.pdf

DIN cable (2010) http://caledonian-cables.com/product/Technical%20Lib/DIN%20cable%20standards.htm

DIN VDE 100 (2010) http://www.baunetzwissen.de/standardartikel/Elektro_DIN-VDE-100_153140.html

DIN VDE 800 (2010) http://de.wikipedia.org/wiki/DIN-VDE-Normen_Teil_8

Engelhardt RG (2010) STRATEGIC FACILITIES Data Centre Standard EN50173-5. http://www.rdm.com/asi/Portaldata/1/Resources/APAC/asia1/documents/Data_Centre_Standard_EN50173-5.pdf

FEMA 461 (2007) Federal Management Emergency Agency, Interim Testing Protocols for Determining the Seismic Performance Characteristics of Structural and Nonstructural Components, FEMA 461, Washington, DC

Horowitz P, Hill W (1989) The art of electronics, 2nd edn. Cambridge University Press, Cambridge

IEC (1991) IEC environmental testing-Part 3: guidance. Seismic test methods for equipment. http://webstore.iec.ch/webstore/webstore.nsf/Artnum_PK/610

IEEE (1987) IEEE Standards Association. http://standards.ieee.org/findstds/standard/1008-1987.html

IEEE (1998) IEEE Standards Association. http://standards.ieee.org/findstds/standard/829-1998.html

IEEE (2005) IEEE Std. 802.1Q-2005, virtual bridged local area networks. http://standards.ieee.org/getieee802/download/802.1Q-2005.pdf

IEEE (2008) IEEE 802.3™: CSMA/CD (Ethernet) Access method. http://standards.ieee.org/findstds/errata/802.3-2008_Cor1.pdf

ISO (2008) International Organisation for Standardisation. http://www.iso.org/iso/iso_9001_2008

ITU-R (2010) Series BS: organization of the work of ITU-R. www.catr.cn/cttlcds/itu/itur/itur.jsp?docplace=BS

Marazzi F, Molina FJ (2009) EFAST inquiry. Joint Research Centre, Institute for the Protection and Security of the Citizen. http://efast.eknowrisk.eu/EFAST/images/docs/JRC_SR_EFASTQ.pdf

National Research Council (2005) Summary of a workshop on using information technology to enhance disaster management. The National Academies Press, Washington, DC, p 2

NEES (2004a) Formally known as the George E. Brown Jr. Network for earthquake engineering simulation, University of California at Berkeley. In: Proceedings of the second annual meeting of the NEES consortium, Catamaran Resort Hotel, San Diego, 19–22 May 2004, pp 60–66. http://www.curee.org/conferences/NEES/annual_mtg2/docs/proceedings2.pdf

NEES (2004b) Formally known as the George E. Brown Jr. Network for earthquake engineering simulation, University of California at Davis. In: Proceedings of the second annual meeting of the NEES consortium, Catamaran Resort Hotel, San Diego, 19–22 May 2004, pp 67–72. http://www.curee.org/conferences/NEES/annual_mtg2/docs/proceedings2.pdf

NEES (2004c) Formally known as the George E. Brown Jr. Network for earthquake engineering simulation, University of California at Lehigh. In: Proceedings of the second annual meeting of the NEES consortium, Catamaran Resort Hotel, San Diego, 19–22 May 2004, pp 39–43. http://www.curee.org/conferences/NEES/annual_mtg2/docs/proceedings2.pdf

NEES (2004d) Formally known as the George E. Brown Jr. Network for earthquake engineering simulation, University of Texas at Austin. In: Proceedings of the second annual meeting of the NEES consortium, Catamaran Resort Hotel, San Diego, 19–22 May 2004, pp 111–116. http://www.curee.org/conferences/NEES/annual_mtg2/docs/proceedings2.pdf

NIST (2010) National Institute of Standards and Technology. http://www.nist.gov/index.html

NXP (2007) I2C-bus specification and user manual. http://www.standardics.nxp.com/support/documents/i2c/pdf/i2c.bus.specification.pdf

Pearlman L et al (2004) Distributed hybrid earthquake engineering experiments: experiences with a ground-shaking grid application, high performance distributed computing. In: Proceedings of the 13th IEEE international symposium, Honolulu, HI, pp 14–23

Pegon P, Dietz M, Martinez IL (2010) Work package WP2, deliverable D2.2, guidelines for implementing a telepresence node. http://www.series.upatras.gr/userfiles/file/Public_Deliverables/Deliverable_2_2.pdf

Rao RR, Eisenberg J, Schmitt T (eds) (2007) National Research Council, Committee on using information technology to enhance disaster management, improving disaster management: the role of IT in mitigation, preparedness, response, and recovery. The National Academies Press, Washington, DC

SEESL (2011) Structural Engineering and Earthquake Simulation Laboratory Publications. http://seesl.buffalo.edu/publications/

SMPTE 370 M (2006) Television – data structure for DV-based Audio, data and compressed video at 100 Mb/s 1080/60i, 1080/50i, 720/60p, 720/50p. Society of Motion Picture and Television Engineers. http://www.techstreet.com/standards/SMPTE/370M_2006?product_id=1624785

Stevens WR (1994) TCP/IP illustrated, vol 1: The protocols. Addison-Wesley Professional, Reading, MA

Taylor SG et al (2010) Multi-scale wireless sensor node for impedance based SHM and long-term civil infrastructure monitoring. In: Proceedings of the fifth international conference on bridge maintenance safety and management. CRC Press, Philadelphia

Telcordia (2010) Telcordia, GR-326, generic requirements for single-mode optical connectors and jumper assemblies. http://telecom-info.telcordia.com/site-cgi/ido/docs.cgi?ID=SEARCH&DOCUMENT=GR-326&

Telcordia (2011) NEBS™ testing facilities and expertise. http://www.telcordia.com/services/testing/nebs/facilities.html#vibration

Uptime (2010) Uptime Institute – TUI. http://upsitetechnologies.com/

Wikipedia (2010) Universal serial bus. http://en.wikipedia.org/wiki/Universal_Serial_Bus

Yang YS et al (2007) ISEE: a platform for internet-based simulation for earthquake. Engineering Part I, Earthq Eng Struct Dyn. Wiley InterScience. http://www.ncree.org.tw/iwsccc/PDF/31%20-%20Wang.pdf

Zaluzec NJ (2001) NEESgrid telePresence system overview, white paper. www.neesgrid.org/documents/TPMOverview20020306.pdf

Chapter 11
Use of Large Numerical Models and High Performance Computers in Geographically Distributed Seismic Tests

Ferran Obón Santacana and Uwe E. Dorka

Abstract Although major improvements have been made in the field of hybrid simulation, the numerical models used in the tests are fairly simple, reaching only an order of ten Dynamic Degrees of Freedom (DDOF). However, the presence of large computational facilities along with the implementation of a Platform for Geographically Distributed Seismic Tests (PGDSTs) provides the possibility to use large and complex numerical models, maybe consisting of a couple of thousand DDOF, within the context of continuous hybrid simulations that work with acceptable time scale factors. However, the use of these facilities requires some approach from both parts in order to solve major issues such as the operating mode and thread to transfer data between facilities, or adapting the substructure algorithms to work in a parallel fashion through the use of special libraries and specifications among others. This paper presents ongoing work within SERIES to assess the extensibility of the PGDSTs to use HPC facilities as well as adapting the substructure algorithm developed by Dorka, which has been used successfully not only in Earthquake Engineering but also in aerospace applications.

11.1 Introduction

The numerical models that are being used in the field of hybrid simulation are fairly simple, reaching an order of ten Dynamic Degrees of Freedom (DDOF) due to hardware limitations. The requirement to improve this aspect is quite simple: use more

---

F. Obón Santacana (✉) · U.E. Dorka
Steel and Composite Section, Department of Civil and Environmental Engineering,
University of Kassel, Kassel, Germany
e-mail: ferran.obon@uni-kassel.de; uwe.dorka@uni-kassel.de

M.N. Fardis and Z.T. Rakicevic (eds.), *Role of Seismic Testing Facilities in Performance-Based Earthquake Engineering: SERIES Workshop*, Geotechnical, Geological and Earthquake Engineering 22, DOI 10.1007/978-94-007-1977-4_11,
© Springer Science+Business Media B.V. 2012

powerful computers. With the availability of multi-core processors the computational power of common personal computers is increasing more and more and, without doubt, the order of the numerical models could be increased without having to use a more extended time scale. This however does not solve the entire problem since one question still arises. What about those extremely large and complex numerical models with thousands of DDOF? Could actual computers handle them within a reasonable time step? The answer is clearly no, common personal computers would not be able to handle them; there is still need for more powerful machines! Therefore the only way to succeed in this aspect could be the use of Supercomputers or the less powerful Linux Clusters that most universities have.

Unfortunately, the use of parallel computers requires some additional software or *tools* to take advantage of multiple processing units. Therefore, the substructure algorithms need to be adapted, or better said, *parallelised*, before being able to run on this kind of machines.

11.2 Parallel Computers

There are different types of parallel computers, but a common way to classify them is known as Flynn's Taxonomy (Flynn 1972): SIMD and MIMD.[1] Single Instruction Stream/Multiple Data Stream (SIMD) is a concept usually associated with vector processors or Graphic Card Units (GPU). In SIMD all the processors perform the same operation simultaneously but using as an input different data, i.e., a vector. Therefore, taking advantage of effective SIMD instructions is usually referred as *vectorisation*. It is good not to confuse this concept with the *data parallelism* (or loop parallelism) concept that usually goes under the name of SPMD (Single Program/ Multiple Data Stream) where the same code is executed on all the processors.

Despite the advantages of SIMD machines, the super-computing migrated however to the more flexible model Multiple Instruction Stream/Multiple Data Stream (MIMD). In this category, multiple instructions are executed on different processors to operate on different data. MIMD computers can have *shared* or *distributed* memory. In the first case, each processor can access a memory that is being shared among the other processors (Fig. 11.1), while in the second case; the processors have access only to a "private" memory and communicate with the other in the network using message passing (Fig. 11.2). It is also possible to combine both methods in order to get the best of the two worlds: the cluster model (Fig. 11.2) The importance of these concepts will be clarified later on, when reviewing the available "tools" to parallelise the substructure algorithms.

---

[1] There are two more categories: SISD (Single Instruction Stream/Multiple Data Stream) and MISD (Multiple Instruction Stream/Single Data Stream) but they are not relevant in the context of this paper.

11 Use of Large Numerical Models and High Performance Computers...

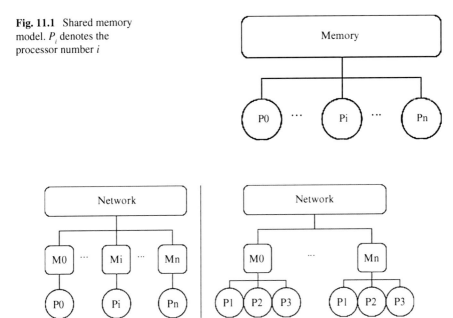

Fig. 11.1 Shared memory model. P_i denotes the processor number i

Fig. 11.2 The distributed memory model (*left*) and the cluster model (*right*). M denotes memory and P processor

Two major obstacles have been found when trying to integrate high performance computers into PGDSTs: *batch mode* and security threat. The *batch mode* is the way supercomputers operate and it can be defined as a *series of execution of programs without manual intervention*. This means that the researcher does not know when the job one has submitted will start running; it could take just a few seconds or an entire week at any time. The second obstacle mentioned refers to the fact that supercomputers have a very restrictive policy when sharing data across the network to prevent being used illegally by third parties. In fact, some of these computers grant access to outside researchers only to a machine that is used only for saving the results of the simulation without any chances to interact with the "numerical part" of the supercomputer. Both problems have been successfully overcome using, in this case, the Linux cluster at University of Kassel.

- **Batch mode**: Some processors will be booked and nobody else will have access, granting the possibility to start the test when desired.
- **Security**: The cluster is behind the main firewall of the university campus and is accessed using the local network and a specific port.

Using the facilities that are "next door" also grants some other advantages like a small time to access it from the laboratory due to the short distance between facilities (Fig. 11.3). However, there are some issues that still have to be addressed, like the peaks in the network every once and a while that in some cases reached around 400 ms. This can of course destroy a test (Fig. 11.3).

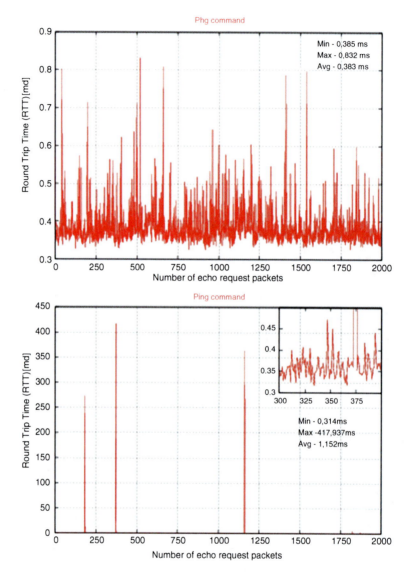

Fig. 11.3 Acceptable (*top*) and unacceptable (*bottom*) Jitter in the round trip time between the laboratory and the Linux cluster

11.3 Adapting the Algorithm to Run in High Performance Computers

In order to be successful in adapting the algorithm to run in a parallel way and achieve high performance, some requirements have to be met:

- The algorithm must be able to run in a wide variety of machines and still run efficiently.

11 Use of Large Numerical Models and High Performance Computers...

- Possibility to divide the matrices into different sub-matrices that will be scattered across the processors so that the linear algebra operations can be performed in parallel.
- Possibility to run different parts of the algorithm simultaneously to speed-up the whole process.

These requirements can be met easily if a look at the tools already available is taken. Most of the available software for supercomputers has been specifically designed to fulfil these requirements and, at the same time, simplify and fasten the development of these kinds of algorithms. There are several layers of software and complexity when designing parallel programs.

11.3.1 Message Passing Interface

Message passing interface (Gropp et al. 1999a, b; Karniadakis and Kirby 2003; Petersen and Arbenz 2004), or by its acronym MPI, is one of the standard API available when writing programs that will run in PC-Clusters or supercomputers. The primary goal was to design an application interface that does not compromise among efficiency, portability and functionality. That means that with MPI one can write a truly portable program that will benefit from specialised hardware and specifically tuned software offered by parallel computer vendors (efficiency and portability). Also, the programmer will have at their hand an extensive set of operations (functionality).

As the key concept behind parallel computing is the *divide-and-conquer* strategy, the whole issue of how to write a parallel code can be summarised as how to partition the problem so that *all the processors are busy and none remain idle*. This is what MPI offers: A set of functions that will enable the programmer to split the work into different parts to finish it in the least amount of time using the single key concept "Send and Receive Messages". How should an algorithm be implemented using MPI? Consider the dot product (Eq. 11.1) and a common implementation in C language.

$$\vec{x} \cdot \vec{y} = \sum_{1}^{n} x_n y_n = x_1 y_1 + x_2 y_2 + \cdots + x_n y_n \qquad (11.1)$$

```
/* Calculate the dot product */
  for ( i = 0; i < N; i = i + 1 ){
    dot = dot + x[i]*y[i];
  }
```

It has been said that the key concept of writing parallel programs was *divide-and-conquer* so a simple way to parallelise this program would be to divide the vector into different parts. Supposing P processes, then the process labelled as 0 will perform the dot product of the elements from 0 to $N/P - 1$, process 1 from N/P to

N/P · 2–1 and so on. At the end all the values are added together in process 0 to get the result of the dot product. To accomplish this, some additional lines of code should be added to the previous example[2]:

```
MPI_Init( &argc, &argv );                    // Initialise MPI
MPI_Comm_size( MPI_COMM_WORLD, &NumNodes ); // Get the number of
                                               nodes
MPI_Comm_rank( MPI_COMM_WORLD, &Iam ); // Get which process I am
/* Initialise variables */
...
/* Calculate the dot produt of the elements between the 0 and
 * Interval=N/NumNodes */
for ( i = 0; i < Interval; i = i + 1 ){
  dot = dot + x[i]*y[i];
}
/* Collect the information from the nodes. All the processes
    except 0 will send a message to process 0 and Process 0 will
    receive a message from all other processes */
if ( Iam != 0 ){
    MPI_Send( &dot, 1, MPI_DOUBLE, 0, 1, MPI_COMM_WORLD );
} else {
  for ( i = 1; i < NumNodes; i = i + 1 ){
    MPI_Recv( &partial, 1, MPI_DOUBLE, i, 1, MPI_COMM_WORLD,
              &status );
    dot = dot + partial;
  }
}
...
MPI_Finalize( );    // End MPI
```

All the MPI programs begin with the statement of `MPI_Init ( )` and end with `MPI_Finalize ( )` and what is in between is executed by all the processes defined at execution time, which means that each process receives an identical copy of the statements to be executed (this also implies that each process has a local copy of all the variables) but that does not mean that all the processes will execute all the lines. This is accomplished using *if* statements to verify the process label ($0, 1,..., P$) and can be seen in the first *if* statement in which all the processes except process 0 are sending while process 0 is receiving. Notice that process 0 has as many receive calls as messages sent to it. The last is an important concept since the program may become idle, and thus not do anything, if some of the processes are sending while

---

[2] This example has been designed only for illustrative purposes.

11 Use of Large Numerical Models and High Performance Computers... 205

there are none waiting to receive the message! In fact, the operation of reducing the values on all the processes to a single value is so common that a specific routine, `MPI_Reduce( )`, was designed to simplify the code. In this case, gathering the information would look like:

```
/* Collect the information from the nodes. The over
 * all dot product is stored in the variable dot */
MPI_reduce( &partial, &dot, 1, MPI_DOUBLE, MPI_SUM,MPI_COMM_WORLD );
```

Note that since in MPI each node has a private copy of the variables, the size of the local vectors can be defined just as a portion of the overall vector (global vector). It is also worth saying that although the previous examples make use only of the global communicator (imagine it as a global work group) `MPI_COMM_WORLD`, it is possible to define communicators that encompass only a bunch of nodes to effectively create some sort of work groups.

11.3.2 Open Multiprocessing: OpenMP

Open Multiprocessing or OpenMP (Chapman et al 2008; Petersen and Arbenz 2004) is a shared memory API that seems to have become a standard. It is relatively easy to use and implement and it is also widely adopted. This comes with a price; OpenMP does not grant the control and fine granularity that other similar libraries, such as the POSIX standard known as *pthreads,* offer. Similar to MPI, OpenMP directives can be called within Fortran and C/C++ but, instead of giving a way to send/receive messages, they use threads. That is, OpenMP divides a program (master thread) into several smaller running entities (slave threads) in order to share, not only work, but also variables (every thread can modify any variable that is being shared). However, to effectively share the work, there is a need of private variables or a way to protect some of them. Returning to the example of the dot product:

```
/* Get the number of processors and create as many
 * threads as processors */
nthreads = omp_get_num_procs( );
omp_set_num_threads( nthreads );
 * Initialise other variables */
...
/* Calculate the dot product */
#pragma omp parallel for
for ( i = 0; i < N; i = i + 1 ){
  dot = dot + x[i]*y[i];
}
```

To start OpenMP one must specify the number of threads that will be created when a parallel region, identifiable by the `#pragma omp` statement, is found.

When exiting this region the threads are destroyed automatically. When programming with OpenMP, one must be careful with the concept of *variable scope*, whether it is shared among the threads or it is private and accessible only by one of the threads. This concept has a major importance if one tries to understand why the previous code gives a bad result; while the counter i is a private variable by definition (each thread has its own copy), the dot variable is shared and it is accessed in an asynchronous way by all the threads! There are different ways of solving this problem but, since this is a common operation, only substituting the statement `#pragma omp parallel for` with `#pragma omp parallel for reduction( +: dot )` will give the best performance (Petersen and Arbenz 2004) and the correct result.

OpenMP features additional functions to parallelise a code, such as *sections* to easily define different parallel regions. For an in depth overview of OpenMP, the reader may refer to Chandra et al. (2001), Chapman et al. (2008) and Quinn (2003).

11.3.3 Hybrid Programming

Up to now, two different models for parallel programming have been presented but it is difficult to say which one is better. Each of them has its own advantages and disadvantages. For example, while OpenMP is easier to implement and debug than MPI, it only scales within one node (the processors that share the memory). On the other hand, MPI grants scalability and private copies of the variables in all the processors, but one must specify the communication explicitly. An interesting thing, however, is that these two APIs can be combined together: use MPI to communicate between nodes using TCP/IP and then use OpenMP within that node to take advantage of shared memory programming and reduce the communication overhead. The dot product code shown under MPI could therefore be rewritten as[3]:

```
/* Initialise MPI */
  . . .
/* Get the number of processors and create as many threads as
processors */
  . . .
/* Calculate the dot product */
#pragma omp parallel for reduction( +: dot )
for ( i = 0; i < 1000; i++ ){
  dot = dot + x[i]*y[i];
}
/* Perform MPI communication and end MPI */
  . . .
```

---

[3] If the number of cores differs, the number of threads will also vary, resulting in load imbalances.

11.3.4 Numerical Software Libraries

It may give the impression that parallelising the substructure algorithms is straight forward. Truly one can have a rough implementation of it just using a few concepts, but the performance will decrease. And besides, synchronising the different processors using message passing or clearly identifying the scope of the variables in more complex algorithms (i.e. matrix inversion) can be a pretty hard job if the complexity of the algorithm increases. This is where numerical software libraries such as ScaLAPACK, PETSc and PLAPACK appear. Although they are different projects, they share some points:

- They hide much of the MPI-level complexity from the programmer.
- The building blocks are the local libraries BLAS and LAPACK. The parallel libraries rely on calling local implementations of the routines.

11.3.4.1 ScaLAPACK

In this paper the name of NETLIB libraries is used to reference a whole family of software that can be freely accessed in Dongarra and Grosse (2010). It comprises of: BLACS (Basic Linear Communication Subprograms) (Chor et al 1995; Dongarra and Whaley 1995), BLAS (Basic Linear Algebra Subprograms) (Lawson et al 1979), LAPACK (Linear Algebra PACKage) (Anderson et al 1999; Dongarra and Van de Geijn 1991), PBLAS (Parallel BLAS) (Chor et al 1995; Petitet 1996) and ScaLAPACK (Scalable LAPACK) (Blackfort et al. 1997). All the components in the zone labelled as *Local* in Fig. 11.4 are called on a single processor and use the data stored in its memory. On the other hand, the components in the area labelled as *Global* are synchronous parallel routines and deal with vector and matrices distributed across multiple processors.

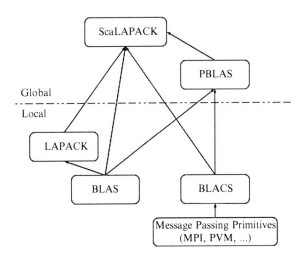

Fig. 11.4 The hierarchy of ScaLAPACK

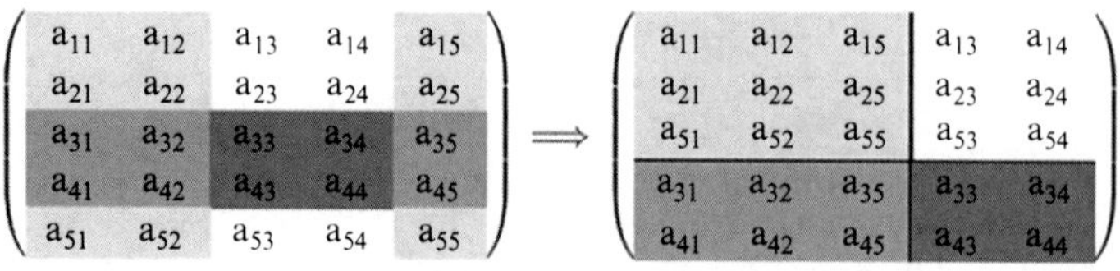

$$
\begin{pmatrix}
a_{11} & a_{12} & a_{13} & a_{14} & a_{15} \\
a_{21} & a_{22} & a_{23} & a_{24} & a_{25} \\
a_{31} & a_{32} & a_{33} & a_{34} & a_{35} \\
a_{41} & a_{42} & a_{43} & a_{44} & a_{45} \\
a_{51} & a_{52} & a_{53} & a_{54} & a_{55}
\end{pmatrix}
\implies
\begin{pmatrix}
a_{11} & a_{12} & a_{15} & a_{13} & a_{14} \\
a_{21} & a_{22} & a_{25} & a_{23} & a_{24} \\
a_{51} & a_{52} & a_{55} & a_{53} & a_{54} \\
a_{31} & a_{32} & a_{35} & a_{33} & a_{34} \\
a_{41} & a_{42} & a_{45} & a_{43} & a_{44}
\end{pmatrix}
$$

Fig. 11.5 Example of a 2×2 cyclic distribution using a 2×2 block size in a 5×5 matrix

ScaLAPACK is a library that comprises some parallelised versions of the LAPACK routines and it will run on any distributed memory computer that supports MPI or PVM. ScaLAPACK relies on a good implementation of the local libraries BLAS and LAPACK to achieve high performance. The algorithms in ScaLAPACK use, like those in PBLAS (the parallelised version of BLAS), a 2D cyclic matrix distribution and the communication library BLACS.

The use of ScaLAPACK can be summarised into four steps (plus initialising and finalising MPI): start BLACS (required for communication between nodes within LAPACK routines) and get a context (a work group), initialise the process grid and vector/matrix descriptors (stores how the vector/matrices are scattered across the processors), call the desired routines and exit BLACS. The second step, along with the 2D cyclic distribution is summarised in Fig. 11.5, and can be understood as dividing the matrix into the different processes (the grid) using blocks. Therefore a 2×2 grid consists of a total of four processes, while a block distribution of 2×2 means that the matrix is split across the processes using 2×2 "sub-matrices".

The following example shows a way to use ScaLAPACK to calculate the dot product. To avoid unnecessary details at this point, only the call to the functions is shown, without showing its arguments.

```
/* Initialise MPI, get the number of nodes... */
...
/* Get a BLACS context */
Cblacs_get( ... );
/* Initialise the process grid and get information about it */
Cblacs_gridinit( ... );
Cblacs_gridinfo( ... );
/* Initialise the vector descriptor */
descinit_( ... );
/* Perform the dot product using ScaLAPACK */
pddot_( ... );
/* Release the process grid */
Cblacs_gridexit( ... );
/* End MPI */
...
```

11.3.4.2 PETSc Library

The Portable Extensible Toolkit for Scientific Computation (PETSc) (Balay et al 1997, 2010, 2011) consists of a variety of object oriented numerical libraries to work in either sequential or parallel way. PETSc builds upon the already introduced BLAS and LAPACK libraries and MPI to provide more advanced functionalities like Krylov subspace methods, non linear solvers and time stepping routines. It is also interesting to point out that PETSc provides routines to handle sparse matrices and also an easy way to show a performance summary of the program (MFlops, time spent, ... of the called PETSc functions). Since PETSc offers an object oriented approach, high level functions are provided that deal with common operations such as memory allocation and vector/matrix distribution, ... that one should specify manually in ScaLAPACK. With PETSc, the dot product has five major steps: initialise PETSC and MPI, create the vectors, call the dot product function, destroy the vectors and end PETSc. Note that each function of PETSc returns an integer that is used to identify different types of errors (CHKERRQ( )).

```
/* Initialise PETSc and MPI */
PetscInitialise( ... );
/* Create Vectors */
ierr = VecCreate( ... ); CHKERRQ( ierr );
...
/* Perform the dot product */
ierr = VecDot( ... ); CHKERRQ( ierr );
/* Destroy the vectors */
ierr = VecDestroy( ... ); CHKERRQ( ierr );
/* End PETSc and MPI */
ierr = PetscFinalize( ); CHKERRQ( ierr );
```

11.3.4.3 PLAPACK

The Parallel Linear Algebra Package (PLAPACK) (Van de Geijn 1997, 2010) is a library for dense linear algebra operations on distributed memory supercomputers developed by the University of Texas at Austin. Unlike ScaLAPACK it uses an object based approach (thus providing a higher level of abstraction) and a different way to distribute the data across the nodes. It is known that using higher level of abstraction introduces some overhead, but as is hinted in Baker et al (1998), the most complex algorithms become more manageable, giving then the possibility to overcome such overhead. Using PLAPACK, the dot product can look like.

```
/* Initialise MPI */
...
/* Initialise PLAPACK */
PLA_Init( ... );
/* Create PLAPACK template */
PLA_Temp_create( ... );
/* Create distributed vectors using the previously
 * defined template */
PLA_Mvector_create( ... );
...
   /* Perform the dot product */
PLA_Dot( ... );
/* Release the objects created with PLAPACK */
PLA_Obj_free( ... );
/* End PLAPACK and MPI */
PLA_Finalize( ); MPI_Finalize( );
```

11.3.5 Achieving High Performance

As has been seen, all the numerical libraries presented in this paper have something in common: they rely on a good implementation of the local libraries BLAS and LAPACK. This means two things: to work with an optimal size of the local matrices to use as efficiently as possible the cache memory and achieve high performance (Karniadakis and Kirby 2003) and to have a good implementation of BLAS and LAPACK (the use of the reference libraries is discouraged).

It is also important to identify and address the bottlenecks using tracers (programs that provide useful information to the user such as idle times, time spent in communication...) and to parallelise the code as much as possible. Gene Amdahl proposed a way to calculate the Speed-up factor known as *Amdahl's Law* (Amdahl 1967).[4] Assuming that a percentage of the code, ε, cannot be parallelised, the Speed-up factor S_p is (see also Fig. 11.6):

$$S_p = \frac{1}{\varepsilon + \dfrac{1-\varepsilon}{P}} \tag{11.2}$$

---

[4] Amdahl's Law has been also criticised for being too pessimistic.

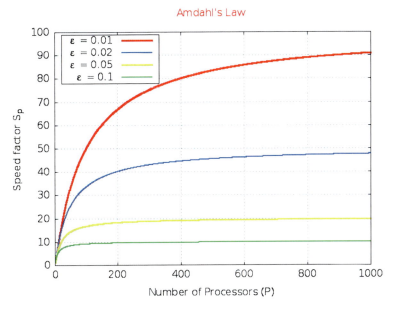

Fig. 11.6 Amdahl's law: speed factor S_p for different percentages of non parallel code

11.4 The Substructure Test

The software tools are there, but now is the turn to use a substructure algorithm that can be scaled and take advantage of parallel computers. The substructure algorithm developed by Dorka has proved to be successful in real time testing, not only in the field of civil engineering but also in aerospace applications using electrodynamic actuators due to frequency requirements (around 20 Hz) (Bayer et al 2005). In this paper only a general overview of the formulation is given. The reader should refer to (Bayer et al 2005; Dorka 2002) for more detailed information.

11.4.1 Basic Formulation of Dorka's Substructure Algorithm

The basics of the substructure algorithm are a discrete formulation of the dynamic equilibrium given by:

$$M\frac{d^2x}{dt} + C\frac{dx}{dt} + Kx + f_r + f_s = p(t) \qquad (11.3)$$

where M, C and K are the mass, damping and stiffness matrices of the numerical model and x, $p(t)$ are the displacement and loading vector respectively. The restoring forces are taken care by the terms f_r (numerical non-linear restoring force) and f_s (force measured at the interface between numerical model and substructure specimen).

Using finite element discretization in the time domain and shape functions with three supporting points[5] (3-step scheme) (Zienkiewicz 1977) for the displacement, lead to a general solution of the previous equation. Nearly all major time stepping algorithms can be obtained if different weighting functions are used when performing the time integration, which results in:

$$u^{n+1} = \left[M + \gamma \Delta t C + \beta \Delta t^2 K \right]^{-1} \cdot$$
$$\left\{ \begin{array}{l} \left[2M - \left(1 - 2\gamma\right)\Delta t C - \left(\frac{1}{2} - 2\beta + \gamma\right)\Delta t^2 K \right] \cdot u^n \\ - \left[M - \left(1 - \gamma\right)\Delta t C + \left(\frac{1}{2} + \beta - \gamma\right)\Delta t^2 K \right] \cdot u^{n-1} \\ + \beta \Delta t^2 f_*^{n+1} + \left(\frac{1}{2} - 2\beta + \gamma\right)\Delta t^2 f_*^n + \left(\frac{1}{2} + \beta - \gamma\right)\Delta t^2 f_*^{n-1} \end{array} \right\}$$

(11.4)

where $f_* = f_r + f_s - p$ and u is the discretized displacement vector in the time domain. The term $[M + \gamma \Delta t C + \beta \Delta t^2 K]^{-1}$ is also known as effective mass matrix M_e (Clough and Penzien 1993) and the super-index **n** denotes the time steps.

A more in depth study of the different parameters (time intervals, integration schemes, …), stability, accuracy and numerical damping, that can be found in (Dorka 2002), identified the so-called *Newmark-β*, with $\gamma = 0,5$ and $\beta = 0,25$, as the only unconditionally stable implicit 3-step algorithm that has no artificial damping and also the least numerical softening.

In Eq. 11.4, the term f_s^{n+1} (the reaction of the subsystem) is only available through measurement. Since iteration will lead to high frequency oscillations in a test, the sub-stepping approach has been applied to deal with this problem. All the implicit algorithms can be expressed in the form of a general linear control equation (Dorka 2002):

$$u^{n+1} = u_0 + G(f_{s^{n+1}} + f_r^{n+1})$$

(11.5)

Where u_0 is the initial vector (updated at the beginning of the time step), G is the gain matrix (constant throughout the test) and f_r, f_s the calculated and measured

---

[5] The formulation can also be extended to four supporting points.

Use of Large Numerical Models and High Performance Computers...

force vectors respectively. These two vectors are updated each sub-step. Therefore, the components of Eq. 11.5 are:

$$
\begin{aligned}
u_0 = & \left[M + \gamma \Delta t C + \beta \Delta t^2 K \right]^{-1} \cdot \\
& \left\{ \begin{aligned}
& \left[2M - (1 - 2\gamma) \Delta t C - \left(\frac{1}{2} - 2\beta + \gamma \right) \Delta t^2 K \right] \cdot u^n \\
& - \left[M - (1 - \gamma) \Delta t C + \left(\frac{1}{2} + \beta - \gamma \right) \Delta t^2 K \right] \cdot u^{n-1} \\
& - \beta \Delta t^2 p^{n+1} + \left(\frac{1}{2} - 2\beta + \gamma \right) \Delta t^2 f_*^n + \left(\frac{1}{2} + \beta - \gamma \right) \Delta t^2 f_*^{n-1}
\end{aligned} \right\}
\end{aligned}
$$

(11.6)

$$
G = \beta \Delta t^2 \left[M + \gamma \Delta t C + \beta \Delta t^2 K \right]^{-1}
$$

(11.7)

Although the number of sub-steps k is an important factor for stability and accuracy (when $k \to \infty$ Eq. 11.6 gives the exact value), there can be other sources that can make the test unstable: errors. In order to deal with them (positioning error of the actuators,...) and avoid a tendency of destabilising the test, a PID compensator is used to minimise the error:

$$
f_e^n = -P \left[e^n + I \Delta t \sum_i^n e^n + \frac{D}{\Delta t} (e^n - e^{n-1}) \right]
$$

(11.8)

where e is the equilibrium error defined as a sum of all dynamic forces, whether they are internal or external, at the end of the time step. P, I and D are the adequately chosen proportional, integral and derivative constants of the error compensator while f_e^n is the compensation force. It is added at the beginning of the next step as a load to the system. Figure 11.7 summarises how the algorithm runs. Finally it is worth mentioning that the algorithm can easily be formulated for displacement, velocity or acceleration control.

11.4.2 A Word About Parallelization

The algorithm presented above gives the appearance of satisfying the scalability requirements to run in a parallel fashion. The idea behind the parallelisation is to use as much as possible the numerical software libraries presented and to use the combination of MPI and OpenMP to deal with the parts of the code that are not covered by them. A nice feature of this substructure algorithm is that all the operations performed during the substructure test are vector addition or scaling (a matrix can be seen also as a vector for this operations) and matrix-vector multiplication. The operations that are more time-consuming, such as matrix inversion and the

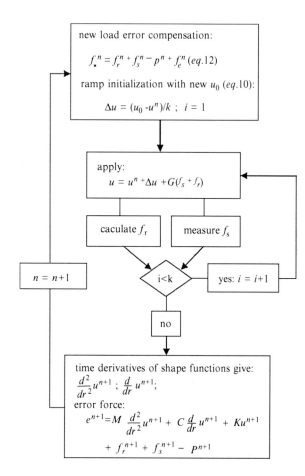

Fig. 11.7 Scheme of the algorithm (Dorka 2002)

operations that involve the addition of the mass, damping and stiffness matrices, can be performed before the start of the test.

It is also worth mentioning that the sub-stepping loop in Fig. 11.7 is performed where the physical substructure is located. This means that data has to be exchanged, u_0 and the result of the sub-stepping u, between the two facilities (using standard communication protocols), but only the part of the vector that belongs to both sub-structures, i.e. the coupling nodes.

11.4.3 Test Setup and Numerical Models

The overall setup comprises of a numerical facility (the PC-Cluster), with large numerical models, and the sub-structure facility, where the test setup shown in Fig. 11.8 is located. Constant communication between the PC-Cluster and ADwin (a system capable of multitasking in charge) is required since the last one requires

11 Use of Large Numerical Models and High Performance Computers... 215

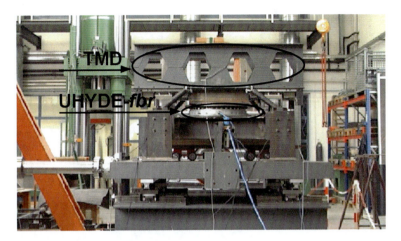

Fig. 11.8 Test setup at University of Kassel

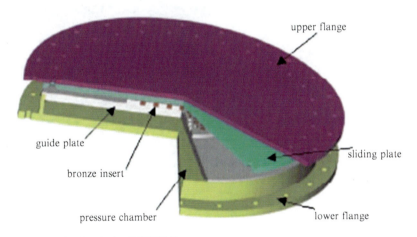

Fig. 11.9 The friction device UHYDE-fbr

u_0 each time step to start with the sub-stepping process. In the sub-structure facility, two actuators, with a capacity of *400 kN* and controlled in displacement by ADwin (by Jäger GmbH 2010), will be responsible to apply the displacements coming from the PC-Cluster to the shake table. Since the tests will not be targeted initially to be real-time, the Tunned Mass Damper (TMD) should be deactivated, leaving the friction device UHYDE-*fbr* (Dorka 1995) as the only component in charge of dissipating energy. The movement of the TMD will be prevented using some bracings.

The friction device dissipates energy due to solid friction and its name is an acronym of Uwe's Hysteretic Device, with *f* for friction and *br* for bridges since it is especially suitable for bridges, where large displacements have to be accommodated. It is composed of a set of bronze inserts and tow steel plates, one for guidance of the inserts and the other with a specially prepared surface that is in contact with the inserts (Fig. 11.9).

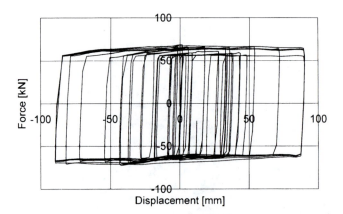

Fig. 11.10 Elasto-plastic behaviour of the friction device (García et al 2004)

Fig. 11.11 ETABS model. Contribution of Middle East Technical University and BLA Associates, Atlanta, GA

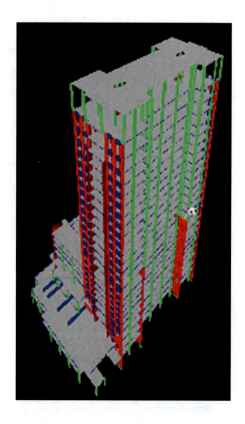

The friction device exhibits an ideal elasto-plastic behaviour (Fig. 11.10), with the friction force easily adjusted varying the gas pressure in the chamber. If this device is made semi-active (a pressure control is implemented) a range of different

Fig. 11.12 Contribution of Technical University of Istanbul

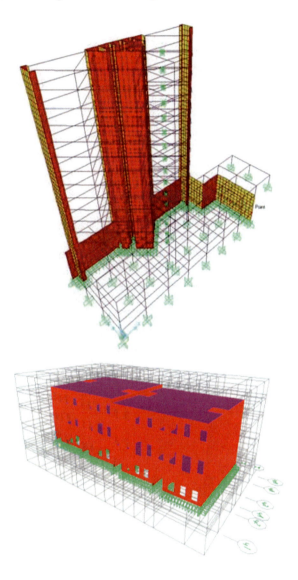

force-displacement characteristics, including viscous damping, may be achieved. A more detailed look at the UHYDE-*fbr* is given in Dorka and Garcia (2005).

The final elements of the test setup are the matrices (M, K and C) of the numerical models shown in Figs. 11.11 and 11.12. These models will need to be slightly modified to integrate them into the test setup and make use of the friction device by creating coupling nodes. The initial idea was to place those nodes (and therefore the friction devices) in the core columns in the floor marked with the yellow line, although a more in depth study needs to be carried out to identify the best solution.

11.5 Conclusions

The use of high complex numerical models seems feasible with the inclusion of supercomputers into the platform for geographically distributed testing since both hardware and software are already there. Parallel computers open a whole new world of possibilities to hybrid simulation but still more work needs to be done. Future work should focus on two general aspects: the computer science part and the substructure study.

- One of the major objectives within this task would be to be able to perform real time tests using local large numerical facilities. This will only be possible if the bottlenecks (communication overhead, load balancing, …) in the parallel implementation of the substructure algorithm are identified and properly addressed. It will be also interesting to see how complex the numerical models can be in common personal computers since the same software tools can be used.
- Reduce the amount of jitter between the two facilities.
- Integrate the large numerical facilities and the substructure algorithm into the communication protocols that will be used in the platform for geographically distributed seismic tests.
- Study the use of novel libraries that in some cases outperform ScaLAPACK and PBLAS such as: Elemental (Poulson 2010) and Libflame (Van Zee 2009).
- Analyse the behaviour of large and complex numerical models during a test and identify and solve possible issues that can affect stability (i.e. the appearance of high frequency modes).

References

Amdahl G (1967) The validity of the single processor approach to achieve large scale computing capabilities, AFIPS conference proceedings, vol 30, Atlantic City, pp 483–485
Anderson E, Bai Z et al (1999) LAPACK users' guide. Society for Industrial and Applied Mathematics, Philadelphia
Baker G, Gunnels J et al (1998) PLAPACK: high performance through high level abstraction. In: Proceedings of ICPP'98, Minneapolis
Balay S, Gropp W et al (1997) Efficient management of parallelism in object numerical software libraries. In: Bruaset A, Arge E, Langtangen HP (eds) Modern software tools in scientific computing. Birkhäuser, Boston, pp 163–202
Balay S, Gropp W et al (2010) PETSc users manual Revision 3.1. Available via http://www.mcs.anl.gov/petsc
Balay S, Gropp W et al (2011) PETSc home page. http://www.mcs.anl.gov/petsc. Accessed 1 Feb 2011
Bayer V, Dorka UE et al (2005) On real-time pseudo dynamic sub-structure testing: algorithm, numerical and experiment results. Aerosp Sci Technol 9:223–232
Blackfort LS, Choi J et al (1997) ScaLAPACK users' guide. Society for Industrial and Applied Mathematics, Philadephia
Chandra R, Menon R et al (2001) Parallel programming in OpenMP. Morgan Kaufmann, San Francisco
Chapman B, Jost G et al (2008) Using OpenMP: portable shared memory parallel programming. The MIT Press, Cambridge

11 Use of Large Numerical Models and High Performance Computers...

Chor J, Dongarra J, Ostroughchov S et al (1995) A proposal for a set of parallel basic linear algebra subprograms. LAPACK working note 100, University of Tennessee, Knoxville

Clough RW, Penzien J (1993) Dynamics of structures. McGraw-Hill, New York

Dongarra J, Grosse E (2010) Netlib home page. http://www.netlib.org. Accessed 1 Feb 2011

Dongarra J, Van de Geijn R (1991) Two dimensional basic linear algebra communication subprograms. LAPACK working note 37. Technical report, University of Tennessee, Knoxville

Dongarra J, Whaley RC (1995) A user's guide to the BLACS v1.1. LAPACK working note 94. Technical report, University of Tennessee, Knoxville

Dorka UE (1995) UHYDE-fbr. US Patent 5456047

Dorka UE (2002) Hybrid experimental – numerical simulation of vibrating structures. In: International conference WAVE2002, Okayama

Dorka UE, Garcia J (2005) Seismic qualification of passive mitigation devices. LNEC Laboratório Nacional de Engenharia Civil, Lisboa

Flynn MJ (1972) Some computer organizations and their effectiveness. IEEE Trans Comput C-21(9):948–960

García J, Dorka UE, Magonette G et al (2004) Testing of algorithms for semi-active control of bridges. EU-ECOLEADER TASCB Report

Gropp W, Lusk E, Skjellum A (1999a) Using MPI: portable parallel programming with the message-passing interface. The MIT Press, Cambridge

Gropp W, Lusk E, Skjellum A (1999b) Using MPI2: portable parallel programming with the message-passing interface. The MIT Press, Cambridge

Jäger Computergesteuerte Messtechnik GmbH (2010) Rheinstrasse 2–4, D-64653 Lorsch, Germany

Karniadakis GE, Kirby RM (2003) Parallel scientific computing in C++ and MPI. Cambridge University Press, New York

Lawson CL, Hanson RJ et al (1979) Basic linear algebra subprograms for Fortran usage. ACM Trans Math Softw 5:308–325

Petersen WP, Arbenz P (2004) Introduction to parallel computing – a practical guide with examples in C. Oxford University Press, Oxford

Petitet A (1996) Algorithmic redistribution methods for block cyclic decompositions. Dissertation, University of Tennessee, Knoxville

Poulson J (2010) Elemental home page. http://code.google.com/p/elemental/. Accessed 1 Feb 2010

Quinn M (2003) Parallel programming in C with MPI and OpenMP. McGraw Hill Education, New York

Van de Geijn RA (1997) Using PLAPACK: parallel linear algebra package. The MIT Press, Cambridge

Van de Geijn RA (2010) PLAPACK home page. http://www.cs.utexas.edu/plapack. Accessed 1 Feb 2010

Van Zee FG (2009) Libflame: the complete reference. www.lulu.com

Zienkiewicz OC (1977) The finite element method. McGraw-Hill/Maidenhead, England

Chapter 12
Shaking Table Testing of Models of Historic Buildings and Monuments – IZIIS' Experience

Veronika Shendova, Zoran T. Rakicevic, Lidija Krstevska, Ljubomir Tashkov, and Predrag Gavrilovic

Abstract The cultural-historic heritage is the key element for the history and the identity of society, contributing to its economic and general well-being. Damages caused to historic buildings and monuments by earthquakes that occurred in the past are irreversible and these lost "documents" cannot be retrieved. In providing the protection of these structures in a manner that requires the least intervention and the greatest care to preserve authenticity, the experts are permanently challenged by the fast development and the improved performance of new materials and techniques. However, the implementation of particular strengthening methodology depends on the extent it has been investigated as well as its analytical and experimental verification. This paper presents the most important seismic shaking table investigations of models of historic buildings and monuments carried out in the IZIIS' Dynamic Testing Laboratory, the main goal of which was assessment of the vulnerability of structures as well as testing and experimental verification of different methodologies for seismic strengthening.

12.1 Introduction

Historic buildings and monuments provide the most tangible legacy of our past civilization and in some cases they speak clearer than any remaining manuscripts. They are usually severely damaged during strong earthquakes due to their stiff and

---

V. Shendova(✉) • Z.T. Rakicevic • L. Krstevska • L. Tashkov • P. Gavrilovic
Institute of Earthquake Engineering and Engineering Seismology, IZIIS,
SS Cyril and Methodius University, Salvador Aljende 73, PO BOX 101,
1000 Skopje, Republic of Macedonia
e-mail: veronika@pluto.iziis.ukim.edu.mk; zoran_r@pluto.iziis.ukim.edu.mk;
lidija@pluto.iziis.ukim.edu.mk; tashkov@pluto.iziis.ukim.edu.mk;
gavrilovicpredrag@yahoo.com

M.N. Fardis and Z.T. Rakicevic (eds.), *Role of Seismic Testing Facilities in Performance-Based Earthquake Engineering: SERIES Workshop*, Geotechnical, Geological and Earthquake Engineering 22, DOI 10.1007/978-94-007-1977-4_12,
© Springer Science+Business Media B.V. 2012

brittle structural components. The main reason for damage is a lack of ductility that prevents a structure from being able to sustain the displacements and distortions caused by severe earthquakes. The goal should then be to strengthen these structures in a manner that requires the least intervention and the greatest care to preserve authenticity, (Feilden 1982, 1987).

Since the beginning of the twentieth century, the traditional techniques and typology of construction have rapidly been replaced by modern methods and new materials used even in restoration of historic structures. The skill and the culture of construction from the past slowly disappears, while the methods of computation and analysis developed primarily for modern structures have become the only practice of young structural engineers. This practice is particularly emphasized in urgent situations (for instance, during an earthquake) in which new techniques and materials are promoted, whose quality is very often overestimated without essentially knowing their durability, chemical compatibility with the existing masonry and reversibility.

The motivation and the challenge for the structural engineer in specific seismic strengthening of a historic structure, with the purpose of preserving its authenticity, should be to provide an economically justified seismic protection and necessary bearing capacity of the structure for an acceptable risk level during future earthquakes, avoiding as much as possible irreversible interventions, (Kelley and Crowe 1995; Tassios 2010; Oliveira and Costa 2010).

12.2 Earthquake Protection of Historic Structures

The problem of earthquake protection of historic structures is radically different from that of other structures, due to the priority given to preservation of aesthetic, architectonic and historic values instead of keeping the structure operational. The difficulties and the complexity of the problem are due to, first of all, the mode in which these structures are constructed as well as the disturbance of their authenticity in the course of their existence. Repair and strengthening as part of modern protection of structures of historic monuments located in seismically active regions should be planned based on a detailed study of the expected seismic hazard, the local soil conditions, the dynamic characteristics of the structure, the strength and deformability of structural elements and built-in materials, as well as on the dynamic response of the structures under expected seismic motions, (Gavrilovic et al. 1991–1995; Shendova 1998).

The key for selecting materials and techniques is classification of repair and strengthening techniques into two main categories: reversible and irreversible. In selecting materials to be used in reversible interventions, there are usually only a few limitations. The materials used in irreversible interventions do impose two additional limitations: compatibility of new with old materials and their durability.

To define an adequate concept of repair and strengthening, it is necessary to carry out a detailed analysis of the existing structure of the historical building or monument, the type and the physical-mechanical characteristics of masonry, dynamic properties of structure, criteria and expected seismic action. If this analysis proves that the structure has a sufficient bearing and deformability capacity, taking measures for its repair shall be sufficient enough. Otherwise, depending on the vulnerability level, the strengthening should be done to increase the strength of the existing structure or/and its deformability.

The specific character of seismic protection of historical buildings and monuments resulting from the variety of structural systems, built-in materials, periods and techniques of construction, stability criteria and contemporary requirements incorporated in the modern principles of conservation and protection needs systematic and scientific approach to achieve a successful solution, (Kelly 1993; Tomazevic and Lutman 1996; Danieli et al. 2008; Manfredi 2009; Martelli 2009; Tassios 2010).

12.2.1 Needs for Shaking Table Testing

For the structures which are not specially designed with seismic rules and codes as historical buildings and monuments are, there is necessity to perform tests for assessing their safety and for designing appropriate upgrading. Many of these historical buildings are built in regions that have been strongly affected by medium and high magnitude earthquakes in the past. For each significant event, specialists have gathered relevant information with respect to their protection, which calls for verifications based on experimental testing.

The problem of interaction between the "old" and the "new" materials and/or elements that arises in their strengthening requires experimental verification of all techniques that have so far been developed (injection, grouting, jacketing, confining, base isolation). Then, the characteristic structural entity, the variability of the built-in materials, the complex history of successful modifications done in the past, as well as the degree of deterioration, make each historic monument a case for itself, which imposes the need for development of a scientifically based methodology for conservation, restoration, repair and seismic strengthening of historic structures in seismically active regions. All the above-mentioned aspects are strongly related to the need for laboratory testing of models on shaking tables, (Modena et al. 1992; Magenes and Cavi 1994; Carydis et al. 1996; De Canio et al. 2008).

To obtain the experimental values of the main parameters (physical-mechanical and chemical characteristics of the built-in material, strength and deformability characteristics, ductility capacity and energy dissipation capacity of the structural elements and structures as whole), different testing techniques are applied in practice. These can be divided into three main groups:

1. Experimental in-situ tests of previously separated wall fragments of the existing structure, which are particularly justified in case of existence of a larger number of similar structures or structures built at about the same time;
2. Static experimental tests of samples of materials (stone, brick, mortar), wall elements or complete structures, known as quasi-static (pseudo-dynamic) tests;

3. Dynamic experimental tests of the main dynamic characteristics of structures (ambient and forced vibration technique), as well as testing of the dynamic response of their models on seismic shaking table by simulation of earthquakes.

The experimental investigation of models on a seismic shaking table is the most corresponding way of investigation from the aspect of dynamic structural behaviour during real earthquakes. Applying an appropriate modelling technique and according to the similarity laws, the models can be designed to different scales and tested under various seismic inputs. Based on experimental data obtained from site testing of the monuments, as well as the results of the seismic hazard analysis, the laboratory testing of models gives very reliable data on the seismic behaviour and stability, pointing out the weak points of the structures. These data are of great importance for further analysis and for the development of an appropriate methodology for seismic strengthening of monuments.

12.3 IZIIS' Experience in Seismic Protection of Historic Buildings and Monuments

Within the framework of IZIIS' research activities, in addition to seismic design of modern structures, particularly noteworthy is also the experience gathered in the field of protection of structures pertaining to the cultural historic heritage. During a period of more than 30 years of activities in this field, the Institute has realized important scientific research projects involving experimental and analytical research, field surveys of historic structures and application of knowledge during earthquake protection of important cultural historic structures and monuments.

Extensive research activities have been performed by IZIIS for the purpose of evaluation of a procedure for repair and strengthening of valuable historic monuments. It can be said that an integrated approach to seismic protection of extraordinarily important cultural historic structures has been adopted by the Institute. This approach that complies with all the restoration and conservation requirements as well as procedures and legislative regulations for high category structures, should encompass the following:

- Definition of expected seismic hazard;
- Definition of soil conditions and dynamic behaviour of soil media;
- Determination of structural characteristics, and bearing and deformability capacity of existing structures;
- Definition of criteria and selection of concept for repair and/or strengthening;
- Definition of structural methods, techniques, materials and types of excitation;
- Analysis of dynamic response of repaired and/or strengthened structures and verification of their seismic stability;
- Definition of field works, execution and inspection.

Although the above stated seems to be the "normal procedure", it is the only way of providing high quality protection of cultural heritage. This task is certainly much more than simply listing what is to be done since it requires a lot of knowledge and efforts.

A particularly important part of IZIIS' experience in the field of earthquake protection of cultural heritage are the numerous shaking table tests on models in order to investigate the structural behaviour of historic buildings and monuments, as well as methodologies for their repair and seismic strengthening, that have been carried out in IZIIS' Dynamic Testing Laboratory. MTS biaxial programmable shaking table with 5.0×5.0 m in plane and a payload of 40.0 ton is used for generation of motions in horizontal direction. The shaking table is supported by four vertical hydraulic actuators which enable a motion in vertical direction. Two horizontal actuators enable motion in horizontal direction.

The considered structures have relatively low levels of axial stresses at the base which justifies the adoption of a model with neglected gravity forces, i.e. the "gravity forces neglected" modelling principle, using the same materials as in the prototype structures. The testing procedure consists of two main phases:

1. Tests for definition of dynamic characteristics of the model, before and after performing seismic tests at each phase, in order to check stiffness degradation of the model produced by micro or macro cracks developed during the tests;
2. Seismic testing by selected earthquake record until collapse. The tests are performed in several steps, increasing the input intensity of the earthquake in order to obtain the response in linear range, as well as to define the initial crack state, development of failure mechanism and collapse of the model.

The most important seismic shaking table investigations of models of historical buildings are presented further.

12.3.1 Shaking Table Testing of Historic Buildings

12.3.1.1 Old Towns Along the Mediterranean Coast

In the course of 1986, a dynamic shaking table test was performed on a model of a single storey masonry house constructed of stone in lime mortar for the purpose of defining the bearing capacity and deformability of stone masonry, characteristic of the Montenegro coastal area (Gavrilovic et al. 1987). It was shown that repair by injection retrofits the structure into a state of being capable of sustaining the seismic loads by a modest improvement of its dynamic characteristics.

Namely, the old towns along the Mediterranean coast (Budva, Kotor, Dubrovnik) were severely damaged due to the 1979 Montenegro earthquake. In the process of renovation of earthquake-affected old towns, extensive studies were performed along with the experimental and analytical investigations. Considering the cultural value of most of the buildings in the old town of Budva, which are typically stone masonry structures with one or several stories, investigations for the purpose of searching for the optimum conditions and methods for reconstruction, repair and strengthening of structures were performed.

Fig. 12.1 Original and repaired model of a typical building in the old town of Budva

Considering the fact that injection was the most frequently applied method and that there is a lack of results in order to evaluate the obtained effects from injection on the basis of bearing and deformability parameters, a 1:2 scaled model of a typical single story building was constructed of the same original stones and mortar and tested on the seismic shaking table in IZIIS. For different levels of seismic excitations and using representative record of the 1979 Montenegro earthquake in Petrovac, the model was brought to the state when large cracks appear. The model was then repaired by injection and tested again, in which case the failure mechanism was not modified, but the model suffered less damage, Fig. 12.1. After these experimental and analytical investigations, rationalization of construction was made and the input data were verified. The whole old town of Budva was repaired at the same time (structural repair and repair of facades) and the interior of each individual building was repaired separately.

12.3.1.2 Historic Adobe Structures in California

Since 1990, the Getty Conservation Institute (GCI) carried out a multi-year, multi-disciplinary project, the Getty Seismic Adobe Project (GSAP), including a survey of existing historic adobe buildings in California, performance of dynamic testing of scaled model at the Stanford University shaking table, and the preparation of an Engineering Guide for designing seismic retrofit measures for adobe buildings. As an extension of GSAP, tests were conducted on two large-scale models (1:2 scale) on the seismic shaking table in the IZIIS' dynamic laboratory (Gavrilovic et al. 1996; Tolles et al. 2000). The two models were of the same design as the previously tested models, i.e. a typical south-western American design that includes floor and roof systems, and highly vulnerable gable end walls. The first model was a control model while the second one was retrofitted with combination of horizontal and vertical straps, center cores, and partial plywood diaphragms.

Unretrofitted and retrofitted tapanco-style models were built to a scale of 1:2 and tested under selected intensities of the Taft earthquake in order to investigate the linear and nonlinear model behaviour and the behaviour in the heavily damaged state.

Unretrofitted Adobe Model   UAB   Retrofitted Adobe Model – RAB

Fig. 12.2 The gable end wall of the UAB and RAB model

In these tests, the gable end wall of the unretrofitted building collapsed, while in the retrofitted building, the straps and, especially, the center core rods proved very effective in improving stability and preventing collapse, Fig. 12.2. Since an increase in the size of the scale models didn't change the test results, gravity does not appear to be a significant factor.

Using a survey of 19 historic adobes in the Los Angeles area that was conducted after the 1994 Northridge earthquake, the behaviour of real and model structures was compared. It was found that many of the structures suffered damages similar to those seen in the test of unretrofitted models.

12.3.2 Shaking Table Testing of Monuments

12.3.2.1 Shaking Table Testing of Models of Byzantine Churches

Within the framework of the scientific research projects realized at the Institute in the period 1990–2000 for the purpose of developing appropriate methods for repair and strengthening of Byzantine monuments in general, and particularly the Byzantine churches located within Macedonia, shaking table testing of a church model in a realistic geometrical scale was performed for the first time in the world. 1:2.75 scaled model of St. Nikita church was constructed and tested on the seismic shaking table in the IZIIS laboratory in its original state, strengthened state by use of "ties and injection", and as a base isolated model.

Seismic Strengthening and Repair of Byzantine
Churches in Macedonia

Experimental and analytical investigations were performed to verify an original methodology that was developed for the repair and seismic strengthening of Byzantine

Fig. 12.3 The church of St. Nikita, v. Banjani

churches. This work was part of the long-term research project realized jointly by IZIIS, the Republic Institute for Protection of Cultural Monuments, Skopje and the Getty Conservation Institute, Los Angeles, (Gavrilovic et al. 1991–1995, 1996, 1999, 2004a; Shendova 1998).

The churches dating from the Byzantine period located in Macedonia are important architectural structures and contain extraordinary collections of highly important frescoes. Based on certain criteria four representative churches were selected for investigations which involve field and analytical tests for definition of seismic parameters, main dynamic characteristics and analysis of seismic resistance. The church of St. Nikita in the village of Banjani has been selected as a prototype representative of the Byzantine churches in Macedonia, Fig. 12.3.

Ample field and analytical studies, in situ and laboratory tests were performed for the existing structure of the St. Nikita church for the purpose of defining the physical-mechanic and chemical characteristics of the built-in materials, the dynamic characteristics of the structure and the seismicity of the terrain. A preliminary analysis of the seismic stability of the existing structure points to nonsufficient ultimate bearing and deformability capacity according to the design criteria on seismic safety. The proposed concept of strengthening consists of incorporating horizontal and vertical steel ties and filling the area around them with a corresponding material for the purpose of increasing the bearing capacity and the deformability of the existing structure.

To experimentally verify the proposed methodology for repair and seismic strengthening, a model of the existing structure of the prototype church (M-SN-EXIST) was constructed and tested on the seismic shaking table in the Dynamic Testing Laboratory

M-SN-EXIST   M-SN-STR

Fig. 12.4 Models on shaking table

of IZIIS (Fig. 12.4). The geometrical scale of the model church was selected on the basis of the characteristics of the seismic shaking table and the precisely defined objectives of testing, i.e. realistic reproduction of nonlinear behaviour and the failure mechanisms. Satisfying these criteria and adopting the "gravity forces neglected" modelling principle, the following main scales were adopted: geometrical scale $L_r = 1:2.75$, scale for the bulk density of the material $\rho_r = 1$, scale for the stresses $E_r = 1$.

The main targets of the experimental tests performed on the church model can be summarized into two groups: (i) assessment of the vulnerability of structures of interest, and (ii) selection of most appropriate procedures for repair and strengthening of damaged structures in post-earthquake protection. To that effect, a programme of experimental tests was adopted by gradual increase of the intensity of input earthquake excitations aimed at monitoring the progressive development of cracks and the failure mechanism, the modification of the dynamic characteristics as well as the phases of dynamic behaviour of the model, i.e. defining the elasticity limit (occurrence of the first cracks).

The investigation was performed by simulating two main types of earthquake: the 1976 Friuli earthquake (Breginj record) as a local earthquake, and the 1979 Montenegro (Petrovac record) and 1940 El Centro earthquakes as earthquakes from distant foci.

From the general behaviour of the model, it was concluded that it behaved as a rigid body in the elastic range, while at occurrence of the first larger cracks, there was separation of the bearing walls and development of damages up to a state close to complete failure. This was proved by the decrease in natural frequency from 11. to 6.6 Hz.

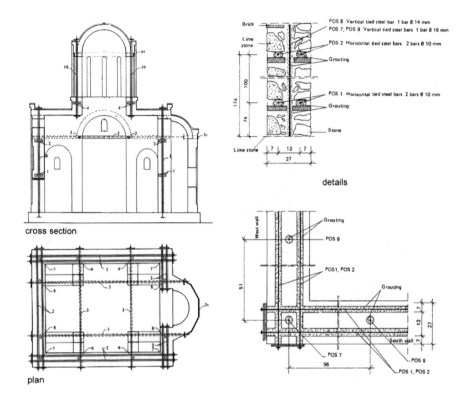

Fig. 12.5 Applied method of strengthening

The damaged model was structurally strengthened in accordance with the proposed methodology (Fig. 12.5), that is implementation of horizontal and vertical steel reinforcement into the wall mass and filling the area by an appropriate injection mixture that enabled contact with the surrounding existing masonry.

The repaired and strengthened model (M-SN-STR) was subjected to the same series of dynamic tests for the purpose of proving the efficiency of the applied method of strengthening. However, due to the high resistance of the strengthened model, the tests were continued under higher intensities. The response of the strengthened model was considerably different from that of the original model. Characteristic was the increased elasticity limit and reduction of displacements at the top. Although there was a considerable deterioration in bearing capacity under maximum seismic effect, the complete stability of the model structure was not disturbed due to the presence of ductile elements, while the damage was such that it was repairable. The applied methodology for repair and strengthening increases the bearing capacity and deformability of the structure up to the level of the designed protection which is proved by comparing the results from the experimental tests on the original and the strengthened model (Table 12.1).

Table 12.1 Comparison between the experimental results of three models

| | | Output acceleration (in g) for the church models | | | | | |
| | | M-SN-EXIST | | M-SN-STR | | M-SN-BIC | |
Earthquake	Input acc (g)	Level 1	Level 2	Level 1	Level 2	Level 1	Level 2
El Centro	0.17	0.29	0.55	0.20	0.47	0.07	0.10
El Centro	0.30	–	–	0.65	1.10	0.10	0.14
El Centro	0.49	–	–	0.91	1.59	0.23	0.31
El Centro	0.54	–	–	0.77	1.41	0.35	0.68
El Centro	0.63	–	–	–	–	0.42	0.82
Petrovac	0.19	0.39	0.76	0.27	0.48	0.09	0.15
Petrovac	0.40	–	–	0.77	1.36	0.15	0.28
Breginj	0.17	0.22	0.52	0.30	0.55	0.07	0.10
Breginj	0.28	–	–	0.20	0.40	0.10	0.16
Breginj	0.38	–	–	0.34	0.79	0.14	0.23

Level 1 – Base of the tambour; level 2 – Top of the dome

Earthquake Protection of Byzantine Churches Using Seismic Isolation

As a continuation of previous activities, experimental and analytical investigations as well as shaking table testing of base isolated church model were performed to develop a methodology for application of seismic isolation as a way of seismic protection of a large number of similar cultural monuments. These investigations were realized at IZIIS within the framework of the joint US-Macedonian research project and PHARE Cultural development program (Gavrilovic et al. 2001b, 2003, 2004b; Shendova et al. 2006).

For the needs of testing the isolated church model, eight seismic isolators of the type of rubber bearings were specially designed and produced in R. Macedonia (Fig. 12.6). The system clearly distinguishes three main elements:

- Laminated rubber bearing element for receipt and transfer of vertical gravity forces and limited displacement (insulation) in horizontal direction;
- Steel plate damper which has the role of a damper in the form of hysteretic behaviour but only after the linear behaviour of the laminated rubber bearings is exhausted.
- In conditions of specific behaviour and requirements for historic monuments regarding their protection in seismic conditions, one of the elements is "limited displacement". This criterion is satisfied by designing the third element of our system – the "stopper element".

In this way, the main requirements are satisfied as to the linear behaviour of the system under slight and moderate earthquakes and its limited but controlled displacements by activation of elements two and three under catastrophic earthquakes. This enables complete protection of the structure and the valuable objects within the structure like frescos and other elements. The base isolated model (M-SN-BIC, Fig. 12.7) was placed on a specially designed steel structure for connection with the isolators and the shaking table.

Fig. 12.6 Laminated rubber bearing

Fig. 12.7 Base isolated model on shaking table

Table 12.1 shows the comparison between experimentally obtained results for all three models. They pointed to a decrease of input acceleration in the model structure by 50–60% and to a completely different failure mechanism. For all testing up to input acceleration of 0.60 g, the base isolated model behaved as a rigid body without any visible cracks. It did not suffer damage under low and moderate earthquake intensities and the damages under maximum expected accelerations with a return period of 1,000 years were minimal and repairable.

The tests undoubtedly proved that the new technology of seismic base isolation of historic monuments offers absolute safety and protection and that its application should become an imperative in earthquake protection of monuments in the future.

Fig. 12.8 Mustafa Pasha Mosque, prototype and model

12.3.2.2 Shaking Table Testing of Monuments Within the PROHITECH Project

In the period 2006–2009, within the framework of FP6 PROHITECH project "Earthquake Protection of Historical Buildings by Reversible Mixed Technologies", experimental shaking table tests on the models of three important historical monuments (mosque, cathedral and church) were carried out in IZIIS: The results have shown the efficiency of all the applied retrofitting systems.

Representative Mosque Structure – Mustafa Pasha Mosque in Skopje

The Mustafa Pasha's Mosque is one of the biggest and the best preserved monuments of the Ottoman sacral architecture in Skopje and the Balkans (Fig. 12.8). It was built at the end of fifteenth century as one-dome structure consisting of massive double-leave walls with a thickness of 2 m. The mosque had one slender minaret with spiral staircases (Mazzolani et al. 2007; Krstevska et al. 2008b).

The model, designed according to the "gravity forces neglected" modelling principle, was constructed in the IZIIS Laboratory (Fig. 12.8). The seismic shaking table testing was performed in three main phases: (1) Testing of the original model under low intensity level, with the aim of provoking damage to the minaret only; (2) Testing of the model with strengthened minaret, with the aim of provoking its collapse and damage to the mosque (Fig. 12.9); and (3) Testing of the strengthened mosque model until reaching heavy damage. The seismic investigation of the model was performed by simulating the 1979 Montenegro Petrovac earthquake,

Fig. 12.9 Model with strengthened minaret

Fig. 12.10 Damage of the model (phase 3)

selected as a representative one and applied in horizontal direction. According to the similitude requirements, the original earthquake record was scaled by 6.

The strengthening solution consisted in forming a horizontal belt course around the bearing walls by CFRP rods as well as around the tambour and at the base of the dome by CFRP wrap (Fig. 12.10.) Push-over curves were constructed based on the

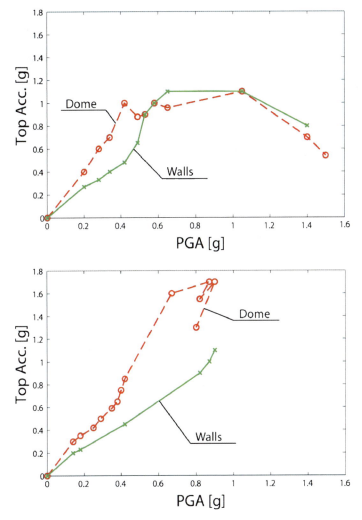

Fig. 12.11 Response of the model: acceleration measured at the top of the shear walls and of the dome as a function of input acceleration, phase 2 and phase 3

different intensity level of the repeated seismic test, within an input acceleration of 0.05–1.50 g (Fig. 12.11).

Shaking table testing for both the original and strengthened model showed that the strengthening of the minaret by application of a CFRP wrap enabled stiffening and increasing of its bending resistance. The mosque model's behaviour, after strengthening, was evidently different in respect to that of the original model. During the intensive tests, the failure mechanism was transferred to the lower zone of the bearing walls where typical diagonal cracks occurred due to shear stress. Comparing the obtained results it can be concluded that the applied strengthening technique has significantly improved the seismic resistance of the monument.

Fig. 12.12 The Fossanova Cathedral, Italy, prototype and model

Representative Gothic Cathedral – Fossanova Abey, Italy

The Fossanova Gothic cathedral dating back to fifteenth century is located in Priverno, Lazio, Italy (Fig. 12.12). In the first phase dynamic characteristics of the prototype were measured applying the ambient vibration testing method (Krstevska et al. 2008a). The model for seismic shake table testing was designed as a true replica, to a scale of 1:5.5. A very detailed model has been conceived in order to take into account the complexity of the prototype and to keep correctly the scale factor in terms of its dynamic response. Equivalent materials have been considered, which have the same weight but a reduced strength according to the adopted scale factor.

The seismic investigation of the model was performed by simulating the Calitri Earthquake selected as a representative one and applied in horizontal direction only. After the testing of the original model and the development of the failure mechanism, the cracks were repaired by an expansive aluminium-cement mortar.

The proposed methodology for strengthening consists in incorporating pre-stressed vertical and horizontal carbon fibre ties at several levels of the model (Fig. 12.13). During the final test, with intensity of 0.4 g, the strengthened model was severely damaged with completely developed failure mechanism (Fig. 12.14).

The main conclusion is that the integrity of the model is significantly improved by the applied strengthening. For the original model the critical input intensity was 0.14 g, while for the strengthened model 0.40 g. These parameters clearly show the effectiveness of the applied strengthening, enabling confining of the structure.

Representative Church Structure – St. Nikola Church Macedonia

The St. Nikola church, dating from the Byzantine period, was built in the mid-fourteenth century. It is about 150 km from Skopje, at Psacha village near Kriva Palanka. The model of the church was designed to the length scale of 1:3.5 according to the "gravity force neglected" modelling principles (Fig. 12.15) (Tashkov et al. 2008).

Fig. 12.13 Strengthening of the model

Fig. 12.14 Damage of the model after final test

The shake-table testing was performed in two main phases: testing of the base-isolated model with the ALSC floating-sliding system and testing of the original fixed-base model. The model with the sliding base was tested first because it was expected that no damage would occur. This was confirmed even under the maximum capacity of the shaking table (1.5 g). The second phase was realized with the fixed base model, which simulates the original structure on site. Testing of the model was performed by simulating the 1979 Montenegro Petrovac earthquake. According to the "gravity force neglected" modelling principle, the original earthquake record was scaled by 3.5 in time domain (compressed).

Fig. 12.15 St. Nikola Church, Psacha, Macedonia, prototype and model

The model was tested by different levels of input acceleration up to 1.45 g but the maximum response acceleration of the sliding plate in all the cases was about 0.2–0.3 g. This was actually a limitation of the transmissibility of the forces from the basin to the sliding plate. This level of acceleration didn't produce any cracks in the model, except an increased relative sliding displacement between the basin and the sliding plate. In the testing phase 2 the model was transformed into a classical fixed-base structure. The input acceleration was changed in several tests from 0.1–0.7 g. The amplification of the response was 1.5–2.0 for the top of the wall and 3–4 for the top of the dome.

The comparative test between the base-isolated model and the fixed base model clearly shows the superior behaviour of the ALSC floating-sliding base-isolation system. The role of the springs, controlling the lateral motion was effective, allowing the structure to slide in the desired and controlled range.

12.4 Implementation of Developed Methodologies for Seismic Strengthening of Monuments

After the realization of these projects, which accumulated unique and incomparable knowledge, IZIIS became partner of the Republic Institute for Protection of Cultural Historic Monuments of R. Macedonia, which enabled direct application of the gained knowledge in actual conditions and for specific historic monuments. Presented further are the three most characteristic examples of application of the developed methodologies for seismic upgrading of monuments.

12.4.1 Reconstruction of St. Athanasius Church in Leshok

On August 21, 2001, during the armed conflict in R. Macedonia, the monastic church of St. Athanasius in Leshok experienced strong detonation, which

Fig. 12.16 The church after detonation

resulted in its almost complete demolition (Fig. 12.16). Based on previous knowledge (Sect. 12.3.2.1), two approaches to the reconstruction of the church were taken: (i) repair and strengthening of the existing damaged part and (ii) complete reconstruction of ruined part with strengthening elements (Shendova and Gavrilovic 2004; Shendova and Stojanoski 2004, 2008).

The solution for repair and structural strengthening of the damaged existing part of the structure anticipates injection of all the cracks and incorporation of horizontal and vertical RC strengthening elements on certain levels. For the demolished part of the structure, a concept of complete reconstruction by maximum possible use of selected material has been adopted, whereat elements for structural strengthening for providing the designed level of seismic safety have also been anticipated: (i) RC belt course below the floor level, in the existing foundation walls, for anchoring the vertical strengthening elements, (ii) vertical strengthening steel elements at the ends of the massive walls and around the openings, (iii) vertical strengthening steel elements into the tambour columns, (iv) horizontal steel elements along the walls, in the base of the tambour and the dome. Due to the different treatment of the structural units constituting the integral structure, an expansion joint between them is anticipated to be constructed.

The church was reconstructed according to the designed methodology during 2003–2004 (Fig. 12.17). The results from the analysis show that both structural units constituting the integral structure possess a sufficient bearing and deformability capacity up to the designed level of seismic protection.

Fig. 12.17 The church after reconstruction

12.4.2 Reconstruction of St. Pantelymon Church in Plaoshnik, Ohrid

In the process of conservation and rebuilding of the St. Panteleymon Church in Ohrid, having in mind the importance and specific nature of the structure (it represents a historic monument classified in the first category), it was necessary to design a building structure that would satisfy the stability conditions in the process of application of the conservation principles regarding shape, system and materials (Gavrilovic et al. 2001a; Shendova et al. 2008). The principal structural system of the church consists of massive stone and brick masonry in lime mortar. Seismic strengthening was provided in accordance with the previously developed and verified methodology (Sect. 12.3.2.1), i.e. the horizontal and vertical steel ties were proportioned and a solution for consolidation of the foundation was given (Fig. 12.18). The church was reconstructed in the course of 2001 (Fig. 12.19).

12.4.3 Repair and Seismic Strengthening of Mustafa Pasha Mosque in Skopje

Respecting the modern requirements in the field of protection of historical monuments, as is the application of new technologies and materials, reversibility and invisibility of the applied technique, the concept of repair and strengthening involving

Fig. 12.18 The church during reconstruction

Fig. 12.19 The rebuilt St. Panteleymon church

the use of composite materials was adopted for the seismic upgrading of the Mustafa Pasha Mosque in Skopje (Shendova et al. 2007, 2008, Mazzolani et al. 2009). It has been selected based on investigations of the: characteristics of the built-in materials, main dynamic characteristics, shaking table testing of the mosque model (Sect. 12.3.2.2);

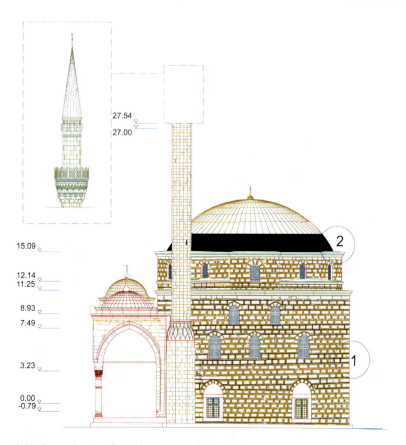

Fig. 12.20 Strengthening of the Mustafa Pasha Mosque

investigations of the soil conditions as well as detailed geophysical surveys for definition of geotechnical and geodynamic models of the site.

The accepted solution of structural strengthening (Fig. 12.20) consists in incorporating CFRP wrap in a layer of epoxy glue along the perimeter of the dome base, placing CFRP bars in an epoxy mortar layer in horizontal joints of bearing walls, and constructing an RC wall along the perimeter of the foundation walls, below the terrain level. Strengthening of the mosque structure in accordance with the designed system started in the fall of 2008.

The design process of the consolidation of the Mustafa Pasha Mosque in Skopje was accurately developed to the greatest possible extent according to the modern principles of seismic protection of historical buildings. The delicate problem of proving the effectiveness of the selected consolidation system has been successfully overcome by using the methodology of design assisted by testing. This methodology, which has been recently codified in all Eurocodes, represents a very powerful tool especially when the object of design is a complex structure, which is difficult, and therefore unsafe, to analyze by using traditional methods.

12.5 Conclusion

Shaking table tests on models in a realistic geometric scale are an excellent method for research investigation of dynamic properties and bearing characteristics, verification of technology and repair and strengthening, as well as for investigation of new technologies in historical monuments. The experimental evidence is helpful in proving the suitability of the design choices and the effectiveness of the applied consolidation system.

The knowledge gained through shaking table testing is unique and incomparable and hence necessary for seismic strengthening of individual important cultural-historic structures where it is important to have an insight into the effect of the interventions upon the authenticity of the monument.

The developed and experimentally verified methodologies were successfully implemented in the structures of the real historic monuments. The design process, which has been followed as shown in this paper, should emphasize the suitable way to follow when facing the complex problem of protection and conservation of a cultural heritage construction.

References

Carydis PG, Mouzakis HP, Papantonopoulos C, Papastamatiou D, Psycharis N, Vougioukas EA, Zambas C (1996) Experimental and numerical investigations of earthquake response of classical monuments. In: 11WCEE. Paper No. 1388, Acapulco

Danieli M, Bloch J, Ribakov Y (2008) Retrofitting heritage buildings by strengthening or using seismic isolation. In: Seismic engineering conference, MERCEA08, Reggio Calabria, Italy

De Canio G, Muscolino G, Palmeri A, Poggi M, Clemente P (2008) Shaking table tests validating two strengthening interventions on masonry buildings. In: Seismic engineering conference, MERCEA08, Reggio Calabria

Feilden BM (1982) Conservation of historic buildings. Butterworth Scientific, London

Feilden BM (1987) Between two earthquakes. ICCROM/GCI joint publication, Rome/Marina del Rey

Gavrilovic P, Stankovic V, Bojadziev M (1987) Experimental investigation of a model of masonry building on seismic shaking table. VIII Congress of Structural Engineers, YU

Gavrilovic P, Ginell W, Sendova V et al (1991–1995) Seismic strengthening, conservation and restoration of churches dating from Byzantine period in Macedonia. Joint research project, IZIIS – Skopje, RZZSK – Skopje, GCI -LA; Reports IZIIS 500-76-91, vols 1–12

Gavrilovic P, Sendova V, Tashkov Lj, Krstevska L, Ginell W, Tolles L (1996) Shaking table tests of adobe structures. Report IZIIS. IZIIS, Skopje, pp 96–36

Gavrilovic P, Shendova V, Ginell W (1999) Seismic strengthening and repair of Byzantine Churches. J Earthq Eng (Imperial College, London) 3–2:199–235

Gavrilovic P, Necevska–Cvetanvska G, Apostolska R (2001a) Consolidation and reconstruction of St. Panteleymon Church in Ohrid. IZIIS Report 2001

Gavrilovic P, Shendova V, Kelley S (2001b) Earthquake protection of Byzantine Churches using seismic isolation. Macedonian – US joint research. Report IZIIS 2001, SS. Cyril and Methodius University, Skopje

Gavrilovic P, Kelley S, Shendova V (2003) A study of seismic protection techniques for the Byzantine Churches in Macedonia. J Assoc Preser Tech APT Bull XXXIV(2–3):63–71

Gavrilovic P, Ginell W, Shendova V, Sumanov L (2004a) Conservation and seismic strengthening of Byzantine churches in Macedonia. The Getty Conservation Institute, GCI Scientific Program Reports, ISBN 0-89236-777-6, J. Paul Getty Trust

Gavrilovic P, Shendova V, Kelley S (2004b) Seismic isolation: a new approach to earthquake protection of historic monuments. In: IV international seminar on structural analysis of historical constructions, SACH, Padova, pp 1257–1264

Kelley SJ, Crowe TM (1995) In: Kelley SJ (ed) The role of the conservation engineer, ASTM STP 1258: standards for preservation and rehabilitation. ASTM, Philadephia

Kelly J (1993) The application of seismic isolation for the retrofit of historic buildings, earthquake engineering research center, University of California, Berkeley

Krstevska L, Tashkov L, Gramatikov K, Kozinakov D (2008a) Shaking table test of Fossanova model. FP6-PROHITECH, Final Report IZIIS 2008

Krstevska L, Tashkov Lj, Gramatikov K, Mazzolani F, Landolfo R (2008b) Shaking table testing of Mustafa Pasha Mosque Model. FP6-PROHITECH Project, DIII-Datasheet No.4.1.2

Magenes G, Calvi G (1994) Shaking table test on brick masonry walls. In: 10ECEE, Vienna

Manfredi G (2009) Composites for seismic protection of cultural heritage. In: Protection of Historical Buildings, PROHITECH09, Italy

Martelli A (2009) Development and application of innovative anti-seismic systems for seismic protection of cultural heritage. In: Protection of Historical Buildings, PROHITECH09, Italy

Mazzolani FM, Krstevska L, Tashkov Lj, Gramatikov K, Landolfo R (2007) Shaking table testing of Mustafa Pasha Mosque model. FP6-PROHITECH Final Report. IZIIS, Skopje

Mazzolani F, Shendova V, Gavrilovic P (2009) Design by Testing of Seismic Restoration of Mustafa Pasha Mosque in Skopje. In: Protection of Historical Buildings, PROHITECH09, Italy

Modena C, La Mendola, Terrusi A (1992) Shaking table study of a reinforced Masonry building Model. 10WCEE, vol 6, Madrid, pp 3523–3526

Oliveira C, Costa A (2010) Reflections on the rehabilitation and the retrofit of historical constructions. Earthquake engineering in Europe, 14ECEE, Skopje

Shendova V (1998) Seismic strengthening and repair of Byzantine churches in Macedonia. Doctoral dissertation, SS. Cyril and Methodius University, Skopje (in Macedonian)

Shendova V, Gavrilovic P (2004) Implementation of a methodology using "ties and injection" developed for repair and strengthening of historic monuments. In: IV international seminar on structural analysis of historical constructions, SACH 2004, Padova

Shendova V, Stojanoski B (2004) Main Project on Repair, Strengthening and Reconstruction of St. Athanasius Church in Leshok. IZIIS Report 2004. SS. Cyril and Methodius University, Skopje, Republic of Macedonia

Shendova V, Stojanoski B (2008) Quasi-static tests on wall elements constructed during the reconstruction of St. Athanasius Church in Leshok. 14WCEE, Beijing

Shendova V, Rakicevic Z, Gavrilovic P, Jurukovski D (2006) Retrofitting of Byzantine Church using passive base control system. In: 4th world conference on structural control and monitoring, San Diego

Shendova V, Stojanoski B, Gavrilovic P (2007) Main project on repair and strengthening of the Mustafa Pasha Mosque in Skopje. IZIIS Report 2007–41, vol 1–3, SS. Cyril and Methodius University, Skopje

Shendova V, Gavrilovic P, Stojanoski B (2008) Integrated approach to repair and seismic strengthening of Mustafa Pasha Mosque in Skopje. In: Seismic engineering international conference, MERCEA09, Reggio di Calabria

Tashkov Lj, Krstevska L, Gramatikov K (2008) Shaking table test of model of St. Nicholas church to scale 1/3.5. FP6-PROHITECH, Final report IZIIS, IZIIS, Skopje

Tassios TP (2010) Seismic engineering of monuments. Earthquake Engineering in Europe, 14ECEE, Skopje

Tolles L, Kimbro E, Webster F, Ginell W (2000) Seismic stabilization of historic adobe structures. Final report of the Getty Seismic Adobe Project, The Getty Conservation Institute, Los Angeles

Tomazevic M, Lutman M (1996) Seismic behaviour of masonry walls: modeling of hysteric rules. J Struct Eng 122(9):1048–1054

Chapter 13
Dynamic Behaviour of Reinforced Soils – Theoretical Modelling and Shaking Table Experiments

Jean Soubestre, Claude Boutin, Matt S. Dietz, Luiza Dihoru, Stéphane Hans, Erdin Ibraim, and Colin A. Taylor

Abstract The dynamic response of soil-pile-group systems are modelled both analytically, using homogenisation theory, and physically, using a shaking table to excite a soft elastic material periodically reinforced by vertical slender inclusions. A large soil/pile stiffness contrast is shown to lead to full coupling in the transverse direction of the bending behaviour from the piles and the shear behaviour from the soil. Analytically derived performance predictions capture important characteristics of the experimentally observed response that are missed when using alternative analytical modelling approaches. The homogenisation theory approach to modelling of generalised media is valid.

13.1 Introduction

The dynamic response of structures founded on groups of piles involves complex interactions between the different system components: piles, soil and structure. Reconnaissance evidence suggests that such systems can perform well during earthquakes. In 2004, the vast majority of structures founded on piles on soft muddy soils

---

J. Soubestre (✉) • C. Boutin • S. Hans
Ecole Nationale des Travaux Publics de l'Etat (ENTPE), Université de Lyon,
FRE 3237 CNRS, 3, rue Maurice Audin, 69120 Vaulx-en-Velin, France
e-mail: jean.soubestre@entpe.fr; claude.boutin@entpe.fr; stephane.hans@entpe.fr

M.S. Dietz • L. Dihoru • E. Ibraim • C.A. Taylor
Department of Civil Engineering, University of Bristol, Queen's Building, University Walk,
Bristol BS8 1TR, UK
e-mail: M.Dietz@bristol.ac.uk; Luiza.Dihoru@bristol.ac.uk; Erdin.Ibraim@bristol.ac.uk;
colin.taylor@bristol.ac.uk

M.N. Fardis and Z.T. Rakicevic (eds.), *Role of Seismic Testing Facilities in Performance-Based Earthquake Engineering: SERIES Workshop*, Geotechnical, Geological and Earthquake Engineering 22, DOI 10.1007/978-94-007-1977-4_13,
© Springer Science+Business Media B.V. 2012

in the city of Pointe-à-Pitre (Guadeloupe) did not experience significant damage under significant earthquake loading despite a lack of seismic protection provisions. To capitalise on this circumstantial evidence, validated models must be developed which can predict the dynamic response of such systems.

The performance of the global behaviour of pile reinforced soil when subjected to lateral ground motions has been a topic of research interest in recent decades (e.g. Makris and Gazetas 1992; Mylonakis and Gazetas 1999; Koo et al 2003). Numerical finite element studies of pile-reinforced soils are generally conducted. However, the resulting models are complex and require significant computing time due to the fine mesh needed to account for the heterogeneities in the medium. In fact, the problem is ill-conditioned because of the high number of piles and the high contrast between the mechanical properties of the soil and the piles. Another limitation lies in the purely numerical form of the result. An understanding of the interactions and the effective influence of the pile parameters can only be extracted by statistical back-analysis of numerous simulations.

An alternative approach models the macroscopic equivalent behaviour analytically. To this end, Sanchez-Palencia's (1980) homogenisation of periodic heterogeneous media using asymptotic expansion techniques has been found to be sufficiently rigorous. Herein, the key assumption lies in the separation of scale between the size of the microstructure (i.e. the distance between the piles) and the scale of evolution of the phenomena (i.e. the macroscopic deformation of the whole system). Postel (1985) recognised that this condition is satisfied for pile-reinforced soils and formulated the problem as a composite of parallel lengths of reinforcement – the piles – periodically arranged in a matrix – the soil. The macroscopic behaviour was that of a strongly anisotropic elastic medium. His approach enabled the large stiffness increase along the reinforcement axis and the moderate increase in the perpendicular direction to be quantified as a function of the reinforcement concentration and material properties.

Sudret and De Buhan (1999) argued that the slenderness of the embedded reinforcement should ensure that the response will involve bending, the classical assumption in earthquake engineering practice. Assuming a sparse reinforcement concentration and a large reinforcement/matrix stiffness ratio, they developed a 'multiphase model' that accounts for the bending effect that remains at the macroscopic scale. Note, however, that this modelling approach is not self-contained because an additional interaction term is required. Numerical simulations based on this model (De Buhan and Hassen 2008) show that the theory enables a better understanding of the kinematics of the reinforcement under static lateral loading.

An analytical model linking the previous approaches and based on homogenisation was proposed by Boutin and Soubestre (2011) to characterize the dynamic behaviour of pile-reinforced soils. Making a systematic use of scaling and of the 1D geometry of the reinforcement, the homogenisation is performed considering different orders of magnitude of matrix-reinforcement stiffness contrast. In accordance to the configuration of the problem, their model can evaluate the contributions made to the global kinematics by both internal shear and internal bending. Both homogenised and multiphase models belong to the framework of generalized elastic continua in which the integration of the bending is related to a scale effect.

The aim of this paper is to provide an experimental validation of the homogenised model proposed by Boutin and Soubestre (2011). First, the main aspect of the theoretical modelling are summarised. In the second half of the chapter, the preliminary findings of an experimental campaign conducted at the University of Bristol (UK) and under the auspices of the European Commission's SERIES project in order to validate the analytical modelling approach are described.

13.2 Theoretical Homogenisation Modelling

This section lays out the conditions for which an elastic material matrix periodically reinforced by linear, slender, elastic lengths of reinforcement can behave as a generalized continuum and introduces the pertinent homogenisation modelling approach.

13.2.1 System Description

The study is conducted considering a medium constituted by a soil (elastic matrix, index m) in which a periodic lattice of reinforcement (parallel, identical, homogeneous, straight beams, index p) is embedded with perfect contact (Fig. 13.1a). The dimension H along the beam axis is significantly larger than the lateral dimension l of the period (Fig. 13.1b). The typical size of the beam section h is of the same order than l (Fig. 13.1b) so that the reinforcements are in finite concentration. The geometry naturally introduces a distinction between the axial direction (unit vector $\underline{e}_l$) and the directions in the plane of the section (unit vectors $\underline{e}_\alpha$ with $\alpha = 2, 3$). Herein, Greek indices run from 2 to 3 and Latin from 1 to 3. The scale parameter of the problem as

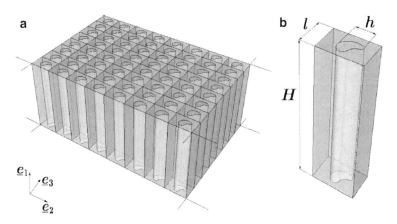

Fig. 13.1 Pile-soil composite material: (**a**) periodic lattice of parallel identical homogeneous lengths of reinforcement embedded in a matrix (**b**) period geometry and dimensions

Fig. 13.2 Notations of the composite material period section

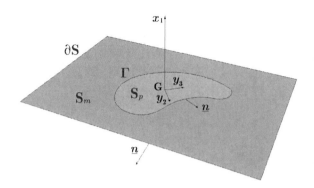

used in the asymptotic expansions is defined as $\varepsilon = l/L$, where the macroscopic length L is much larger than l. The notations related to the period section are displayed on the Fig. 13.2. The relevant dimensionless space variables are $(x_1/L, x_2/l, x_3/l)$ and the appropriate physical space variables are (x_1, y_2, y_3), where $y_\alpha = (L/l)\, x_\alpha = \varepsilon^{-1} x_\alpha$.

The soil and the piles are assumed to be isotropic linear elastic. For soils in geophysics this assumption is justifiable for sufficiently small strains, which is the operational case for deep foundation systems or under small amplitude earthquakes. The two constituents are characterized by their Lame coefficients λ_q and μ_q ($q=m, p$) or, equivalently, by their Young's modulus E_q and Poisson's ratio v_q. The contrast between the elastic properties of the matrix and reinforcement plays a crucial role. Without the matrix, the reinforcement lattice is governed by bending; if the matrix and the reinforcement materials are identical, one has a homogeneous medium governed by shear.

13.2.2 Basic Idea

The homogenised model is derived in accordance with the following reasoning. The transverse static equilibrium and constitutive laws for a matrix of elastic material (ante exponent m) in the plane direction (for instance e_2) are:

$$\frac{d\,^m\sigma_{12}}{dx_1} = 0; \quad ^m\sigma_{12} = \mu_m \frac{d\,^m u}{dx_1} \tag{13.1}$$

Conversely, for reinforcement governed by bending (ante exponent p) they are:

$$\frac{d\,^p T_2}{dx_1} = 0; \quad ^p T_2 = \frac{d\,^p M}{dx_1} \quad \text{and} \quad ^p M = -E_p I_p \frac{d^2\,^p u}{dx_1^2} \tag{13.2}$$

Full coupling between the reinforcement beam behaviour and the matrix shear behaviour emerges when the transverse forces in both constituents are of the same order of magnitude. Multiplying the stress in the matrix $^m\sigma_{12}$ by the matrix section of the period, S_m, to get an action effect, gives:

$$E_p I_p \frac{u_p}{L^3} = O\left(\mu_m S_m \frac{^m u}{L}\right) \qquad (13.3)$$

Considering that $I_p = O(l^4)$, $S_m = O(l^2)$ and that the displacement in the reinforcement and the matrix are of the same order of magnitude ($^p u = O(^m u)$), for coupling to occur the shear modulus contrast μ_m/μ_p (equivalent to the modulus contrast μ_m/E_p because $E_p = O(\mu_p)$) has to be of the order of magnitude of ε^2:

$$\frac{\mu_m}{\mu_p} = O\left(\frac{u_m}{E_p}\right) = O\left(\frac{l^2}{L^2}\right) = O\left(\varepsilon^2\right) \qquad (13.4)$$

13.2.3 Homogenised Model

The displacement, strain and stress fields of the reinforcement and the matrix can be expanded asymptotically according to the powers of ε. The stiffness contrast is integrated in the asymptotic process. For stiffness contrast $\mu_m = O(\varepsilon^2 \mu_p)$ and for a bi-symmetric period geometry (as is usual in practice when piles are arranged in a square or a hexagonal grid) the resolution, detailed in Boutin and Soubestre (2011), shows that:

– axially, under a macroscopic vertical motion $U_1(x_1)$, the system is driven by the piles subjected to kinematic beam compression:

$$\frac{d\langle\sigma_{11}\rangle}{dx_1} = 0; \quad \langle\sigma_{11}\rangle = E_p S_p \frac{dU_1}{dx_1} \qquad (13.5)$$

– Transversely, (for instance in the direction e_2), the macroscopic behaviour of the system submitted to a macroscopic horizontal motion $U_2(x_1)$ is:

$$\frac{d\langle\sigma_{12}\rangle}{dx_1} = 0 \qquad (13.6)$$

$$\langle\sigma_{12}\rangle = C_{12} \frac{1}{2}\frac{dU_2}{dx_1} - \frac{E_p I_{P2}}{|S|}\frac{d^3 U_2}{dx_1^3} \qquad (13.7)$$

Here, the shear coefficient C_{12} is equal to the soil coefficient $(2\mu_m)$ corrected by a form function κ that takes into account the presence of the pile.

$$C_{12} = 2\mu_m\left(1+\kappa\right) \qquad (13.8)$$

κ can be evaluated either through a finite-element-method simulation for complex period geometry or approximated by a self-coherent estimation given by Hashin and Rosen (1964) for small reinforcement concentrations.

$$\kappa = \frac{2c}{\dfrac{2}{\mu_p / \mu_m - 1} + (1 + c)} \quad \text{for } c \ll 1 \tag{13.9}$$

Equations 13.6 and 13.7 define the static macroscopic behaviour of the unloaded reinforced matrix (without body forces or dynamic loading) under macroscopic transverse motion $U_2(x_1)$. The macroscopic constitutive law includes:

- a classical shear contribution related to the distortion $U2, x_1$. The elastic coefficient C_{12} is the matrix coefficient $2\ \mu m$ corrected by a small term κ to take into account the presence of the reinforcement. It coincides with that given by the usual homogenization approach in the case of infinitively rigid reinforcement (Léné 1978).
- a bending contribution related to the derivative of the curvature $U_2, x_1 x_1 x_1$. The bending inertia parameter is exactly equivalent to that of the reinforcement (divided by the period section). Quite unlike other composite models where higher gradient terms appear as correctors (see e.g. Boutin 1996), here the bending effect arises at the leading order.

This general solution is shown to degenerate either into the usual shear behaviour of elastic composite media when $\mu_m/\mu_p \geq O(\varepsilon)$, or into the usual Euler-Bernoulli bending behaviour when $\mu_m/\mu_p \leq O(\varepsilon^3)$.

The macroscopic behaviour differs from the description of composites usually derived by homogenisation (Léné 1978; Sanchez-Palencia 1980; Postel 1985). Here we obtain a generalized inner bending continuum in which the macroscopic variable is the translation $U_2(x_1)$ and the mean transverse stress combines at the same order local and non-local terms related respectively to the strain tensor (shear) and to the derivative of the curvature (bending). Note that, as the rigid reinforcement 'imposes' its motion on the soft matrix, the stiffness of the reinforcement in bending and of the matrix in shear are combined 'in parallel'. Thus, the internal mechanism drastically differs from the one of Timoshenko beams. Actually, the present description enables one to avoid the introduction of the unknown interaction term of the phenomenological model developed by Sudret and De Buhan (1999). It provides a generalization of the mathematical analysis of energy of Bellieud and Bouchitté (2002) and of the work of Pideri and Seppecher (1997) who consider infinitesimal concentration of cylindrical fibres with extremely high modulus.

The dynamic behaviour of the system in the harmonic regime is derived by introducing inertia terms $-\rho_q \omega^2 \,^q\underline{u}$ $(q = m, p)$ into the equilibrium of both constituents. Provided that the scale separation condition is satisfied, it can be demonstrated that they can be introduced in the model as volume forces without modifying

13 Dynamic Behaviour of Reinforced Soils – Theoretical Modelling... 253

the macroscopic constitutive law obtained in static regime. Then the reinforced soil behaves macroscopically in a dynamic regime transversally:

$$\frac{d\langle \sigma_{12} \rangle}{dx_1} = -\frac{\left(\rho_p |S_p| + \rho_m |S_m| \right)}{|S|} \omega^2 U_2$$

$$\langle \sigma_{12} \rangle = C_{12} \frac{1}{2} \frac{dU_2}{dx_1} - \frac{E_p I_{p2}}{|S|} \frac{d^3 U_2}{dx_1^3} \qquad (13.10)$$

13.2.4 Energy and Boundary Conditions

The higher order of differentiation in the constitutive law modifies the nature of the usual boundary conditions which can be identified through the formulation of the elastic and kinetic energy at the macroscale. For an infinite layer of reinforced soil of height H along $\underline{e}_1$, taking the product of the equilibrium equation by a test field U_2 and integrating over the height, one obtains:

$$\int_0^H \left(C_{12} \frac{1}{2} \frac{dU_2}{dx_1} \frac{dU_2}{dx_1} + \frac{E_p I_{p2}}{|S|} \frac{d^2 U_2}{dx_1^2} \frac{d^2 U_2}{dx_1^2} \right) dx_1$$

$$- \int_0^H \frac{\left(\rho_p |S_p| + \rho_m |S_m| \right)}{|S|} \omega^2 U_2 U_2 dx_1 = \left[\langle \sigma_{12} \rangle U_2 \right]_0^H - \left[\frac{1}{|S|} {}^P M_2 \frac{dU_2}{dx_1} \right]_0^H \qquad (13.11)$$

where ${}^P M_2 = - E_p I_{p2} U_{2,x1x1}$ is the reinforcement momentum. The elastic energy (LHS of Eq. 13.11) accounts for inertial energy and both shear and bending deformations. It balances the work (RHS of Eq. 13.11) produced at the boundary ($x_1 = 0$ and $x_1 = H$) by the mean stress vector $\langle \sigma_{12} \rangle$ submitted to the motion U_2 on the one hand, and by the momentum ${}^P M_2$ submitted to the pile section rotation $U_{2,x1}$ on the other. Hence, according to the fourth degree differential equation, two boundary conditions must be specified at each extremity: one in terms of displacement or stress as for continuous media, and one in terms of rotation or momentum as for beams. By construction of the macroscopic modelling, the interpretation of these latter conditions is directly linked to the actual conditions imposed on the reinforcement.

13.2.5 Application to a Reinforced Soil Layer

Consider the transverse mode of an infinite lateral extension of reinforced matrix. The period is constituted by circular section of reinforcement centrally positioned within a square section of matrix. Writing:

$$U_2 = U; \quad x_1 = x; \quad I_{p2} = I_p; \quad C_{12}/2 = G; \qquad (13.12)$$

the equilibrium condition verified by the mean displacement U is:

$$G|S|\frac{d^2U}{dx^2} - E_p I_p \frac{d^4U}{dx^4} + \left(\rho_p |S_p| + \rho_m |S_m|\right)\omega^2 U = 0 \qquad (13.13)$$

The dimensionless parameter $C = (G|S|H^2)/(E_p I_p)$ 'weights' the bending effects compared to the shear effects. Bending predominates when C is small, shear when C is large. The general solution of the fourth order differential equation is:

$$U(x) = a\cosh\left(\delta_2 \frac{x}{H}\right) + b\sinh\left(\delta_2 \frac{x}{H}\right) + c\cos\left(\delta_1 \frac{x}{H}\right) + d\sin\left(\delta_1 \frac{x}{H}\right) \quad (13.14)$$

with

$$\begin{cases} \delta_1^2 \delta_2^2 = \dfrac{(2\pi f)^2 \left(\rho_p |S_p| + \rho_m |S_m|\right)H^4}{E_p I_p} \\[4mm] \delta_1^2 - \delta_2^2 = \dfrac{G|S|H^2}{E_p I_p} = C \end{cases} \qquad (13.15)$$

The boundary conditions must be specified to get access to the modal characteristics of the system. For an encastré condition at the lower end of the reinforcement and a free condition at the top (i.e. unrestrained translation and rotation), the boundary conditions in terms of displacement are:

$$\begin{cases} U(0) = 0 \\ U'(0) = 0 \\ {}^p M_2(H) = 0 \\ \langle \sigma_{12} \rangle(H) = 0 \end{cases} \Rightarrow \begin{cases} U(0) = 0 \\ U'(0) = 0 \\ U''(H) = 0 \\ GU'(H) - E_p I_p /|S| U'''(H) = 0 \end{cases} \qquad (13.16)$$

Substituted into the general solution (13.14), we obtain the modal equation:

$$\frac{C\cos(\delta_1)}{\delta_1^2 \delta_2^2} + \frac{\tanh(\delta_2)\sin(\delta_1)}{\delta_1 \delta_2} + \frac{2}{C}\left(\cos(\delta_1) + \frac{1}{\cosh(\delta_2)}\right) = 0 \qquad (13.17)$$

The modal equation can be solved numerically as a function of C for coefficients δ_{1i} and δ_{2i} corresponding to the mode number i. The frequency of the ith mode is:

$$f_i = \frac{\delta_{1i}\delta_{2i}}{2\pi H^2}\sqrt{\frac{E_p I_p}{\left(\rho_p |S_p| + \rho_m |S_m|\right)}} \qquad (13.18)$$

Figure 13.3 shows the evolution of the eigen frequencies ratios $f_i/f_1 = \delta_{1i}\delta_{2i}/(\delta_{11}\delta_{21})$ varying according to C for the three first modes. For small values of C the reinforced soil has the same frequency distribution as a fixed-free bending beam

13 Dynamic Behaviour of Reinforced Soils – Theoretical Modelling...

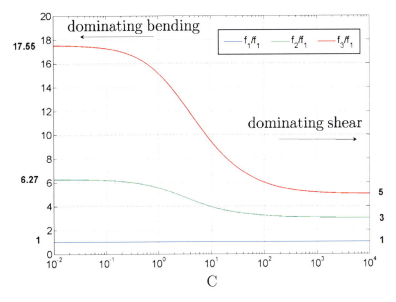

Fig. 13.3 Evolution of eigen frequency with C

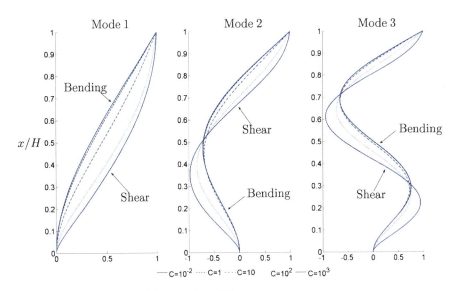

Fig. 13.4 Modal shapes for different values of C

(1, 6.27, 17.55, ...) and for large values of C the ratios are close to those of a shear layer (1, 3, 5, ...). The three first modal shapes (normalized to 1 at $x_1 = H$) of the reinforced soil (each modal shape is plotted on a "slice" of the system as the displacement U is the same in both constituents) are displayed in Fig. 13.4 for different values of the C parameter. The curves vary from a bending to a shear shape as C increases.

13.3 Experimental Validation

An experimental campaign was carried within the framework of the SERIES project to validate the theoretical model. Tests were conducted at the shaking table of the University of Bristol's Earthquake and Large Structures (EQUALS) Laboratory. The objective was to identify and quantify the actual bending effect due to the reinforcement under transverse motions by analysing the spectral response of a reinforced matrix subjected to transverse excitation.

13.3.1 Experiment Design

Since the main objective was phenomenological identification, the physical model was constructed from analogue materials consistent with the basic assumptions of the theoretical modelling, namely: large stiffness contrast between the matrix and reinforcement, homogeneous, linear-elastic materials and a perfect adherence at their interface. Thus, the inherent difficulties of using soil – a vertical gradient of mechanical properties and lateral inhomogeneities close to the inclusions – were avoided. Moreover, the small linear region of natural soil would otherwise impose small deformation restriction resulting in less accurate measurements.

An overview and schematic of the model is provided as Figs. 13.5 and 13.6 respectively. The matrix was a polyurethane foam block, 2.13 by 1.75 by

Fig. 13.5 Overview of physical model

13 Dynamic Behaviour of Reinforced Soils – Theoretical Modelling...

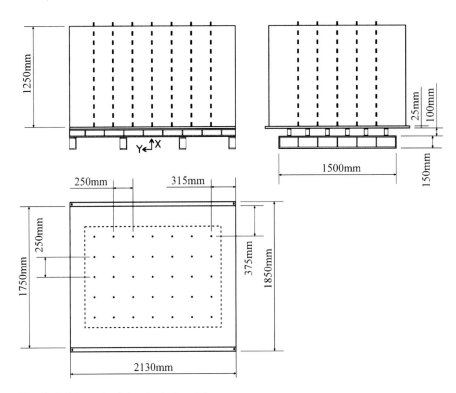

Fig. 13.6 Schematic of the physical model

1.25 m tall, and of density $\rho_m = 48$ kg/m^3. The foam was subjected to preliminary characterisation tests using standard laboratory equipment in order to obtain its mechanical properties. The foam exhibited linear elastic behaviour up to 4–5% of axial strain and thereafter became non-linear. A Young's modulus $E_m = 54$ kPa, Poisson's ratio $\nu_m = 0.11$ and shear modulus $G_m = 24.3$ kPa were derived. The experiments on the reinforced foam block were conducted with a global distortion level of about 0.1% to ensure that the foam remained within its linear elastic range.

The reinforcement was round, seamless, mild steel tube with 12.7 mm outside diameter and 3.25 mm wall thickness. The mechanical properties were, classically, Young's modulus $E_p = 210$ GPa, Poisson's ratio $\nu_p = 0.3$ and density $\rho_p = 7800$ kg/m^3. To reflect the pile-group attributes commonly seen in practice, thirty-five 1.3 m lengths of reinforcement were used on a seven by five grid at 250 mm centres (Fig. 13.6). An array of seven by five holes at 250 mm centres was bored through the 1.25 m deep block of foam in order to take the lengths of reinforcement. To ensure that the inclusions maintained good contact with the foam during testing, the bore diameter was 1 mm less than the diameter of the inclusions.

To accurately represent the boundary conditions of the analytical model the lengths of reinforcement were bolted to a base plate secured to the shaking table. The block of foam was adhered to the base plate. The model was positioned so that

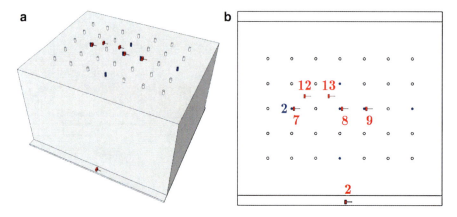

Fig. 13.7 Test instrumentation layout featuring accelerometers (*red*) and strain gauges (*blue*): (**a**) orthographic projection (**b**) in plan

its principal axes aligned with the principal axes of the shaking table and so that the long section was in the principal direction of shaking (Y).

Mesurands were acceleration and strain. Accelerometers were mounted on the shaking table, on the uppermost surface of the foam, and on the 50 mm free length of reinforcement protruding from the top of the foam. The longitudinal strains generated in 6 of the 35 lengths of reinforcement were monitored using strain gauges. All strain gauge cabling was fed into the interior of the reinforcement through small holes ($\approx$1 mm diameter) drilled through the wall. The instrumented lengths of reinforcement were each fitted with six strain gauges. The strain gauges were deployed in pairs with each pair having a distinct ordinate: one pair at the reinforcement bottom (38.5 mm from the base), one at the middle (625 mm from the base) and one at the top (1211.5 mm from the base). One of each pair faced in the +Y direction, the other in the −Y direction. The measurement axes of the strain gauges were aligned with the axis of the reinforcement to allow the measurement of bending strains. Sensor positions are displayed in Fig. 13.7.

Different magnitudes of random (white noise) excitation with frequency content between 1 and 30 Hz were used to drive the shaking table in the Y direction. Harmonic sinusoidal waveforms were also used to excite the model at its eigen frequency for accurate mode shape determination.

13.3.2 Theoretical/Experimental Results Comparison

Comparison between experiment and theory is based on the response of the first mode of the system. The C parameter of the experimental model can be evaluated:

$$C = \frac{G|S|H^2}{E_p I_p} = 9.42 \tag{13.19}$$

13 Dynamic Behaviour of Reinforced Soils – Theoretical Modelling...

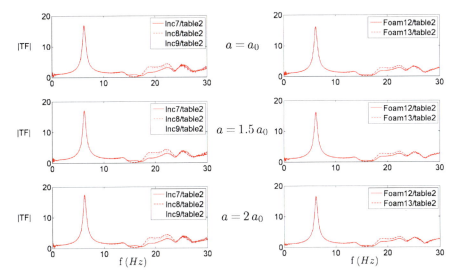

Fig. 13.8 Transfer function modulus between accelerometers located on inclusions/table (*left*) and foam/table (*right*) for white noise excitations of different mean amplitudes: a_0 (*top*), $1.5a_0$ (*middle*) and $2a_0$ (*bottom*)

The theoretical first eigen frequency of the system is then $f_{1,th} = 5.88$ Hz and the theoretical mode shape is close to the one corresponding to $C = 10$ in Fig. 13.4.

13.3.2.1 White Noise Response

Figure 13.8 shows the acceleration response of the system to three 0–30 Hz horizontal white noise excitation tests of different mean amplitudes: a_0, $1.5a_0$ and $2a_0$. The linear response of the system is confirmed with the observation that the magnitude of the transfer function (TF) modulus does not depend of the excitation level. Moreover the coincidence of TFs derived using accelerometers placed on different components of the system means that the system has an in-plane (geometrically) homogeneous kinematic for its first mode. The measured eigen frequency $f_{1,exp} = 5.95$ Hz is close to the prediction of the shear/bending theoretical homogenised model ($f_{1,th} = 5.88$ Hz) and clearly different from the pure shear (unreinforced media: $f_{1,th} = 4.51$ Hz) or pure bending (negligible foam effect: $f_{1,th} = 6.56$ Hz) response.

13.3.2.2 Harmonic Forcing Response

Figure 13.9 presents the time response of the six strain gauges located on the length of reinforcement numbered 2 (c.f. Fig. 13.7) to harmonic sinusoidal forcing at its first eigen frequency. Figure 13.10 presents the first mode shape (top) and curvature (bottom) of response if dominated by bending (left), dominated by shear (right) and

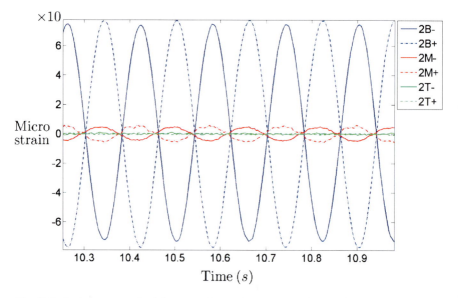

Fig. 13.9 Strain response of reinforcement bar no. 2 to harmonic loading at the first mode

dominated shear/bending coupling (centre). Note that for a system excited at its first mode the curvature of the first mode shape $\Phi_1''(x_1)$ is related to the normal strain on the side of the inclusion $|{}^p e_n(x_1, d_p/2)|$. Indeed, the normal strain in the inclusion ${}^p e_n(x_1, y_2)$ is proportional to the curvature of the transverse displacement $U(x_1)$ which is equal to the mode shape $\Phi_1(x_1)$ when the system responds at first mode:

$$ {}^p e_n(x_1, y_2) = -\frac{d^2 U(x_1)}{dx_1^2} y_2 = -\frac{d^2 \Phi_1(x_1)}{dx_1^2} y_2 \tag{13.20} $$

Considering that $\Phi_1(x_1)$ is normalized to $U(H)$ at the top (i.e. $x_1 = H$) and since during an harmonic forcing test $a_r(H) = (2\pi f_1)^2 U(H)$, the normal strain is related to the relative acceleration at the top of the system $a_r(H)$. As the relative acceleration $a_r(H)$ is the difference between the total acceleration measured at the top $a_T(H)$ and the acceleration of the shaking table a_{base}, the curvature of the normalized first mode shape is proportional to the normal strain measured on the side of the inclusion of diameter d_p:

$$ \Phi_1''(x_1) = |{}^p e_n(x_1, d_p/2)| \frac{2}{d_p} \frac{(2\pi f_1)^2}{a_T(H) - a_{\text{base}}} \tag{13.21} $$

Since the bending moment in the reinforcement is given by ${}^p M(x_1) = -\Phi_1''(x_1)/(E_p I_p)$, measurements of the strain gauges provide a direct evaluation of the inner momentum in the reinforcement.

Figure 13.9 shows the system response to harmonic forcing at its first mode and indicates that there is no strain at the reinforcement top (*T*). At the bottom (*B*) and

13 Dynamic Behaviour of Reinforced Soils – Theoretical Modelling...

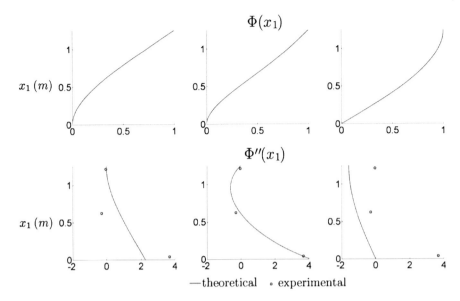

Fig. 13.10 Mode shape (*top*) and curvature of the mode shape (*bottom*) of reinforced materials dominated by bending (*left*), dominated by shear (*right*) and dominated by shear/bending coupling (*centre*, C=9.42)

at the middle (*M*) of the reinforcement we observe equal amplitude phase opposition for opposite (+/−) strain gauges. Extension on one side of the inclusion and compression (of the same amplitude) on the other is clear evidence of bending. Moreover, the gauges 2B+ and 2M+ (or 2B− and 2M−) are in phase opposition. This signifies that there is an inflexion point between the heights *B* and *M* in the first mode shape of the system and an inversion of sign of the momentum. As presented in Fig. 13.10, this observation is in accordance with the theoretical homogenised model. Both the classical composite model (with only macroscopic shear) and the pure bending model fail to capture the experimentally observed momentum distribution. Furthermore, both the amplitude and the sign of the experimentally measured strains (red points) coincide with the shear/bending coupling model. Neither the pure shear nor the pure bending models can demonstrate an inflexion point in the first mode shape because the associated second derivatives do not change in sign. These experimental observations provide a first validation of the theoretical modelling of generalized media derived by homogenisation.

13.4 Conclusion

An experimental programme has been conducted to validate an analytical approach to modelling pile-reinforced soils based on homogenisation theory. The validated model provides new insights into the dynamic behaviour of pile-reinforced soils in the elastic range.

The formulation of the analytical model is straight-forward and contains a few parameters. The model can be used to perform parametric studies of pile-reinforced soil systems that cannot be achieved using the finite element method due to the problem being numerically ill-conditioned. The homogenised model shows that a pile-reinforced soil system is highly anisotropic. In the vertical direction, the response is simply evaluated and dominated by the pile. In the transverse direction, and where there is a high pile/soil stiffness contrast, the shear response of the soil may be coupled with the bending response of the pile.

A physical model of a pile-reinforced soil system was constructed using analogous materials and tested using the shaking table at the University of Bristol's Earthquake and Large Structures Laboratory. In accordance with the homogenised model, the experimental observations give a clear evidence that an atypical transverse behaviour (shear/bending coupling) can occur.

Future work will expand the scope of the analytical model by considering soil/pile interface laws and poro-elastic soil behaviour of the soil. The influence of boundary conditions on the dynamic response of the model will also be investigated.

Acknowledgments The research leading to these results has received funding from the European Community's Seventh Framework Programme [FP7/2007-2013] under grant agreement n° 227887 for the SERIES project. The Authors would also like to thank Frederic Sallet from the University of Lyon ENTPE/CNRS Laboratory and David Ward and Edward Skuse from the EQUALS Laboratory for their key contribution in realising the experiments. They also would like to thank Marek Lefik and Marek Wojciechowski from the Technical University of Lodz, Jonas Snaebjornsson and Ragnar Sigbjornsson from the University of Iceland and Loretta Batali and Horatiu Popa from the Technical University of Civil Engineering of Bucharest for their participation.

References

Bellieud M, Bouchitté G (2002) Homogenization of soft elastic material reinforced by fibers. Asymp Anal 32(2):153–183

Boutin C (1996) Microstructural effects in elastic composites. Int J Solids Struct 33(7):1023–1051

Boutin C, Soubestre J (2011) Generalized inner bending continua for linear fiber reinforced materials. Int J Solids Struct 48:517–534

De Buhan P, Hassen G (2008) Multiphase approach as a generalized homogenization procedure for the macroscopic behaviour of soils reinforced by linear inclusions. Eur J Mech A Solids 27:662–679

Hashin Z, Rosen B (1964) The elastic moduli of fiber-reinforced materials. ASME J Appl Mech 31:223–232

Koo K, Chau K, Yang X, Lam S, Wong Y (2003) Soil-pile-structure interaction under SH wave excitation. Earthq Eng Struct Dyn 32:395–415

Léné F (1978) Comportement macroscopique de matériaux élastiques comportant des inclusion rigides ou des trous répartis périodiquement. CR Acad Sci IIB 286:75–78

Makris N, Gazetas G (1992) Dynamic pile-soil-pile interaction part ii: lateral and seismic response. Earthq Eng Struct Dyn 21:145–162

Mylonakis G, Gazetas G (1999) Lateral vibration and internal forces of group piles in layered soil. J Geotech Geoenviron Eng 125(1):16–25

Pideri C, Seppecher P (1997) Un résultat d'homogénéisation pour un matériau élastique renforcé périodiquement par des fibres élastiques de très grande rigidité. CR Acad Sci IIB 324(7):475–481

Postel M (1985) Réponse sismique de fondations sur pieux. PhD thesis, Ecole Centrale de Paris, Châtenay-Malabry

Sanchez-Palencia E (1980) Non homogeneous media and vibration theory, lectures notes in physics, vol 127. Springer, Berlin

Sudret B, De Buhan P (1999) Modélisation multiphasique de matériaux renforcés par inclusions linéaires. CR Acad Sci IIb 327:7–12

Chapter 14
Evaluation and Impact of Qualification of Experimental Facilities in Europe

Maurizio Zola and Colin A. Taylor

Abstract The paper addresses the need of qualification of large seismic research infrastructures operating in Europe and how this may be achieved. Questions about current limits to the free circulation of the products and services of the European Industry in earthquake engineering are discussed, with emphasis given on the issues – of technical, quality or commercial relevance – which constitute obstacles to the mutual accreditation. Results are reported relating to the general requirements for the qualification, both technical competence and quality assurance, as they are identified by large testing facilities, accreditation bodies, standardization institutions and industry.

14.1 Introduction

One of the aims of the SERIES project of the European Community's 7th Framework Programme is to enhance the services provided by the research infrastructures, transcending their current extreme fragmentation. This is to pursued, through:

- Common European standards and protocols for similar research infrastructures and
- Qualification criteria for European research infrastructures in earthquake engineering.

M. Zola (✉)
Mechanical Testing Department, Consultant of P&P LMC, Via Pastrengo 9,
24068 Seriate, BG, Italy
e-mail: maurizio.zola@gmail.com

C.A. Taylor
Department of Civil Engineering, University of Bristol, Queen's Building, University Walk,
Bristol BS8 1TR, UK
e-mail: colin.taylor@bristol.ac.uk

M.N. Fardis and Z.T. Rakicevic (eds.), *Role of Seismic Testing Facilities in Performance-Based Earthquake Engineering: SERIES Workshop*, Geotechnical, Geological and Earthquake Engineering 22, DOI 10.1007/978-94-007-1977-4_14,
© Springer Science+Business Media B.V. 2012

The present paper addresses the need of qualification of large seismic research infrastructures operating in Europe. Questions about the present limits to the free circulation of the products and services of the European Industry in earthquake engineering are raised and discussed, with emphasis given on the issues – of technical, quality or commercial relevance – which constitute obstacles to the mutual accreditation. Results are reported relating to the general requirements for the qualification, both technical competence and quality assurance, as they are identified by large testing facilities, accreditation bodies, standardization institutions and industry.

A companion paper (Kurç et al. 2012) in this Volume presents complementary information.

14.2 The Activity Towards Qualification of Research Infrastructures

14.2.1 Objectives

The activity in the framework of the SERIES aims at creating the conditions leading to the qualification of Structural Testing Laboratories specialising in earthquake engineering and equipped for large scale testing.

The Qualification process after ISO 9000 (2000) is a "Process to demonstrate the ability to fulfil specified requirements". For the qualification of the laboratories the combination of two main requirements is necessary:

- technical competence and
- quality assurance.

The final objective is to guarantee the reliability of testing in each laboratory. Reliability of experimental testing means repeatability and reproducibility:

- Repeatability (ISO 3534-1 1993) is the principle that experimental activities repeated on the same specimen in the same laboratory lead to the same results, and
- Reproducibility (ISO 3534-1 1993) is the principle that experimental activities repeated on the same specimen in different laboratories lead to the same results.

Besides establishing the general reliability of structural testing in Europe, a common platform for qualification will significantly enhance the expertise of testing facilities, as a result of the continuous benchmarking of similar laboratories.

The assessment of testing and instrumentation management procedures is in progress and it is expected that it will most likely lead to a critical analysis of the requirements imposed by official standardization and accreditation organisations, National and European.

It is expected that the mutual acknowledgement of European research infrastructures in earthquake engineering through the accreditation will enhance their standing with respect to their American or Japanese counterparts, promoting a unified EU policy on acceptance criteria for products and techniques.

14.2.2 General Description of Work

The activities towards qualification of Seismic Testing Laboratories have the aim to address the assessment criteria for technical competence of the beneficiary laboratories – and of similar ones in future – and to develop the basis for their mutual accreditation. This is focused on seismic experimentation through real-time (shaking table), pseudo-dynamic (reaction wall) or even quasi-static testing and on-site testing and monitoring. The activity is directly involving the Structural Research Infrastructures performing seismic experimentation and equipped for large scale testing.

Furthermore the activities involve the other players, i.e. the users of the research infrastructures (Scientific and Technological Community and Industry), which will ultimately draw the benefits from the qualification of laboratories for Products Certification.

The activity is broken down in four tasks:

1. Evaluation and impact of qualification of experimental facilities in Europe.
2. Assessment of testing procedures and standards requirements.
3. Criteria for instrumentation and equipment management.
4. Development and implementation of a common protocol for qualification.

As far as the two main requirements for the qualification of the laboratories, quality management system and technical competence, a sound experience is available in the application of international standards (EN ISO 9000 2000; EN ISO 9001 2008; ISO/IEC 17025 2005). Then it was decided to conduct the activities with reference to those standards, to receive a contribution from European Accreditation and Standardisation Organisations and to pursue agreement to issue a final Common Protocol for the qualification of research infrastructures in earthquake engineering.

From a technical point of view the mutual accreditation is a tool to guarantee the reliability of testing laboratories, i.e. repeatability and reproducibility of test results.

Repeatability and reproducibility can be easily achieved if common standard test procedures are used. Moreover a common language can be established on the basis of international standards which can help avoid mismatches and misunderstandings.

14.2.3 Evaluation and Impact of Qualification of Experimental Facilities in Europe

This task includes:

- a critical analysis of the problems that limit the free circulation of the products of the European Industry, the solution of which will be promoted by the qualification of structural laboratories, and
- a study of the issues – of technical, quality or commercial relevance – which in fact constitute obstacles to the mutual accreditation.

Final Users (Industry) and Accreditation or Standardisation Organisations have been directly involved, through questionnaires and meetings. This will lead to a detailed roadmap towards the common protocol for mutual accreditation among the laboratories.

The task has included the following activities:

- Identification of European Large Testing Facilities Laboratories;
- Identification of national accreditation bodies to be contacted;
- Identification of standardization organizations, both national and international;
- Identification of industrial organizations involved in the seismic qualification of products and in the seismic research activities;
- Preparation of a questionnaire to be circulated among European Large Testing Facilities Laboratories in order to collect information about:

  - The state of the certification of the quality systems of Laboratories;
  - The state of the accreditation of Laboratories for seismic testing;
  - The state of testing, personnel, instrumentation and equipment management procedures of the laboratories;
  - The evaluation of the time and cost which a laboratory should spend to fulfil the accreditation requirements;
  - The identification of the Standards adopted for the performance of experimental seismic testing;
  - The identification of the Standards adopted for the management of experimental facilities, measuring equipment, data acquisition systems;
  - The identification of the Standards adopted for data acquisition and processing, including frequency analysis, statistics, modal analysis, etc.;
  - The demand of qualification by structural laboratories;

- Preparation of a questionnaire for circulation among accreditation organizations of European countries to collect information about:

  - The general requirements for the accreditation of Laboratories with Large Test Facilities in the seismic field;
  - The evaluation of main activities which should be performed by a testing laboratory to get the accreditation to conduct seismic testing;

14 Evaluation and Impact of Qualification of Experimental Facilities in Europe

- The evaluation of the time and cost which a laboratory should spend to perform the formal activities requested by the accreditation organization, such as meetings, inspection visits and other related activities.

- Preparation of a questionnaire for circulation among standardization organizations in order to collect information on:

  - Standards for the performance of experimental seismic testing;
  - Standards for the management of experimental facilities, measuring equipment, data acquisition systems;
  - Standards for data acquisition and processing, including frequency analysis, statistics, modal analysis, etc.;

- Preparation of a questionnaire for circulation among industry organizations to collect information about:

  - Problems that limit the free circulation of the products of the European Industry;
  - Demand of qualification of structural laboratories to facilitate the free circulation of the products;
  - Standards for data acquisition and processing, including frequency analysis, statistics, modal analysis, etc.;

- Study of the issues – of technical, quality or commercial relevance – which in fact constitute obstacles to the mutual accreditation.

The questions were mainly based on EN ISO/IEC 17025 Standard; this choice was done in order to collect information covering all the aspects that are internationally recognized to be necessary for the certification of the technical competence of an experimental laboratory.

After the collection of the filled questionnaires, the responses were processed with reference to the specific parts dealing with the relevant objectives.

The outcome is a final report:

- collecting all the data from the questionnaires
- drawing the conclusions of the study of technical, quality and commercial relevance problems related to the accreditation of Large Testing Facilities Laboratories
- drawing a roadmap towards a common protocol for mutual accreditation.

14.3 Analysis of the Responses to the Questionnaires

14.3.1 Introduction

The questionnaires were prepared, distributed and collected filled by March 2010. In the following the information collected from the European Institutions are reported.

14.3.2 European Large Testing Facilities Laboratories

Thirty-six Institutions replied to the questionnaire.

As far as the state of the certification of the quality system, the situation is not uniform: 40% of the laboratories have a quality system and 60% do not, but most all laboratories (66%) lack any certification of the quality system nor any accreditation for the performance of experimental tests.

The same is the situation for the management procedures for personnel (40% having one), instrumentation (66% with one) and equipment (60% having a management procedure). Note that if these procedures are not identified then it has been assumed that the Laboratory has internal procedures and not standard procedures.

For the implementation of the accreditation requirements the laboratories consider as reasonable a time ranging from 1 month to 2 years and a cost varying from 1,500 to one million euros.

A similar percentage of laboratories, about 50%, adopts Standards for the performance of experimental seismic testing, but the percentage drops to less than 20% regarding the standards for the management of forcing/excitation equipment and measuring instrumentation.

Forty-six percent of the laboratories state that they have a demand for qualification, but only a few (29%) indicate a reference standard, mainly ISO 9001.

Most laboratories (more than 80%) seems to cover the general and technical requirements coming from the EN ISO/IEC 17025 standard, which has been taken as a reference document to list those requirements.

14.3.3 National Accreditation Bodies

Six Institutions replied to the questionnaires.

One of the main issues as far as the "General requirements for the accreditation of Laboratories" has been the necessity to take into account the REGULATION (EC) No 765/2008 of the European Parliament and of the European Council of 9 July 2008 (EC No 765/2008 2008), setting out the requirements for accreditation and market surveillance; this regulation entered into force on January 1st, 2010.

The Regulation lays down rules for the organization and operation of accreditation of conformity assessment bodies performing conformity assessment activities and lays down the general principles of the CE marking. It states that accreditation is part of an overall system, including conformity assessment and market surveillance, designed to assess and ensure conformity with the applicable requirements. Moreover the particular value of accreditation lies in the fact that it provides an authoritative statement of the technical competence of bodies whose task is to ensure conformity with the applicable requirements.

Accreditation, though so far not regulated at the European Community level, is carried out in all EU Member States. The lack of common rules for that activity has resulted in different approaches and differing systems throughout the Community,

14 Evaluation and Impact of Qualification of Experimental Facilities in Europe 271

the result being that the degree of strictness applied in the performance of accreditation has varied between Member States. Regulation 765 was issued in order to develop a comprehensive framework for accreditation and to lay down at Community level the principles for its operation and organization. A uniform national accreditation body is then established.

A system of accreditation which functions by reference to binding rules helps to strengthen mutual confidence between Member States, as regards the competence of conformity assessment bodies and consequently the certificates and test reports issued by them.

In order to avoid multiple accreditation, to enhance acceptance and recognition of accreditation certificates and to carry out effective monitoring of accredited conformity assessment bodies, these latter bodies should request accreditationzby the national accreditation body of the Member State in which they are established.

Regulation 765 provides for the recognition of a single organization at European level in respect of certain functions in the field of accreditation. The European cooperation for Accreditation (the EA) manages a peer evaluation system among national accreditation bodies from the Member States and other European countries. Therefore EA is the first body recognized under the Regulation and Member States should ensure that their national accreditation bodies seek and maintain membership of the EA for as long as it is so recognized.

Sectorial accreditation schemes should cover the fields of activity where general requirements for the competence of conformity assessment bodies are not sufficient to ensure the necessary level of protection where specific detailed technology or health and safety-related requirements are imposed. The EA should be requested to develop such schemes, especially for areas covered by Community legislation.

As far as the certification and accreditation of an experimental test facility the only applicable standards are respectively ISO 9001 (2008) and EN ISO/IEC 17025 (2005).

It is the accreditation body that will assess the procedures of the applicant laboratory as required by ISO 17025 for the management of:

- personnel
- instrumentation
- equipment
- acquisition, recording, storage and processing of data.

The same criterion is applied to the procedures for assessing experimental testing performance.

Any objective test may be accredited on the base of ISO 17025 which is covering all the technical aspects for the management of a Laboratory. It seems there are no specific standards to cover:

- the management of experimental facilities (shaking table, actuators, hydraulic hoses, servo-controllers, pumps, power supply units);
- the management of the measuring equipment (measurement transducers, cables, signal conditioners and amplifiers, filters);
- the management of data acquisition systems;

- data acquisition, recording and storage;
- data processing, including frequency analysis, statistics, modal analysis, etc.

As far as the evaluation of the time and cost which a laboratory is requested to spend to perform the formal activities requested by the accreditation organization, rough estimates are the following:

- time: 1 month for informal pre-assessment; 3 months for initial assessment and 1 month for clearance, assuming a laboratory compliant for the assessment;
- cost: ranging from 2,300 to 6,000 €/year.

ISO 17025 Standard covers also the accreditation of a facility that conducts non-standard testing and research at large scale on the basis of a flexible scope.

14.3.4 Standardization Organizations

Four Institutions replied to the questionnaires.

No specific standards were identified covering qualification and accreditation of seismic tests facilities; the outcome was that reference should be made to the general approach to certification of quality system by ISO 9001 and to the accreditation of laboratories by EN ISO/IEC 17025.

The following Standards covering seismic testing were identified with specific reference to:

- Nuclear Power Plants
- Electro-technical Equipment
- Telecommunication Equipment.

No specific standards were identified for the seismic testing of Civil Structures and Buildings.

Some standards were found for the measurement of vibrations on buildings and evaluation of their effects; some standards cover the seismic testing of components for civil structural applications.

As far as standards covering the management of experimental facilities and measuring equipment, some ISO standards were identified. No standards were found specifically devoted to the management of data acquisition systems.

ISO standards cover data acquisition and processing, including frequency analysis, statistics and modal analysis. Moreover specific ISO standards cover the accuracy (trueness and precision) of measurement methods and results.

14.3.5 Industry

Twelve industrial institutions replied to the questionnaires.

As far as the free circulation of the products, the industry requires facilities with a quality management system, if possible with certification and with the accreditation for seismic tests performance. Moreover, reference should be made to the certification of quality system by ISO 9001 and to the accreditation of laboratories by EN ISO/IEC 17025.

A special request is the outcome of the telecommunication industry, that is relaying to QuEST Forum who has developed the TL 9000 quality management system (QMS). Built on ISO 9001 and the eight quality principles, TL 9000 is designed specifically for the communications industry.

Legally independent facilities are preferred and facility's policies and procedures to ensure the protection of its customer's confidential information and proprietary rights are strongly required.

Standardized test methods are preferred and the uncertainty of test measurements should be addressed.

Opinions and interpretations of a test should be included the test report only upon request by the client.

14.4   Conclusions

14.4.1   General Requirements

As far as the general requirements for the qualification of experimental facilities the following conclusions can be drawn:

- Most of the testing facilities declare to have a Quality Management System. However, they do not have any official certification;
- The facilities give very rough estimates of the time and cost for the accreditation, because, not being certified yet, then they do not have experience about what is needed and how much time it takes to do what is necessary for the certification; by contrast, the accreditation bodies give a time estimate of less than a year and a maximum cost of 6,000 €/year to perform the formal activities requested by the accreditation organization;
- The laboratories report that clients request them to be certified. Laboratories which are not certified do claim to fulfill most of the requirements for the accreditation after EN ISO/IEC 17025;
- The free circulation of the products of the European Industry is ruled by REGULATION (EC) No 765/2008, setting out the requirements for accreditation and market surveillance;
- Regulation (EC) No 765/2008 is relaying for the accreditation to the European cooperation for Accreditation (the EA) and to the national accreditation bodies from the EU Member States;
- Regulation (EC) No 765/2008 states that the only applicable standards for the certification and the accreditation are respectively ISO 9001 and EN ISO/IEC 17025; anyway, any objective test may be accredited on the basis of ISO 17025;
- No specific standards cover the qualification of facilities for the performance of seismic tests;

- There are Standards covering seismic testing of equipment for Nuclear Power Plants, electro-technical and telecommunication applications; no specific standards cover seismic testing of Civil Structures and Buildings, both on site and on models, but some standards cover seismic testing of structural components;
- Some ISO standards cover the management of both experimental facilities and measuring equipment and data acquisition and processing;
- Industry requires qualified facilities with reference to ISO 9001 and to EN ISO/IEC 17025;
- Industry also requires Standardized test methods and estimates of the uncertainty in the test measurements.

Irrespective of the activity, research or services to industry, the testing facilities are requested to be qualified.

14.4.2 Specific Technical Requirements

As far as the specific technical requirements, the two main experimental procedures that should be dealt with are:

- Shaking table testing and
- Concentrated load testing.

Shaking table testing mainly entails the base excitation of a specimen, whereas concentrated load testing is referred to static, quasi-static and dynamic excitation by the application of loads through actuators directly connected to the tested specimen.

Regarding shaking table testing, many international standards are available for shock and vibration tests on electrical and mechanical equipment, but there are no specific standards for the testing of Civil Structures and components. Design standards for structures made of concrete, steel or wood are legion, but there are no standards covering their experimental testing.

For the concentrated load testing there are neither standards for equipment nor for structures.

14.4.3 Concluding Remarks from the Workshop

During the concluding panel session of the International Workshop on the Qualification of Large Testing Facilities in Ohrid (MK), the following worthy remarks were made:

- To qualify a laboratory requires demonstrating its competence to conduct experimental testing. To guarantee international recognition, the demonstration should be done by an "accredited Third Party".
- Basic methods for the determination of repeatability and reproducibility are given by international standards. To enable high repeatability and reproducibility,

common standard test methods should be used. Nevertheless, this is only one of the requirements for the qualification.

- For the management of testing equipment, common standard procedures are not required and each laboratory may adopt its own procedures.
- As far as product certification, REGULATION (EC) No 765/2008 gives the rules, but neither the Regulation nor the CE marking do pertain to research activities. The CE marking is related to the essential requirements, mainly devoted to safety aspects, of EN standards or of ETA (European Technical Approval) procedures for a specific product to be marketed in EU Member States.
- The qualification of a laboratory could be performed on the base of a specific agreement with each client. However, this procedure does not guarantee the mutual international recognition of the research laboratories qualification.
- Inter-laboratory comparison is a suitable tool both for the validation of internal testing methods and for the assessment of measurements traceability; nevertheless it is only one of the requirements for the qualification.
- Proficiency testing requires time and money which will contribute to the increase of the management costs of the laboratories. This is against the demand from industry for low test pricing.
- The cost of the initial accreditation is about 50,000 €; the maintenance of the accreditation may cost about 10,000 €/year. Such costs need to be covered by a research laboratory through reducing the already scarce research budget. The new trend for the University Laboratories is to use their resources also to offer services to industry, which is requiring a third party certification of the laboratory.
- The funding of research centres may be cut during economically difficult times, irrespective of whether they are public or private. This in turn puts pressure on the costs of test programmes, especially for seismic qualification purposes.
- Current standardized test methods tend to impose a generic set of requirements that are not necessarily relevant for the particular test specimen in question. Thus, unnecessary tests might be conducted, posing unnecessary costs and demands on the specimen. Alternatively, the standardized test programme might not exercise the specimen in a seismic range that it might actually be exposed to in service. The emergence of performance based engineering (PBE) approaches in earthquake engineering offers an opportunity to rethink laboratory test approaches by tailoring the test programme to achieved performance outcomes that are explicitly defined for the particular test specimen. The standardization would be in the generic process that establishes and justifies the performance outcomes and the test programme that achieves these; the test programme could still default to existing standards if appropriate, but this would be come about as the result of an explicit, rational decision. Such a PBE approach opens the door to cost efficiencies and better focused test programmes. It would also be consistent philosophically with the PBE methods that are gradually being adopted for structural design. However, these concepts are still novel and require detailed development and evaluation before they can be promoted and adopted as an alternative to the current standardized approaches. They would also involve wider training in PBE thinking and approaches amongst clients, consultants and test engineers.

Moreover the following remark should be considered:

- The European cooperation for Accreditation (EA) manages a peer evaluation system among national accreditation bodies both from the EU Member States and from other European countries; the International Laboratory Accreditation Cooperation (ILAC) together with the International Accreditation Forum (IAF) manage the mutual agreements at international level. Nevertheless some European Countries do not yet have an accreditation body.

Acknowledgements The research leading to these results has received funding from the European Community's Seventh Framework Programme [FP7/2007-2013] under grant agreement n° 227887 for the SERIES project.

Annex 1: Questionnaire for European Large Testing Facilities Laboratories

The questionnaire circulated among European Large Testing Facilities Laboratories was the following:

- State of the certification of the quality systems of Laboratories:
  - Does the Facility use quality system management?
  - If you have answered 'yes', does the Facility have any certification of the quality system management? Please specify the certification body.
  - Does the Facility have any certification for the performance of experimental tests?
  - If you have answered 'yes', please list the certified tests.

- State of the accreditation of Laboratories for seismic testing:
  - Does the Facility have any accreditation for the performance of experimental tests? If so, please specify the accreditation body.
  - Does the Facility have accreditation for the performance of any experimental tests? If so, please list the accredited tests.

- State of testing, personnel, instrumentation and equipment management procedures of the laboratories:
  - Does the Facility utilise defined procedures to which assure the performance of experimental tests? If so, please list the test procedures.
  - Does the Facility utilise defined procedures for managing personnel? If so, please list the personnel management procedures.
  - Does the Facility utilise defined procedures for instrumentation management? If so, please list the instrumentation management procedures.
  - Does the Facility utilise defined procedures for equipment management? Please list the equipment management procedures.

14 Evaluation and Impact of Qualification of Experimental Facilities in Europe

- Evaluation of time and cost which a laboratory should spend to fulfil the accreditation requirements:

  - What amount of time is deemed reasonable for the implementation of the accreditation requirements?
  - How large is the workforce that could be dedicated towards the implementation of the accreditation requirements?
  - Please indicate an estimate of the maximum financial cost deemed reasonable for the implementation of the accreditation requirements.

- Identification of the adopted Standards for the performance of experimental seismic testing:
- Identification of the adopted Standards for the management of experimental facilities, measuring equipment, data acquisition systems:

  - Does the Facility adopt Standards for the performance of experimental seismic testing?
  - If the Facility adopts Standards for the performance of experimental seismic testing, please list the adopted Standards.
  - Does the Facility adopt Standards for the management of experimental facilities (shaking table, actuators, pipelines, servo-controllers, pumps, power supply units)?
  - If the Facility adopts Standards for the management of experimental facilities, please list the adopted Standards.
  - Does the Facility adopt Standards for the management of measuring equipment (measurement transducers, cables, signal conditioners and amplifiers, filters)?
  - If the Facility adopts Standards for the management of measuring equipment, please list the adopted Standards.
  - Does the Facility adopt Standards for the management of data acquisition systems?
  - If the Facility adopts Standards for the management of data acquisition systems, please list the adopted Standards.
  - Does the Facility adopt Standards for data acquisition, recording and storage?
  - If the Facility adopts Standards for the management of data acquisition, recording and storage, please list the adopted Standards.

- Identification of the adopted Standards for data acquisition and processing, including frequency analysis, statistics, modal analysis:

  - Does the Facility adopt Standards for data processing, including frequency analysis, statistics, and modal analysis?
  - If the Facility adopts Standards for data processing, please list the adopted Standards.

- Demand of qualification by structural laboratories:

  - Do you have a demand for qualification?
  - Have you perceived a demand for qualification from someone else (industry, university, and state owned institutions)?
  - If so, are any particular Standards in demand? If so, please specify.

Moreover some specific questions were asked dealing with the management requirements:

- Organization:

  - Is the Facility legally independent or is it a part of an independent institution?
  - Can the Facility be held legally responsible?
  - Does the Facility have the freedom to carry out its activities (e.g. testing, reporting) without seeking the approval of other parties (e.g. holding companies)?
  - Please indicate which kind of activities are performed in the Facility:
  - Research
  - Services to industry (tests and/or consultancy)
  - Development tests for the holding company.
  - Do the managerial and technical personnel have the authority and resources needed to carry out their duties?
  - Are the managerial and technical personnel free from any undue internal and external commercial, financial and other pressures and influences?
  - Does the Facility have policies and procedures to ensure the protection of its customer's confidential information and proprietary rights, including procedures for protecting the electronic storage and transmission of results?
  - Has the Facility specified the responsibility, authority and interrelationships of all personnel?
  - Does the technical management have overall responsibility for the technical operations and the provision of the resources?
  - Does the Facility have a quality manager who, irrespective of other duties and responsibilities, has defined responsibility and authority for ensuring that the quality management system is implemented and followed at all times?

- Management system:

  - Does the Facility have a quality management system manual?
  - Has the top management of the Facility issued a quality policy statement?
  - Are the roles and responsibilities of the technical management and the quality manager defined in the quality manual or in another system document?

- Document control:

  - Does the Facility have procedures to control all documents that are part of its management system?

14 Evaluation and Impact of Qualification of Experimental Facilities in Europe

- Has the Facility defined the responsibilities for the issue, review and approval of the documents?
- Does the Facility uniquely identify a document's:
  - Date of issue
  - Revision identification
  - Page numbering
  - Total number of pages
  - Issuing authorities?

- Review of requests, tenders and contracts:
  - Does the Facility have a procedure for the review of requests, tenders and contracts?
  - Does the Facility assure that:
    - The requirements of the customer are adequately defined, documented and understood?
    - It has the capability and resources to meet the requirements?
    - The appropriate test method is selected and that the Facility is capable of meeting the customer's requirements?

- Subcontracting of tests and calibrations:
  - Does the Facility subcontract work?
  - Does the Facility advise the customer of the subcontract and does it obtain the approval of the customer?

- Purchasing services and supplies:
  - Does the Facility have procedures for the purchasing and acceptance of:
    - Equipment and instrumentation?
    - Consumable materials?
    - Services such as maintenance, calibration, consultancy?

- Service to the customer:
  - Is access allowed by the Facility to the customer during the tests?
  - Does the Facility seek feedback from the customers to improve the management system, testing and calibration activities and customer service?

- Complaints:
  - Does the Facility have a policy and procedures for the resolution of complaints received by the customers or other parties?

- Control of nonconforming testing and/or calibration work:
  - Does the Facility have the policy and procedures to deal with any aspect of its testing work that does not conform to the Facility's procedures or the agreed requirements of the customer?

- Improvement:
  - Does the Facility continually improve the effectiveness of its management system?
- Corrective action:
  - Does the Facility have a procedure for implementing corrective actions?
- Preventive action:
  - Does the Facility have a plan to monitor test quality to allow corrective actions to be taken in advance when necessary?
- Control of records:
  - Does the Facility have procedures for:
    - Identification
    - Collection
    - Indexing
    - Access
    - Filing
    - Storage
    - Maintenance
    - and Disposal of quality and technical records?
  - Does the Facility have a procedure such that documents and recordings of the test activities allow traceability of equipment, materials, personnel, environmental conditions and performed controls?
- Internal audits:
  - Does the Facility conduct internal audits of its activities to verify the compliance with the requirements of the management system?
- Management reviews:
  - Does the top management of the Facility conduct a periodic review of the management system and of the testing activities?

Annex 2: Questionnaire for Certification and Accreditation Organizations

The questionnaire circulated among Certification and Accreditation Organizations of the EC Countries was the following:

- General requirements for the accreditation of Laboratories with Large Test Facilities in the seismic field;
  - Are there any Standards that cover the certification of a Facility's quality management system? If so, please specify.

14 Evaluation and Impact of Qualification of Experimental Facilities in Europe

- Are there any Standards that cover the accreditation of Facilities? If so, please specify.
- Are there any recognised procedures for assessing experimental testing performance? If so, please specify.
- Are there any recognised procedures for personnel management? If so, please specify.
- Are there any recognised procedures for instrumentation management? If so, please specify.
- Are there any recognised procedures for equipment management? If so, please specify.
- Are there any recognised procedures for data management (acquisition, recording, storage and processing)? If so, please specify.

- Evaluation of main activities which should be performed by a Testing laboratory to get the accreditation to conduct seismic testing;

  - Are there any Standards for assessing the performance of experimental seismic testing? If so, please list the existing Standards.
  - Are there any Standards for the management of experimental facilities (shaking table, actuators, hydraulic hoses, servo-controllers, pumps, power supply units)? If so, please list the existing Standards.
  - Are there any Standards for the management of the measuring equipment (measurement transducers, cables, signal conditioners and amplifiers, filters)? If so, please list the existing Standards.
  - Are there any Standards for the management of data acquisition systems? If so, please list the existing Standards.
  - Are there any Standards for data acquisition, recording and storage? If so, please list the existing Standards.
  - Are there any Standards for data processing, including frequency analysis, statistics, modal analysis? If so, please list the existing Standards.

- Evaluation of time and cost which a laboratory should spend to perform the formal activities requested by the accreditation organization, such as meetings, inspection visits and other related activities;

  - Please provide an estimate of the time deemed necessary for the implementation of the accreditation requirements.
  - Please indicate an estimate of the total expected costs for the implementation of the accreditation requirements.
  - If a Facility has a demand for qualification, the reference standard could be ISO/IEC 17025. Do other Standards cover the accreditation of a Facility that conducts non-standard testing and research at large scales? If so, please specify.

- Additional requirements related to the organization;

  - Must the Facility be a legally independent institution or a part of an independent institution?

- May the Facility perform the following activities: Research? Services to industry (tests and consultancy)? Development tests for the holding company?
- May the Facility share managerial and technical personnel with the holding company or institution?
- May the Facility have a quality manager who covers other duties and responsibilities in either the Facility itself or in the holding company?

• Additional technical requirements related to the personnel:

- May the large testing facility make use of temporarily contracted personnel?

Annex 3: Questionnaire for Standardization Organizations

The questionnaire circulated among Standardization Organizations of the EC Countries was the following:

• Standards for the performance of experimental seismic testing;

- Are there any Standards which cover the qualification of seismic test facilities? If so, please specify.
- Are there any Standards which cover the accreditation of seismic test facilities? If so, please specify.
- Are there any Standards which cover seismic testing? If so, please specify.

• Standards for the management of experimental facilities, measuring equipment, data acquisition systems:

- Does any Standard exist for the management of experimental facilities (e.g. shaking table, actuators, hydraulic hoses, servo-controllers, pumps, power supply units)? If so, please specify.
- Are there any Standards that cover the management of measuring equipment (e.g. measurement transducers, cables, signal conditioners and amplifiers, filters)? If so, please specify.
- Are there any Standards that cover the management of data acquisition systems? If so, please specify.

• Standards for data acquisition and processing, including frequency analysis, statistics, modal analysis:

- Are there any Standards that cover data acquisition, recording and storage? If so, please specify.
- Are there any Standards covering data processing, including frequency analysis, statistics, modal analysis? If so, please specify.

14 Evaluation and Impact of Qualification of Experimental Facilities in Europe 283

Annex 4: Questionnaire for European Industry Organizations

The questionnaire circulated among Industry Organizations of the EC Countries was the following:

- Problems that limit the free circulation of the products of the European Industry:

  - Does the industry require Facilities with quality management system?
  - Does the industry need Facilities having certification of the quality management system?
  - Does the industry need Facilities having certification for the performance of seismic tests?

- Demand of qualification of structural laboratories to ease the free circulation of the products:

  - Does the industry need Facilities accredited for seismic testing?
  - Does the industry recognise any procedures for the seismic testing? If so, please specify.
  - Does the industry employ Standards that cover seismic testing? If so, please specify.
  - Does the industry have a demand for the qualification of Facilities?
  - Does the industry perceive a demand for Facility qualification from an external party (final user, state owned institutions, other)?
  - If the industry has a demand for the qualification of Facilities, please note the reference Standards.
  - Does the industry need legally independent Facilities or Facilities that are part of manufacturing and construction companies?
  - Please indicate the kind of activities which you contract to Facilities:

    - Research
    - Services to industry (testing and/or consultancy)
    - Other (specify)

  - Is it important whether the Facility has policies and procedures to ensure the protection of its customer's confidential information and proprietary rights, including procedures for protecting the electronic storage and transmission of results?
  - Is it important whether the Facility has a quality management system manual?
  - Is it important whether the Facility assures that:

    - the requirements of the customer are adequately defined, documented and understood;
    - it has the capability and resources to meet those requirements;

- the appropriate test method is selected and
- the Facility is capable of meeting the customer's requirements?

- Is it important whether the Facility has well-defined test and calibration methods?
- Is it important whether the Facility has technical procedures on the use and operation of all relevant equipment, and on the handling and preparation of items for testing and calibration?
- Are standardized test methods preferred?
- Does the Industry require the Facility to address the uncertainty of test measurements?
- Please give some examples of specimens that you typically test or would like to test.
- Do you demand that a Facility adopts Standardised test methods? If so, please list the reference Standards.
- What type of tests have you asked a Facility to perform (static test, pseudo-dynamic test, sinusoidal vibrations, multi-frequency vibrations, etc.)?
- What measurements are typically taken during the course of the test?
- What is the measurement range?
- Do you demand that a Facility quantifies the measurement uncertainty in accordance to ENV 13005?
- Do you require that a Facility adopts calibration procedures which state personnel responsibilities and calibration time intervals?
- Do you request a Facility to express its own opinions and interpretations of a test within the test report?

- Standards for data acquisition and processing, including frequency analysis, statistics, modal analysis:

- Does the industry employ Standards for equipment management? If so, please specify.
- Does the industry employ Standards that cover the management of measuring equipment (calibration)? If so, please specify.
- Does the industry employ Standards for data management (acquisition, recording, storage and processing)? If so, please specify.

References

EC No 765/2008 (2008) REGULATION (EC) No 765/2008 of the European Parliament and of the Council of 9 July 2008 setting out the requirements for accreditation and market surveillance relating to the marketing of products and repealing Regulation (EEC) No 339/93. European Commission, Brussels

EN ISO 9000 (2000) Quality management systems – fundamentals and vocabulary. European Committee of Standardisation, Brussels

EN ISO 9001 (2008) Quality management systems – requirements. European Committee of Standardisation, Brussels

14 Evaluation and Impact of Qualification of Experimental Facilities in Europe 285

ISO 3534–1 (1993) Statistics – vocabulary and symbols – Part 1: probability and general statistical terms. International Standards Organisation, Geneva

ISO/IEC 17025 (2005) General requirements for the competence of testing and calibration laboratories. International Standards Organisation, Geneva

Kurç Ö, Sucuoğlu H, Molinari M, Zanon G, Ohrid MK (2012) Qualification of large testing facilities in earthquake engineering research. In: Fardis MN, Radikevic Z (eds) Proceedings of SERIES workshop: role of research infrastructures in performance-based earthquake engineering. Springer, Dordrecht

Chapter 15
Qualification of Large Testing Facilities in Earthquake Engineering Research

Özgür Kurç, Haluk Sucuoğlu, Marco Molinari, and Gabriele Zanon

Abstract The qualification of large testing facilities serving in earthquake engineering research and the management of the instrumentation are the central issues of this paper. The main aspects of instrumentation management and seismic test methods are presented, commented upon and critically assessed based on both the answers to a questionnaire and the outcomes of the SERIES Workshop about the Qualification of Facilities held in Ohrid, MK. On the basis of the results presented herein and the proposals and comments formulated at the Ohrid workshop, a roadmap for the qualification of earthquake research laboratories will be drawn up.

15.1 Introduction

One of the aims of the FP7 Project SERIES (Seismic Engineering Research Infrastructures for European Synergies) is to create the conditions leading to the qualification of Structural Testing Laboratories specializing in earthquake engineering which are equipped for large scale testing. A reference document, recognized as the

---

Ö. Kurç (✉) • H. Sucuoğlu
Department of Civil Engineering, Middle East Technical University,
Inonu Bulvari, Campus, Ankara 06531, Turkey
e-mail: kurc@metu.edu.tr; sucuoglu@ce.metu.edu.tr

M. Molinari • G. Zanon
Department of Mechanical and Structural Engineering, University of Trento,
Via Mesiano 77, 38100 Trento, Italy
e-mail: marco.molinari@ing.unitn.it; gabriele.zanon@ing.unitn.it

M.N. Fardis and Z.T. Rakicevic (eds.), *Role of Seismic Testing Facilities in Performance-Based Earthquake Engineering: SERIES Workshop*, Geotechnical, Geological and Earthquake Engineering 22, DOI 10.1007/978-94-007-1977-4_15,
© Springer Science+Business Media B.V. 2012

best one to follow in order to draw a roadmap for the qualification of Laboratories, is ISO/IEC 17025:2005, to which it is necessary to refer in order to obtain the accreditation of Laboratory for certain tests. This norm is strictly connected to ISO/EN 9001:2008 for aspects relevant to management criteria. In accordance to the aforementioned norm, qualification implies technical competence and quality assurance, necessary to guarantee the reliability of testing in each laboratory. Reliability is translated into practice in terms of:

- *Repeatability*: the principle that experimental activities repeated on the same specimen in the same laboratory lead to the same results, and
- *Reproducibility*: the principle that experimental activities repeated on the same specimen in different laboratories lead to the same results.

Repeatability and reproducibility can be easily achieved if common standard test procedures are used. Moreover a common language can be established on the basis of international standards which could help avoiding mismatches and misunderstandings.

The activities regarding qualification of Structural Testing Laboratories specializing in earthquake engineering fall under four tasks:

1. Evaluation and impact of qualification of experimental facilities in Europe
2. Assessment of testing procedures and standards requirements
3. Criteria for instrumentation and equipment management
4. Development and implementation of a common protocol for qualification

This paper mainly covers all the tasks but the first one, which is covered in a companion paper (Zola and Taylor 2012) in this Volume. It focuses on facilities endowed with a shaking table or a reaction wall, i.e. so-called large seismic testing facilities. Their special character and relatively small number, the differences in technical solutions and the wide variety of testing procedures followed by their operators – often developed and implemented by the personnel – do not lend themselves to fully harmonized approaches. Quasi-static testing is also considered in the scope of the study since a significant portion of the experimental data related to earthquake engineering are still obtained with this method, despite the fact that it lacks the definition of seismic demand. Moreover, problems related to instrumentation management – i.e. storage, calibration, identification in tests – are common to quasi-static and dynamic tests.

The studies carried out for achieving the stated objectives were organized in terms of the following activities:

- Identification of European Large Testing Facilities with shaking tables, reaction walls, on-site testing and large structures monitoring capabilities.
- Collection of information about instrumentation management and testing procedures, checking the compliance of the procedures with the requirements of the applicable standards (ISO/IEC 17025) (ISO/IEC 17025 2005) or reference recommendations.
- Comparative analysis and critical assessment of presented procedures.

Information collected from the European Large Testing Facilities are compiled, analyzed and critically assessed in the following sections. Finally, a roadmap is suggested for the qualification of earthquake engineering testing facilities and instrumentation management.

15.2 Collection of Information

An internationally recognized standard for seismic testing qualification procedures does not exist and a simple inquiry of this aspect is not feasible. Therefore, the first step in view of the definition of a qualification roadmap was the collection of information relevant to testing and management procedures among important European large testing facilities by means of questionnaires. To save time and optimize the compilation, only four questionnaires simultaneously covering the main issues of the abovementioned tasks 1, 2 and 3 were distributed. The collected answers were processed with reference to the specific parts dealing with the relevant tasks.

Answers related to instrumentation management were mainly relevant to appropriateness, maintenance and calibration. Questions related to the testing procedures mainly focused on identifying the employed test, equipment utilized, quantities measured, personnel in charge of testing and the manner in which tests are carried out. The conclusions of the SERIES workshop in Ohrid, MK, of September 2, 2010, highlighted in (Zola and Taylor 2012), were also taken into account.

15.2.1 Identification of European Large Testing Facilities

Thirty-two testing facilities – mostly from European countries, with a high percentage from the most seismic areas – answered the questionnaires. The list of the facilities is presented below:

International Institutions

- EU-JRC – ELSA Laboratory

Belgium

- University of Liège – Laboratory of Mechanics of Materials and Structures

France

- CEA Saclay – Tamaris Laboratory
- LCPC – Laboratoire Central des Ponts et Chaussées

Germany

- University of Kassel – Department of Civil Engineering

Greece

- University of Patras – Department of Civil Engineering, Structures Laboratory
- National Technical University of Athens – Laboratory for Earthquake Engineering
- Aristotle University – Laboratory of Geotechnical Earthquake Engineering and Soil Dynamics
- LABOR S.A. – Certification Laboratory

Italy

- EUCENTRE, Università di Pavia
- CESI – LPS Laboratory – Seriate
- University of Trento – Laboratory for material and structural testing
- University of Naples Federico II – Structural Engineering Department
- Alga S.p.A
- FIP Industriale – Research and Development Department
- PeP LMC Srl
- GEWISS S.p.A

Portugal

- FEUP – Faculdade de Engenharia da Universidade do Porto
- IST – Instituto Superior Técnico
- UA – Universidade de Aveiro
- UM – Universidade do Minho
- FCT – Faculdade de Ciências e Tecnologia da Universidade Nova de Lisboa
- LNEC – Laboratório Nacional de Engenharia Civil

Republic of Macedonia

- IZIIS – Dynamic Testing Laboratory

Romania

- Technical University Gheorghe Asachi of Iasi – Faculty of Civil Engineering
- INCERC Branch of Iasi

Turkey

- Middle East Technical University – Structural and Earthquake Engineering Laboratory
- Istanbul Technical University – Structural and Earthquake Engineering Laboratory
- Bogazici University – Department of Civil Engineering, Structures Laboratory

UK

- University of Bristol
- University of Cambridge
- Cardiff School of Engineering

15.2.2 European Large Testing Facility Questionnaire: Instrumentation Management

The questionnaire circulated among European Large Testing Laboratories included the following items regarding instrumentation management:

Qualification state: state of the procedures

(a) Does the Facility utilise defined procedures to assure the performance of experimental tests? If so, please list the test procedures. What test method is typically employed (standard, internal)?
(b) Does the Facility utilise defined procedures for instrumentation management? If so, please list the instrumentation management procedures. What parameters are typically measured?
(c) Does the Facility utilise defined procedures for equipment management? Please list the equipment management procedures.

Qualification state: standards

(a) Does the Facility adopt Standards for the management of measuring equipment (measurement transducers, cables, signal conditioners and amplifiers, filters)?

Technical requirements: accommodation and environmental conditions

(a) Are the testing and calibration, lighting and environmental conditions appropriate for the correct performance of the tests?

Technical requirements: equipment

(a) Is the deployed equipment uniquely identified as follows?

  - Name and description of the equipment and its software;
  - Name of the manufacturer; model number and serial number;
  - Initial check with respect to the purchasing requirements;
  - Date of receipt and date of start up;
  - The current location, where appropriate.

(b) Is the maintenance of all the employed equipment specified as follows?

  - Manufacturer's instructions;
  - The maintenance plan and maintenance carried out to date;
  - Frequency of maintenance;
  - Log of damage, malfunction, modification or repair.

(c) Is the calibration of all the employed equipment recorded as follows?

  - Dates; results; copies of reports; certificates of all calibrations; due date of next calibration.

15.2.3 European Large Testing Facility Questionnaire: Testing Procedures

The questionnaire circulated among European Large Testing Laboratories included the following items regarding testing procedures:

Identification of the test

(a) What types of items are typically tested?
(b) What test method is typically employed (standard, internal)?
(c) What type of test is typically performed (static test, pseudo-dynamic test, sinusoidal vibrations, multi-frequency vibrations, etc.)?
(d) What parameters are typically measured?
(e) What is the typical measurement range?

Personnel in charge of the test

(a) What are the typical qualifications of personnel (study degree, years of experience)?

Test methods

(a) Is the test method a standard?
(b) Does the facility have its own test procedures?
(c) Does the facility verify its own tests repeatability?
(d) Does the Facility have a procedure for the evaluation of measurements uncertainty according to ENV 13005?

Test performance

(a) Does the Facility verify that the testing equipment comply with the requirements of the test standard or method?
(b) Are the raw data traceable starting from the test report?

Test report

(a) When opinions and interpretations are included, are they clearly marked as such in the test report?

15.3 Analysis of Data from Laboratory Responses: Instrumentation Management

Table 15.1 summarizes, with data in percentages, the responses to questions related to maintenance and calibration of instrumentation. Problems relevant to instrumentation management are common to structural laboratories notwithstanding the different facilities, complexity of performed tests and customers, public or private.

15 Qualification of Large Testing Facilities in Earthquake Engineering Research 293

Table 15.1 Responses to questions related to maintenance and calibration of instrumentation

Question	Yes (%)	No (%)
Does the facility utilise defined procedures which assure the performance of experimental tests?	55	45
Does the facility utilise defined procedures for instrumentation management?	72	28
Does the facility utilise defined procedures for equipment management?	66	34
Does the facility adopt Standards for the management of measuring equipment?	45	55
Are the testing and calibration, lighting and environmental conditions appropriate for the correct performance of the tests?	100	0
Is the measurement equipment appropriate for the type of tests you are conducting?	100	0
Is the deployed equipment uniquely identified as name and description of the equipment and its software?	93	7
Is the deployed equipment uniquely identified as name of the manufacturer, model number and serial number?	86	14
Is the deployed equipment uniquely identified as initial check in with respect to the purchasing requirements?	86	14
Is the deployed equipment uniquely identified as date of receipt and date of start up?	93	7
Is the deployed equipment uniquely identified as current location, where appropriate?	90	10
Is the maintenance of all the employed equipment specified as manufacturer's instructions?	86	14
Is the maintenance of all the employed equipment specified as the maintenance plan and maintenance carried out to date?	83	17
Is the maintenance of all the employed equipment specified as frequency of maintenance?	86	14
Is the maintenance of all the employed equipment specified as log of damage, malfunction, modification or repair?	83	17
Is the calibration of all the employed equipment recorded as dates, results, copies of reports, certificates of calibrations, due date of next calibration?	86	14

15.3.1 Evaluation of Responses

Answers show the perception of large seismic testing laboratories managers to own adequate facilities, instrumentation and procedures for performing a correct testing activity. The calibration of the acquisition instrumentation is periodically checked and the relevant database is maintained and updated. Moreover, most of the laboratories apply procedures for the management of measuring instrumentation and equipment. A lower percentage asserts to possess procedures for assuring the properness of experimental tests: however, only slightly more than half of the laboratories declare the utilized procedure and less than 10% refer to standard procedures.

A contradiction in the answers exists, since on one side tests are performed by experienced scientists of very high level and there are no doubts about the reliability of results; on the other side, only few laboratories refer to written procedures to this aim, and a minimum number to standards. It is interesting to look at the causes, as a result of the discussion at the Workshop in Ohrid.

One major problem is related to the cost of implementation and maintenance of the management quality system, both financial and in terms of the personnel required for this aim. Some researchers believe that only laboratories performing tests for private companies have the possibility to attract the necessary funding, unlike purely research institutions.

With reference to standards, it should be noted that research tests in particular are not easy to standardize, since they often involve numerical methods and instrumentation developed by the Laboratories themselves. Conversely, several reference standards are available for testing electrical equipment and some are applied for mechanical systems and components.

15.4 Analysis of Data from Laboratory Responses: Testing System

The collected data were further analyzed by a categorization with respect to the tested items, testing equipment, type of test, and laboratory personnel. Responses to questionnaires are compiled in a table which is organized in line with the questions listed above.

15.4.1 Tested Items

The first part of the questionnaire about the testing procedures targeted at identifying the items that are being tested at European Large Testing Facilities, as reported in the summary presented below. The items are first classified according to the industry to which they belong and then according to their structural behaviour.

Civil Structures; System Level

- Large/full scale structural systems
- Shelter
- RC Mock-up
- Sub-assemblages
- Frames with/without infill walls

Civil Structures; Component Level

- Beams, columns, slabs and connections
- Composite members
- Steel bracing (bars, metal sheets)

15 Qualification of Large Testing Facilities in Earthquake Engineering Research

- RC and masonry walls
- Composite timber-fiber concrete walls
- Non-structural components

Civil Structures, Special Devices

- Bridge Bearings
- Seismic Isolators

Geotechnical Structures

- Soil Specimens
- Earthworks
- Structures on Soils

Electrical Structures

- Electrical cabinets
- Medium-high voltage electrical equipment
- Electromechanical cubicles
- Telecommunication equipment

Mechanical Structures

- Mechanical Components
- Valves

Other Industrial Systems

- Automotive, nuclear, aeronautical, oil industry, medical industry.

A very wide range of items is tested in the European Large Testing Facilities, including items for the civil engineering industry and components for the electrical, oil and nuclear industry. These items are not only single components such as columns, beams or other devices, but also a combination of various components such as frames and sub-assemblies.

15.4.2 Test Equipment

Another important identification category is the test equipment and testing capabilities related to the employed equipment. The current test equipment at the European Large Testing Facilities can be grouped under five basic categories such as:

- Reaction wall with static/quasi-static test equipment
- Reaction wall with pseudo-dynamic test equipment
- Shake table with uni-axial or multi-axial excitation
- Centrifuge with sinusoidal or random dynamic excitation
- On-site testing with related equipment.

15.4.3 Type of Conducted Tests

Different types of tests are conducted at the facilities in order to understand the seismic behaviour of items. The test types depend on the items and on the experimental capabilities and equipment of the testing facility. According to the responses to the questionnaires, several types of tests were identified, as follows:

- Testing apparatus:

  - Single axis or multi axes shaking table, for controlled excitation through the base;
  - Concentrated forces, for distributed and controlled point excitation (reaction wall tests, both static or dynamic or pseudo-dynamic, on site tests with inertial exciters or with reaction structures);
  - Ambient vibration, for uncontrolled excitation (mainly on site tests).

- Excitation types:

  - Harmonic excitation (fixed frequency sinusoid, sweeping frequency sinusoid, stepped frequency sinusoid);
  - Stationary random excitation (white noise, wide frequency band vibration, narrow frequency band vibration);
  - Transient excitation (seismic-like excitation, sine-beat excitation, impulsive or shock excitation, quick release free oscillation);
  - Fatigue excitation (low cycle fatigue, high cycle fatigue, spectrum fatigue);
  - Static or quasi-static tests.

- Modal testing.

The last part of the questions related to the test methods focused on the measured quantities and their range. The laboratories that perform static and pseudo-dynamic tests mostly acquire force, displacement, strain, crack opening and rotations. Those endowed with real-time dynamic testing capability, such as shake tables, and with on-site testing equipment, acquire accelerations, velocities, and frequencies. Pressure and pore pressure are measured in centrifuges.

15.4.4 Laboratory Personnel

The personnel in charge of testing is another important aspect that affects the quality of tests. According to the questionnaires, highly educated and experienced engineers and academics perform the tests: when laboratories belong to academic institutions, usually academics with a PhD together with graduate students and/or technicians are in charge. The summary of the responses regarding the personnel is presented below.

- Graduate Students
- Engineers – BS Degree

15 Qualification of Large Testing Facilities in Earthquake Engineering Research

- Engineers – MS Degree
- Engineers – PhD Degree (5–20 years of experience)
- Technicians (3–10 years of experience)
- Academics, PhD (10–30 years of experience).

15.5   Response to Questionnaires – Testing Procedures

Table 15.2 summarizes the responses to questions related to test methods, test performance, maintenance, calibration and test reports (presented in percentages).

15.5.1   Evaluation of Responses – Testing Procedures

The summary table (Table 15.2) reveals that most of the experiments carried out in large seismic testing laboratories do not follow standard test methods, but rather that laboratories apply their own test procedures. In this case, they also verify the repeatability of results by comparing the experimental results with corresponding estimations of analytical simulations or by repeating low amplitude tests. Ten percent of the laboratories verify repeatability on a single test specimen only, instead of on all specimens.

Few laboratories only perform tests according to common, internationally accepted procedures or standards. Several testing standards and guidelines exist for commercial products such as electrical relays (EN 60255-21-3 1996) and circuit breakers (IEC1166 1993). Most of seismic testing standards and guidelines for

Table 15.2 Responses to questions related to test methods, test performance, maintenance, calibration and test reports

Question	Yes (%)	Sometimes (%)	No (%)	N/A (%)
Is the test method a standard?	30	24	42	4
Does the facility has its own test procedures?	81	–	19	–
Does the facility verify its own test repeatability?	78	10	12	–
Does the facility have a procedure for the evaluation of measurements uncertainty according to ENV 13005?	–	12	88	–
Does the facility verify that the testing equipment comply with the requirements of the test standard or method?	88	–	12	–
Are the raw data traceable starting from the test report?	85	12	3	–
When opinions and interpretations are included, are they clearly marked as such in the test report?	87	–	13	–

structural components are relevant to specific items such as steel connections or elastomeric isolators and define procedures for a single test type: as an example, ECCS (1996) and ES 2126-02a (2003) define loading protocols for cyclic testing of steel elements and seismic walls of buildings, respectively. Moreover, the majority of the test standards are limited to structural components rather than to systems. Conversely, tests protocols for systems are defined only for electrical and non-structural products. The list of the standards and guidelines for seismic testing of various items is presented in (ECCS 1996; ATC-24 1992; ISO/TC 165 WD 16670 1998; FEMA 461 2007; ES 2126-02a 2003; ANSI/AISC 341-05 2005; ICC-ES AC156 2010; ISO 22762-1 2005; ISO 2762; ISO 22762-3 2005) for structural components, and in (IEC1166 1993; IEC/TS 61463 2000; EN 60255-21-3 1996; EN 61587-2 2001; IEC/TR 62271-300 2006; IEC 60980 1989; IEEE/ANSI Standard 344 2005; ANSI/IEEE C37.98-1987 1987) for electrical products, respectively.

Although measurement uncertainties are not evaluated according to a standard, such as ENV-13005 (UNI CEI ENV 13005 2000), the testing equipment is declared to comply with the test method. Further, the test equipment is regularly maintained and mostly periodically calibrated.

15.6 ISO/IEC 17025 and Answers from SERIES Questionnaires

Whilst a roadmap for the qualification of seismic laboratories hasn't yet been drawn, it was stated at the Workshop held in Ohrid to consider ISO/IEC 17025 as a reference norm for the matter. This does not mean that all requirements of the norm should be fulfilled – there was concern about the excessive cost required for the accreditation, which is also limited to specific tests and does not fit with the need for freedom and innovation required by research laboratories. However, the basic aspects of laboratory qualification are clearly codified: the need for a management system of instrumentation, the requirement to prove reliability and repeatability, the distinction between data results and their interpretation are probably the more important, as commented in this paragraph.

Earthquake engineering laboratories carry out seismic tests with the following motivations, and related sources of funding:

(a) Verification testing of a commercial product with industry funding.
(b) Confirmation of code procedures for providing public safety, with public funding.
(c) Pure research and development related testing, mostly with public funding.

Commercial products include factory-produced or prefabricated structural components such as steel and timber members; non-structural components such as exterior cladding, curtain walls and partitions; and several seismic devices such as seismic isolators, energy absorbers and bridge bearings. They also include electrical and mechanical products. Guidelines and specifications for the certification of these products are well established (ECCS 1996; ATC-24 1992; ISO/TC 165 WD 16670 1998; FEMA 461 2007; ES 2126-02a 2003; ANSI/AISC 341-05 2005; ICC-ES AC156 2010; ISO 22762-1 2005; ISO 2762; ISO 22762-3 2005; IEC1166 1993; IEC/TS 61463 2000; EN 60255-21-3 1996; EN 61587-2 2001; IEC/TR 62271-300 2006; IEC 60980 1989; IEEE/ANSI Standard 344 2005; ANSI/IEEE C37.98-1987 1987) where the producer stands as the client and the laboratory is the service provider. Laboratories are asked by the client to possess an accreditation certificate for eligibility in verification testing. Accreditation does not guarantee the accuracy of the results, but is related to the appropriateness of the testing procedures employed. When accreditation is performed according to 17025, an accuracy assessment is further required through the evaluation of the uncertainty of measurements. However the accuracy implied herein is basically related to the technical quality of the testing procedure employed. The cost of accreditation varies from one country to another; however, it is usually a significant amount which can only be provided by a laboratory which has income from commercial testing.

Commercial motivation does not prevail in the other two categories of testing. Most of the laboratories which conduct seismic testing predominantly for research purposes do not apply to an accreditation agency for accreditation according to ISO/IEC 17025 or any other standard simply because nobody asks for such an accreditation, and probably because they do not have available resources to meet the cost of accreditation.

According to the results of the questionnaires, the existing test procedures in various standards and guidelines usually focused on a single component type (such as beam-column connection test) and a single type of experiment for a specific purpose. On the other hand, laboratories have a wide range and variety of test equipment for performing seismic tests. They perform several tests from simple static tests to more complex dynamic tests. The items that are tested are not only single components such as beams, columns and walls but also sub-assemblies and structures that involve a combination of components. The physical parameters that are measured or investigated also change depending on the type of items that are being tested and the aim of the test. Thus, in view of the responses of many large testing facilities in Europe, a single standard is not sufficient for understanding, defining and predicting the seismic performance of components, devices or structural systems. Because of this reason, the majority of laboratories have their own testing procedures which depend on their testing capabilities and on the items they test.

ISO/IEC 17025 covers most of the issues that are required for the qualification of a testing facility for technical competence. Technical competence guarantees the reliability of tests conducted in the facility. The reliability of testing laboratories can be achieved through repeatability and reproducibility of an experimental activity. Repeatability and reproducibility can be easily achieved if common standard test procedures are used. Moreover, a common language can be established on the basis of international standards which could help avoid mismatches and misunderstandings. For seismic testing, however, it is not possible to have a single standard. For each specific item and for each test technique, a different standard must be produced.

As a long term project, however, internationally accepted standards for each type of seismic test equipment are required for the qualification of laboratories. There are several international standards for quasi-static testing of various components but standards for dynamic tests such as pseudo-dynamic and shake table testing have not been produced yet. As a starting point, a common standard that focuses on the general aspects of dynamic testing for producing reliable test results can be established.

According to the results of the questionnaires, most of the laboratories that conduct their own test methods confirm repeatability of their testing procedures, but reproducibility is not assessed for most of their experimental activity. ISO/IEC 17025 obligates the validation of a non-standard or laboratory designed/developed method. Among various validation techniques that exist in ISO/IEC 17025, two of them have significant importance in terms of seismic testing: comparison of results achieved with other methods, and inter-laboratory comparisons. These two validation techniques are actually enforcing the reproducibility of an experimental activity which is another requirement for qualification besides repeatability. The last one requires the execution of so-called proficiency tests, i.e. tests on identical specimens performed in different Laboratories, also required by ISO/IEC 17025. Proficiency tests have sometimes a significant cost and require significant effort; hence not all laboratories agree to perform them owing to limited available time and resources. Moreover, effective comparisons are limited to elastic tests whilst results in the plastic field are somewhat dependent on the algorithm. Therefore extension to seismic tests under severe excitations appears limited.

ISO/IEC 17025 also requires that methods developed by laboratories be assigned to qualified personnel equipped with adequate resources. Responses to questionnaires revealed that seismic tests are conducted by very well-educated and highly-experienced teams of engineers and academics. The person in charge of tests is usually an engineer with at least a MS degree and 20 years of experience, or an academic with a PhD and a minimum of 10 years of experience. Engineers and technicians that perform the tests have also several years of experience in the field. As a result, all kind of seismic tests performed at European Large Testing Facilities are being conducted by highly competent personnel and experts in the area. Furthermore, the majority of laboratories have testing equipment that complies with the requirements of the test standard or method. Thus, it can be concluded that the laboratories that utilize their own testing procedure most likely comply with the requirements of ISO/IEC 17025.

According to ISO/IEC 17025, all equipment used in testing which may have a significant effect on the accuracy or validity of the results have to be calibrated before being put into service. It is also required that the laboratory have an established program and procedure for the calibration of its test instrumentation. Most of the large testing facilities had calibration procedures stating the calibration protocol and the time intervals of the calibrations. Some of them also perform calibrations between scheduled time intervals. It is mandatory to have a maintenance plan for the testing equipment in order to ensure proper functioning according to ISO/IEC 17025. It was observed that most of the laboratories had a maintenance procedure for the important testing equipment, thus fulfilling the requirements of ISO/IEC 17025. Another requirement of ISO/IEC 17025 is the estimation of uncertainty of measurements, but only a few laboratories have a procedure for evaluating the measurement uncertainty according to ENV-13005.

The way the test reports should be prepared are explicitly defined in ISO/IEC 17025. It is required that the result of each test or series of tests shall be reported accurately, clearly, unambiguously and objectively. In addition to this, when opinions and interpretations are included in the report, the laboratory shall document the basis upon which the opinions and interpretations have been made. Opinions and interpretations shall be clearly marked as such in the test report. The way test reports are prepared generally fulfils the reporting requirements of ISO/IEC 17025. Most of the laboratories include raw data and mark any opinion and interpretation as such in the test report.

15.7 Summary

The main outcomes from questionnaires gathered within the framework of the SERIES research project were presented and discussed in this paper. Answers have shown that few laboratories follow written procedures for testing and far less follow standards, especially in research testing activities which, by nature, are hard to standardize. In a similar way, few laboratories follow written procedures or quality systems for the measurement instrumentation management. Test equipment – pumps, actuators, shaking tables and reaction structures – is believed to affect test reliability to a limited extent. Moreover, it is generally adequately maintained for reliable testing. Therefore, the management of test equipment should not a major point in the management system.

Regarding the option of drawing a roadmap for the qualification of Laboratories, which was discussed during the Ohrid Workshop, a decision will be taken by partners in the following months. However, some common points have already been stated, as follows:

1. Any standard should resemble the frame of ISO/IEC 17025; the management criteria should be confirmed by an accreditation body, depending on financial resources. If they are limited, then ISO/IEC 17025 requirements could be possibly checked and reported by the laboratory itself.
2. Any reference procedure similar to ISO/IEC 17025 should fulfil the main requirements of ISO 9001 on management criteria, which is another important outcome of the questionnaires.
3. Verification of the test results by a third party established by the SERIES Consortium is preferable, but this is open to discussion depending on the duration of the SERIES Consortium and on the acceptance of third party statements by USA, Japan, China or other scientific institutions. One option is to have the SERIES Consortium stay in charge even after the project is finalized.

4. An effort to produce international standards for dynamic testing procedures should be initiated. A common standard that covers the general aspects of dynamic testing can be a good starting point.
5. Test results should be documented through the SERIES distributed database by filling all the related data fields in order to provide enough information on the reproducibility of the test results.

Implementation of the suggested qualification procedure is expected to facilitate the establishment of research consortiums for future seismic research activities, and to serve as a powerful tool for receiving research grants from international research sponsoring organizations.

Acknowledgements The research leading to these results has received funding from the European Community's Seventh Framework Programme [FP7/2007-2013] under grant agreement n° 227887 for the SERIES project.

References

ANSI/AISC 341-05 (2005) Seismic provisions for structural steel buildings, AISC, Chicago, Ill
ANSI/IEEE C37.98-1987 (1987) IEEE Standard Seismic Testing of Relays
ES 2126-02a (2003) Standard test method for cyclic (reversed) load test for shear resistance of walls for buildings. ASTM West Conshohocken, Pennsylvania
ATC-24 (1992) Guidelines for cyclic seismic testing components of steel structures for buildings. Report no. ATC-24, Applied Technology Council, Redwood City
ECCS (1996) Recommended testing procedure for assessing the behavior of structural steel elements under cyclic Loads, ECCS, N°45
EN 60255-21-3 (1996) Electrical relays – Part 21: vibration, shock, bump and seismic tests on measuring relays and protection equipment-Section 3: Seismic tests
EN 61587-2 (2001) Mechanical structures for electronic equipment – Tests for IEC 60917 and IEC 60297 – Part 2: seismic tests for cabinets and racks
FEMA 461 (2007) Interim Testing Protocols for Determining Seismic Performance Characteristics of Structural and Nonstructural Components Through Laboratory Testing, Federal Emergency Management Agency, Washington, DC
ICC-ES AC156 (2010) Seismic qualification by shake-table testing of nonstructural components and systems, ICC Evaluation Service, Approved Criteria, Falls church, VA
IEC 60980 (1989) Recommended practices for seismic qualification of electrical equipment of the safety system for nuclear generating stations
IEC/TR 62271-300 (2006) High-voltage switchgear and control gear – Part 300: seismic qualification of alternating current circuit-breakers
IEC/TS 61463 (2000) Bushings –seismic qualification. International Electrotechnical Commission, Geneva
IEC1166 (1993), First Edition, High-voltage alternating current circuit-breakers – Guide for seismic qualification of high-voltage alternating current circuit-breakers. International Electrotechnical Commission, Geneva
IEEE/ANSI Standard 344 (2005) IEEE Recommended Practice for Seismic qualification of class 1E equipment for nuclear power generating stations
ISO 2762-2 (2010) Elastomeric seismic protection isolators – Part 2: applications for bridges – specifications. ISO, Geneva
ISO 22762-1 (2010) Elastomeric seismic protection isolators – Part 1: test methods. ISO, Geneva
ISO 22762-3 (2010) Elastomeric seismic protection isolators – Part 3: applications for buildings – specifications, ISO, Geneva

15 Qualification of Large Testing Facilities in Earthquake Engineering Research

ISO/IEC 17025 (2005) General requirements for the competence of testing and calibration laboratories. ISO, Geneva

ISO/TC 165 WD 16670 (1998) Timber structures – joints made with mechanical fasteners – quasi static reversed-cyclic test method. Standards Council of Canada

UNI CEI ENV 13005 (2000) Guide to the expression of uncertainty in measurement, UNI – CEI Metrologia Generale, Italy

Zola M, Taylor C (2012) Evaluation and impact of qualification of experimental facilities in Europe. In: Fardis MN, Radikevic Z (eds) Proceedings of SERIES workshop: role of research infrastructures in performance-based earthquake engineering. Springer, Dordrecht

Chapter 16
Performance Based Seismic Qualification of Large-Class Building Equipment: An Implementation Perspective

Jeffrey Gatscher

Abstract This paper examines three aspects of nonstructural seismic qualification that become enablers for implementation of performance-based seismic design (PBSD) principles. These include: (1) identification of a nonstructural "large-class" category to address qualification of physically massive and costly equipment platforms, (2) establishment of a systems design framework for earthquake demand allocation at the subsystem level, and (3) introduction of subsystem testing techniques to qualify large-class equipment platforms by testing at the subassembly level. The fundamental premise prescribed in this paper is that implementation of PBSD approaches must be considered during foundation research investigations. Without consideration of the implementation aspect, PBSD will risk rejection by industry. The natural evolution of nonstructural earthquake protection likely points to PBSD, however understanding how to implement PBSD procedures is critical for eventual industry acceptance.

16.1 Large-Class Nonstructural Systems

The nonstructural category includes everything contained within the building skeleton that makes a building function. The building skeleton includes the beams, columns, walls, lateral bracing and floor slabs. Nonstructural are the architectural components, equipment and distribution systems that are not part of the building structure (i.e., nonstructural); whether inside or outside, above or below grade. Nonstructural systems transform an empty skeleton into a functioning building

---

J. Gatscher (✉)
Fellow Engineer,
Schneider Electric, Nashville, TN 37217, USA
e-mail: jeff.gatscher@schneider-electric.com

M.N. Fardis and Z.T. Rakicevic (eds.), *Role of Seismic Testing Facilities in Performance-Based Earthquake Engineering: SERIES Workshop*, Geotechnical, Geological and Earthquake Engineering 22, DOI 10.1007/978-94-007-1977-4_16, © Springer Science+Business Media B.V. 2012

suitable for human occupancy. The nonstructural category covers a plethora of possible building applications. The diverse types of nonstructural systems can be grouped into four general categories:

- Architectural Components – such as parapets, partitions, facades, soffits, cladding, glazing, ceiling systems, etc.
- Mechanical and Electrical Equipment (also called components) – including building service equipment like pumps, generators, air handlers, compressors, transformers, switchgear, power supplies, and building tenant equipment like medical-related technology and equipment, emergency communication equipment and other specialized process equipment that are anchored during installation.
- Mechanical and Electrical Distribution Systems – such as process piping, fire sprinkler systems, HVAC ductwork, lighting systems, electrical busway, conduit, cable trays, etc.
- Building Occupancy Contents – including bookcases, shelving, office equipment and everything else a building contains that is not permanently anchored to the building structure.

The focus of this paper is on mechanical and electrical equipment and mechanical and electrical distribution systems. The term nonstructural is used to describe both equipment and distribution system categories.

Essential nonstructural systems (also called designated seismic systems) are those which have been determined to have critical functional importance after a design earthquake event. The integrity of these systems must be maintained for continued operation of essential buildings in the aftermath of earthquake recovery. This includes emergency power systems, fire suppression systems, emergency communication systems, medical treatment systems and many other types of essential building applications.

Large-class nonstructural systems are a subset of the nonstructural category. Large-class systems are defined as physically massive equipment platforms that are typically too large for practical testing as a complete unit. In addition, these systems are most often very expensive items. These two qualities make the process of seismic qualification for large-class systems a significant challenge to equipment suppliers and original equipment manufacturers (OEM). The focal point of this paper is on performance-based seismic qualification of large-class designated seismic systems, since this represents the greatest difficulty.

The list of large-class systems that fall into this special category is not inconsequential. In fact, this category constitutes a significant portion of designated essential seismic systems. Here are a few examples:

- Generator Sets (>600 kW)
- Uninterruptible Power Supply (>600 kW)
- Cooling Towers (>1,000 ton)
- Industrial Chillers
- Semiconductor Fabrication Equipment
- Medical Radiation Treatment Equipment
- Medical Imaging Equipment (MRI, CT, X-ray).

16.2 Prescriptive Versus Performance-Based Seismic Qualification

The building application addressed in this paper is seismic qualification to model building code seismic requirements. The current state requirement definition, contained in most building codes used around the world, is prescriptive in nature (i.e., life-safety related). In other words, most present-day equipment qualification involves establishing an equipment seismic withstand capacity that exceeds the project-specific seismic demand for a given building application. The code's performance expectation is to satisfy life-safety concerns. The natural evolution of building code seismic requirements is to transition from prescriptive to performance-based (FEMA 445 2006).

A simple comparison between prescriptive and performance-based approaches is analogous to comparing digital to analog systems. In digital systems it's either functioning or not functioning, and nothing in between. In analog systems it's a continuous spectrum between functioning and not functioning. Prescriptive requirements are digital-like and performance requirements are more analog-like in nature. Thus, in a performance-based paradigm seismic withstand capacity, with regard to nonstructural systems, will adopt "levels of performance." These levels will include various damage states. For example, damage states could include the categories: cosmetic damage, intermittent malfunction and permanent failure. Each performance level will have an associated earthquake demand. The goal in PBSD is to determine the demand level that initiates an identified damage state for a given equipment platform.

Figure 16.1 illustrates the difference between prescriptive qualification and performance-based qualification. The response spectrum plot shown in Fig. 16.1b is a typical approach for prescriptive qualification. There is a predefined (i.e., prescribed) spectral qualification requirement and a seismic simulation test is conducted. If the equipment test item passes at the code prescribed seismic demand, the qualification activity ends. However, in a performance-based paradigm the response spectrum plot in Fig. 16.1b only represents a single time slice out of the performance continuum surface plot shown in Fig. 16.1a. The ultimate goal in performance-based qualification is to establish the fragility surface that corresponds with an identified damage state (e.g., permanent failure). As can be observed, there is a fundamental difference between conducting a single qualification test at a prescribed seismic demand (Fig. 16.1b) and conducting enough tests with enough test specimens to develop a performance continuum profile (Fig. 16.1a).

Dynamic testing, or specifically seismic simulation testing, is the only viable method to validate active operation performance under design-level earthquake demands. This is applicable regardless of which strategy is being adopted in code requirements (i.e., prescriptive or performance approaches). However, seismic qualification is a product design process and implies more than showing up at the test lab with a single test unit. Almost all modern-day nonstructural product offerings are modular product platforms, pre-engineered to support many different end use applications (Ericsson and Erixon 1999).

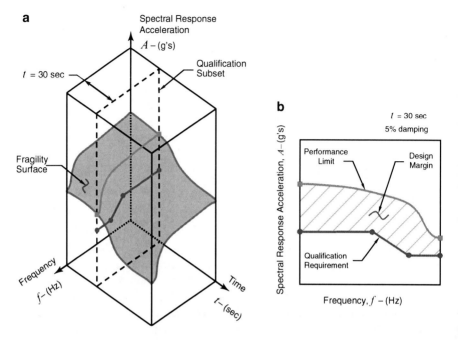

Fig. 16.1 Nonstructural seismic performance: (**a**) Shown as a continuous fragility surface and (**b**) Shown as a discrete response spectrum time slice

The goal during qualification testing is to qualify the entire family of configuration options associated with a given product platform by selecting a few rationalized product samples. The process of selecting the right product configurations for qualification purposes is called product line rationalization (also called type testing). The rationalization objective is to select the fewest number of design configurations that can adequately represent the entire product line family. Thus, by qualifying the rationalized configurations, in effect, the entire product line offering is qualified.

The output of rationalization is directly linked to the complexity of the product line's configuration offerings. For example, if the nonstructural platform to be qualified has minimal configuration options, then only a handful of possible product variations need consideration, resulting in one or two test unit candidates. Conversely, if the platform has hundreds or even thousands of possible configuration variants to consider, then the rationalization process will simply yield more test candidates (typically around five to ten test units). Why is this important? Some perspective is needed to better appreciate the answer.

Testing multiple product samples is difficult enough to justify in today's economic environment via a prescriptive code philosophy. Today, the OEM's seismic qualification process entails judiciously selecting the smallest number of design configurations to represent a specific product line family. Next, the product samples are sent to a test lab and shake tested to a code prescribed earthquake demand level. This is essentially a destructive test, since no customer would ever consider receiving a product order for a production unit that has been previously shaken to high levels. This process is repeated for all different product lines that need seismic qualification. Overall, the cost of present-day equipment qualification is not trivial. The cost can easily exceed several hundred thousand Euros over the span of a couple years.

Eventual transition to a performance-based code philosophy will greatly exacerbate the qualification cost requirement. This is necessitated by the fact that PBSD is a probability based procedure. Seismic performance is expressed as the probable consequences of earthquake damage. Each damage state is defined as a probability function, indicating the likelihood of the loss assessment based on the intensity of the earthquake ground shaking exposure. The need is to test multiple samples of the same design configuration in order to construct probability functions. The literature is not clear in defining how many test units are needed to establish a damage-state probability function. Obviously the greater number of test samples the better the result. A minimum of three test samples has been discussed as a compromise solution (FEMA 461 2007). This translates into a qualification cost level that is at least three times that for present-day equipment qualification.

For example, a low configurability productline that requires two test units would need three samples of each configuration for a total of six test units. And with a highly configurable equipment platform that requires five or more design configurations to cover the product line family, would need three of each for a total of 15 or more test samples. Multiply these by the number of different product lines that require seismic qualification and you quickly reach unreasonable cost levels. In all fairness, not many nonstructural suppliers or OEMs will be willing to invest millions to seismically qualify product offers when testing is conducted at the top-level assembly. This approach is a nonstarter and risks rejection by industry. The primary question addressed here is what PBSD compatible approach can be implemented that is more reasonable from a cost and implementation perspective.

A more practical method to conduct fragility-type testing in support of PBSD, is employing subsystem testing techniques. Device level testing (i.e., subassembly testing) is attractive for several reasons. Table 16.1 highlights the pros and cons for subsystem testing. The primary advantage is that multiple subassembly devices can be appropriated and tested at reasonable cost. Full-scale testing, conducted at the top-level assembly, becomes impractical and too costly for nonstructural suppliers to justify. The concept is to allocate earthquake demands to the nonstructural subsystem. A fundamental prerequisite for conducting subsystem testing is characterization of the "device-level" seismic demand. This requires employing a systems design framework.

16.3 Systems Design Framework

The path traveled by the earthquake shock wave from geologic source to a device located within a piece of equipment that is located somewhere within a building is obviously influenced by many variables. This picture becomes clearer if you think

Table 16.1 The pros and cons for subsystem qualification testing

Top-level system testing		Device-level subsystem testing	
Pros	Cons	Pros	Cons
Simulates end-use building application environment	Top-level assembly is functionally complex	Functional device assembly is functionally less complex	Simulates internal nonstructural functional environment
What you see installed is what you test	Top-level assembly is physically large and heavy	Functional device assembly is physically smaller and lighter	Top-level assembly qualification becomes a composite calculation with uncertainties
No ambiguity with transfer functions	Top-level assembly test unit is expensive	Functional device assembly test unit is less expensive and the cost of qualification testing can be incorporated into the cost of device product development	Internal structural transmission paths (i.e., transfer functions) must be characterized
Mechanical impedance interaction between device and FRS is naturally accounted for	Few test facility options and typically requires the services of large-scale test laboratories where testing cost can be expensive	More test facility options and testing cost is less expensive using small-scale shake-tables	Mechanical impedance interaction between device and FRS is typically ignored
No opportunity for overtesting of functional devices	Difficult to simulate building-level functional inputs during testing	Easier to simulate device-level functional inputs during testing	The possibility for functional device overtesting is high

16 Performance Based Seismic Qualification of Large-Class Building Equipment... 311

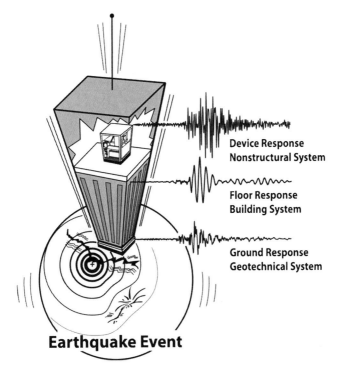

Fig. 16.2 Illustration showing three primary systems involved in nonstructural earthquake protection: (*1*) geotechnical, (*2*) building, and (*3*) nonstructural

of buildings as appendages embedded into the earth's surface and nonstructural items as appendages attached to buildings. Ground motion must traverse a directional path into buildings and then into nonstructural items. As the ground shakes, the two appendages will shake and respond to the base input in series (i.e., ground shakes building and building shakes nonstructural). Figure 16.2 displays an illustration of this concept.

At the core of systems design is the concept of framing problems in a top-down perspective and then to divide and sub-divide until a big problem is transformed into smaller, isolated elements. This process is called system decomposition. The next step is to identify interface relationships between the isolated pieces. For example, Fig. 16.2 can be decomposed into three primary systems: (1) geotechnical, (2) building, and (3) nonstructural. Figure 16.3a shows this systems view representation as a top-level block diagram. This diagram is not very useful as is. Each system element needs further decomposition and the system interfaces need to include functional relationships between elements. Figure 16.3b shows an expanded version of the block diagram with the nonstructural system divided into two subsystems: (1) mechanical-related elements, and (2) active operation-related elements. Mechanical-related elements include a force resisting skeleton (FRS), attachments

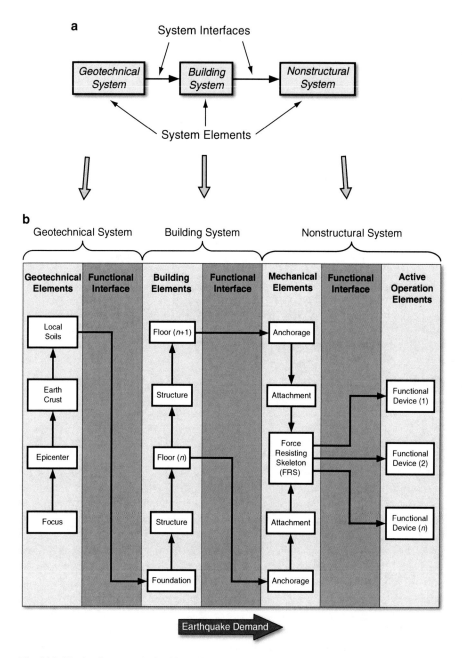

Fig. 16.3 Top level systems design block diagrams: (**a**) basic block diagram concept and (**b**) expanded diagram using subsystem decomposition

16 Performance Based Seismic Qualification of Large-Class Building Equipment... 313

and anchorage. The active operation elements include functional devices. To clarify these terms a generic definition for each subsystem element is provided in Table 16.2.

Now that we have defined the system elements depicted in Fig. 16.3b, the remaining task is to identify the functional interface relationships between the elements. Identification of functional relationships between system elements is an important aspect of systems design. These interface relationships provide the mechanism to extend seismic protection principles beyond the single component philosophy. The objective is to consider an entire building system, which may be composed of multiple equipment items that are functionally connected via a nonstructural distribution chain. This is a key concept. The interface relationships relevant to nonstructural protection and seismic qualification include:

- Structural Transmissibility – characterizes the seismic loading transmission path between system elements. The transmission path is quantified by using structural dynamic transfer functions. Structural transmission paths need to include response reduction modifiers that account for inelastic dynamic response. Transfer functions adjusted for inelastic response are more representative of earthquake dynamics. Linear-based transfer functions are too conservative for implementation purposes.
- Functional Interaction – includes several types of interactions. There are mechanical impedance interactions that represent a structural feedback relationship between the forcing system (e.g., FRS) and the system being driven (e.g., functional device). There are connection interactions, as well, that describe the electrical and mechanical connection interface relationships between system elements.
- Seismic Demand – defines the resultant environmental loading and becomes the input seismic demand requirement for the subsystem elements. Seismic demand includes forces, accelerations, velocities and relative displacements.

In essence, each connecting arrow between system blocks in Fig. 16.3b is replaced with interface elements. Figure 16.4 displays a generalized systems design framework that includes the necessary interface relationships between system elements. This framework depicts the relationships for one nonstructural system containing one active functional device that is installed between building floor elevations n and n + 1. Two attachment options are depicted considering possible attachment designs that could secure to the floor, walls or ceiling of a given floor elevation.

The framework may be readily extended by adding more branch points. Additional branches could be included to represent multiple equipment items in a distribution chain and/or represent multiple functional devices per equipment item in a given building nonstructural system. These branches may be cascaded or in parallel. Furthermore, the generalized framework may be simplified when typical engineering assumptions are made regarding the various system interactions; for example, ignoring mechanical impedance effects between system elements.

The generalized systems framework provides a tool for allocating seismic demands on individual nonstructural system elements regardless if the system represents a

Table 16.2 Nonstructural system element naming convention and description

Nonstructural system element			Definition/description
Mechanical subsystem	Force resisting skeleton(FRS)		Structural members or assemblies of members, including frames, enclosures, struts, rods, panels, etc. FRS assembly members can be joined together using mechanical fasteners or can be weldments. Monocoque construction techniques are also included as FRS types. The nonstructural FRS provides support for subassemblies, modules and internal devices. The FRS also provides overall structural stability for the nonstructural platform. The FRS should be viewed as the nonstructural systems structural skeleton to resist all environmental and operating loads
	Attachments	Operational attachment	Operating elements that connect between the nonstructural FRS and building structure or could insert between two nonstructural systems in a distribution chain. An operational attachment includes both mechanical and electrical elements that are intended to support active operation functions. These attachments are necessary operational umbilicals that are required in order for the nonstructural system to function. Examples of operational attachments include piping, ducting, tubing, cabling, conduit, wiring, grounding, etc
		Bracing attachment	Structural elements that connect between the FRS and building structure. A bracing attachment is purely structural in nature and used to provide a structural link between FRS and building to resist environmental loads. Examples of bracing attachments include brackets, angles or other structural shapes that provide structural support and load resistance to the FRS
		Isolation attachment	Mechanical elements (springs and dampers) that insert between the FRS and building structure to isolate the FRS from the building. Isolation attachments (also called shock and vibration isolators) are mechanical energy absorbers that are intended to attenuate loading between the nonstructural FRS and building. Examples of isolation attachments include compression springs and dashpots, compression springs with snubbers, etc

Anchorage	The final connection points securing the nonstructural FRS to the building structure, with or without the use of attachments. Anchorage includes bolts, concrete anchors, welds or other mechanical fasteners for positively securing the nonstructural FRS to the building without consideration of frictional resistance produced by the effects of gravity. For example, a typical equipment FRS can be anchored to a concrete pad directly with anchor bolts without the use of either bracing or isolation attachments. Or an equipment FRS can be placed on isolation attachments and the isolator is anchored to the pad. Anchorage connection points to the building structure are the elements that cross the line between building and nonstructural systems to react dynamic loading experienced during the earthquake
Functional device	Logical sub-groupings of nonstructural active functions (operational) typically organized and arranged as physical devices, modules, components or subassemblies that mount to the nonstructural FRS. Functional devices can be electrical, mechanical or electro-mechanical in nature. Some devices also function as load bearing members of the FRS in addition to being a functional device. For example, with a piping distribution system the pipe acts as both FRS and functional device with containment of pipe contents being the active function
Active operation subsystem	

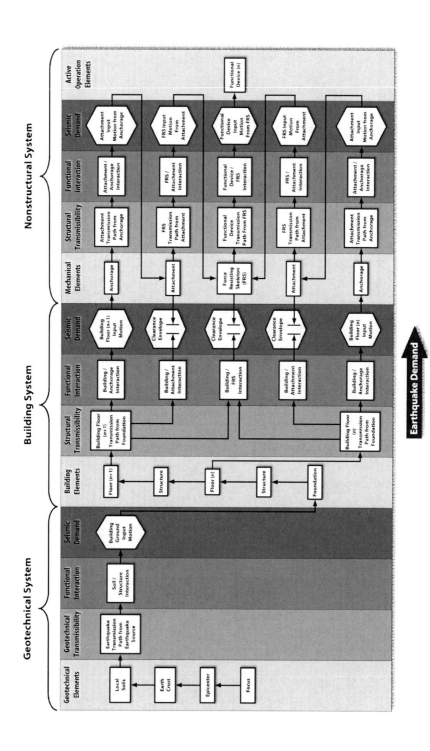

Fig. 16.4 Generalized systems design framework for seismic qualification applications

single equipment item or a distribution chain of equipment. This approach is applicable for specifying demand requirements on a project-specific basis or based on a wide geographic area approach. The project-specific method endeavors to evaluate demand requirements for a known building site location and for a specific building type. In this instance, the exercise is to use geotechnical site survey data for ground input motion in conjunction with using a detailed building model to arrive at project-specific building floor input motions. Project-specific nonstructural demand requirements are based on unique calculated floor level inputs and represent building specific requirements.

The geographic area approach attempts to select nonstructural demand levels that can envelope (i.e., umbrella) a known geographic region. For example, the region could be a single U.S. State, a portion of a state or the entire country. The concept is to establish nonstructural input demands using upper limit code requirements. This approach typically assumes the maximum ground motion requirement for the target region, in conjunction with selecting upper-level building installation heights for maximum amplification effects. Geographic area demand requirements are based on generic floor level inputs in accordance with code default maximums. This represents worst case nonstructural requirements for any building in the specified region and for any floor in a building.

The systems framework introduced here is highly flexible. Depending on the nonstructural application, this framework can contract to a few elements and interfaces or expand to include many application features. When applied to simple nonstructural systems using code default maximums the framework is minimal. When applied to cover complex nonstructural systems using project-specific geotechnical and building input models the framework can expand accordingly. And since functional devices contain unique system interfaces, seismic qualification of functional devices can be pursued as a separate activity. The systems framework introduced here can be viewed as a necessary foundation schema supporting PBSD principles. This approach is directly applicable when the need is to support PBSD qualification. Subsystem testing implements the systems framework to establish device-level seismic demands.

16.4 Subsystem Testing Techniques

The general process is shown in Fig. 16.5 (Gatscher and Bachman 2010). The steps are summarized as follows:

- **Step 1:** The system-level seismic demand is defined with a response spectrum requirement. This spectral demand would originate from the building code or standard that is applicable for the nonstructural application. For example, for most equipment applications in the United States the applicable code is the International Building Code (ICC 2011) and the shake-table test demand is defined in AC156 test protocol (ICC ES 2010). Both horizontal and vertical spectral demands are defined.

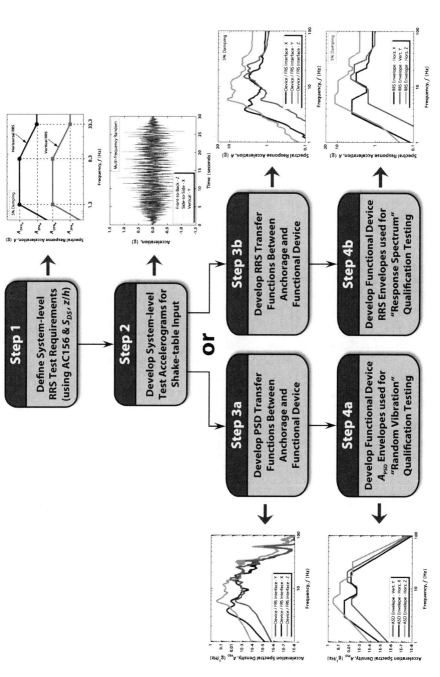

Fig. 16.5 Large-class nonstructural process flow for creating a functional device qualification test requirement used for standalone device testing, supporting either response spectrum or random vibration techniques

16 Performance Based Seismic Qualification of Large-Class Building Equipment... 319

- **Step 2:** Appropriate system-level time history test accelerograms (three independent inputs, two horizontal and vertical) are developed to satisfy the spectral demands defined in Step 1. There are four general characteristics that need to be addressed when developing accelerograms that can be used for seismic qualification: (1) adequate amplitude intensity (including velocity and displacement intensity), (2) adequate frequency content, (3) representative rise and decay rates, and (4) proper phasing interactions.
- **Step 3a:** Used with a random vibration approach. The transfer function is developed between system-level input location (e.g., base anchorage for most equipment applications) and location of functional devices. The power spectral density transfer functions should include response reduction modifiers to account for nonlinear dynamic response. Three transfer functions are required, two horizontal and vertical.
- **Step 3b:** Used with a response spectrum approach. The transfer function is developed between system-level input location (e.g., base anchorage for most equipment applications) and location of functional devices. The response spectra transfer functions should include response reduction modifiers to account for nonlinear dynamic response. Three transfer functions are required, two horizontal and vertical.
- **Step 4a:** Used with a random vibration approach. Test power spectral density envelopes are developed to conduct functional device-level qualification testing. Three test spectra are required, two horizontal and vertical. The functional devices are then separately qualified using the test envelopes as seismic input demand via random vibration testing.
- **Step 4b:** Used with a response spectrum approach. Test response spectra envelopes are developed to conduct functional device-level qualification testing. Three test response spectra are required, two horizontal and vertical. The functional devices are then separately qualified using the test envelopes as seismic input demand via response spectrum testing.

The process described assumes that the system-level demand (Step 1) is the starting point. In most cases, it is easier to test the functional device to its limit state. Once the approximate design limit of the device can be detected under seismic inputs, the process can be reversed to determine the maximum system-level demand that can be satisfied by the functional device given the applicable transfer functions. This approach is more flexible in that a single device may be used in multiple equipment platforms and thus the various transfer functions will be different from one equipment platform to the next. Stated simply, if the device has a known performance limit, the only variable that effects the system-level performance is the specific transfer function for a given equipment platform.

When the functional device is tested separately from the equipment platform, three or more devices can be practically tested for a reasonable cost. In fact, this type of seismic qualification testing of devices can be added to the normal design process and included with other "shock and vibration" design assurance testing conducted during the product development cycle. This approach does require characterization of system-level transfer functions that have been adjusted for inelastic dynamic response. However, it is reasoned that the added effort to characterize system-level transfer functions is minimal compared to the implementation cost of performing performance-based seismic qualification testing at the top-level of system assembly.

16.5 Summary and Conclusions

After decades of secondary treatment, nonstructural research has recently become a topic of interest in the earthquake engineering community. This trend is likely attributed to the recognition that nonstructural systems are paramount for the continued operation of essential buildings. In addition to this, as the science and understanding of building structural performance has increased over the past few decades, there is realization that nonstructural items may cause the greatest loss to life (and cause most injuries) as well as causing a great economic loss in otherwise well prepared/ designed facilities. Earthquakes over the last quarter century have proven the nonstructural vulnerability when it comes to maintaining building functions after the event. Building structural designs have made good progress in mitigating earthquake damage, due in most part to extensive research. While nonstructural systems have lagged behind in earthquake protection effectiveness, with limited research interest. Perhaps the time has come for nonstructural research to move to the forefront.

The evolution of earthquake risk mitigation is moving in the direction of a performance-based philosophy. PBSD procedures are being investigated by researchers world-wide. There is need for consideration of the implementation aspect of PBSD, specifically with regard to performance-based qualification of large-class nonstructural equipment and systems. Large-class equipment platforms are physically massive and most often impractical to test as a complete unit. In addition, these products are highly expensive making any qualification activity difficult to justify in today's economic climate. Performance-based qualification will require testing multiple samples of a test unit to develop necessary damage state fragility functions. This exacerbates the cost perspective. Performance testing of large-class equipment at the highest level of assembly is not a practical approach. However, performance testing of subassembly functional devices is a more practical option from a cost and implementation perspective.

The allocation of seismic demand, to subsystem devices and components, can be achieved when employing systems analysis techniques. Individual subassemblies can be performance tested to their functional limits. This establishes the seismic demand threshold associated with a given damage state (for example, permanent failure). Once a subassembly device has a defined limit state, the system-level equipment platform's performance becomes a function of the applicable device transfer functions. This methodology has the practical benefit of conducting performance testing at a much lower level of product assembly. The seismic performance testing of smaller devices can be incorporated into the normal product development process along with other "shock and vibration" design assurance testing.

The only realistic method to determine active operation performance is testing. This necessitates using multiple samples of the test item to construct the needed probability functions. The only practical method to conduct fragility-type testing is employing subsystem testing techniques. When testing is conducted at the device level, multiple devices can be appropriated and tested at a reasonable cost. When testing is conducted at the top-level assembly, full-scale testing of multiple units becomes impractical and too costly for nonstructural suppliers and OEMs to justify.

The main requirement for subsystem testing is characterization of device-level transfer functions that are adjusted accounting for FRS inelastic response. Subsystem testing also introduces a degree of uncertainty in the calculation of system-level seismic performance. Research is needed to both validate this proposed approach and to better understand the bounds of uncertainty that result.

References

Ericsson A, Erixon G (1999) Controlling design variants: modular product platforms. Society of Manufacturing Engineers, Dearborn, pp 1–15

Federal Emergency Management Agency (2006) Next-generation performance-based seismic design guidelines: program plan for New and Existing Buildings. FEMA 445, Washington, DC

Federal Emergency Management Agency (2007) Interim testing protocols for determining the seismic performance characteristics of structural and nonstructural components. FEMA 461, Washington, DC

Gatscher JA, Bachman RE (2010) Seismic qualification of large-class equipment using combined analysis and testing. In: Proceedings of the 14th European Conference on Earthquake Engineering, Ohrid, MK, pp 1–10

ICC ES (2010) AC156 acceptance criteria for seismic certification by shake-table testing of nonstructural components. International Code Council Evaluation Service, Country Club Hills, IL

ICC (2011) 2012 international building code®. International Code Council, Country Club Hills

Chapter 17
Experimental Evaluation of the Seismic Performance of Steel Buildings with Passive Dampers Using Real-Time Hybrid Simulation

Theodore L. Karavasilis, James M. Ricles, Richard Sause, and Cheng Chen

Abstract An experimental program is presented involving the use of the real-time hybrid simulation method to verify the performance-based seismic design of a two-story, four-bay steel moment resisting frame (MRF) equipped with compressed elastomer dampers. The laboratory specimens, referred to as experimental substructures, are two individual compressed elastomer dampers with the remainder of the building modeled as an analytical substructure. The proposed experimental technique enables an ensemble of ground motions to be applied to the building, resulting in various levels of damage, without the need to repair the experimental substructures, since the damage will be within the analytical substructure. Statistical experimental response results incorporating the ground motion variability show that a steel MRF with compressed elastomer dampers can be designed to perform better than conventional steel special moment resisting frames (SMRFs), even when the MRF with dampers is significantly lighter in weight than the conventional MRF.

T.L. Karavasilis (✉)
Department of Engineering Science, University of Oxford, Oxford OX1 3PJ, UK
e-mail: theodore.karavasilis@eng.ox.ac.uk

J.M. Ricles • R. Sause
Department of Civil and Environmental Engineering, ATLSS Engineering Research Center,
Lehigh University, Bethlehem, PA 18015, USA
e-mail: jmr5@lehigh.edu; rc0c@lehigh.edu

C. Chen
School of Engineering, San Francisco State University, San Francisco, CA 94132, USA
e-mail: chcsfsu@sfsu.edu

M.N. Fardis and Z.T. Rakicevic (eds.), *Role of Seismic Testing Facilities in Performance-Based Earthquake Engineering: SERIES Workshop*, Geotechnical, Geological and Earthquake Engineering 22, DOI 10.1007/978-94-007-1977-4_17,
© Springer Science+Business Media B.V. 2012

17.1 Introduction

Several experimental studies have been performed on steel frames with passive dampers. These studies include mainly shaking table tests similar to those conducted by Chang and Lin (2004) on a full-scale five-story frame, and recently by Kasai et al. (2009) on a full-scale five-story steel building.

Recent research studies have proposed new kinds of dampers. Karavasilis et al. (2011) experimentally evaluated the hysteretic behavior of a new innovative compressed elastomer damper developed by Sweeney and Michael (2006) and used the design procedure of Lee et al. (2005) to design steel MRFs with compressed elastomer dampers.

To demonstrate the full potential of new types of dampers, damper designs and performance-based design procedures for structural systems with dampers need to be experimentally validated. The 2004 NEHRP provisions (BSSC 2004) allow the design of buildings with passive damping systems to experience controlled inelastic deformations associated with typical design drifts limits, e.g. a 2% drift limit. Therefore, to experimentally validate damper designs and performance-based design procedures, inelastic response statistics incorporating ground motion variability should be obtained. Full-scale testing (Chang and Lin 2004; Kasai et al. 2009) is a reliable, but at the same time, a challenging experimental technique. In particular, conducting a series of full-scale tests to obtain response statistics of structural systems under earthquake intensities which produce inelastic deformations may be cost and time prohibitive since the damaged components of the structural system need to be repaired or rebuilt after each test.

This chapter presents an experimental program based on the use of real-time hybrid simulation to verify the performance-based seismic design of a two-story, four-bay steel MRF equipped with compressed elastomer dampers. The MRF was designed using a reduced base shear design force compared to a conventional SMRF, where the dampers are designed to control the drift of the structure. The laboratory specimens, referred to as experimental substructures, are two individual compressed elastomer dampers. The remainder of the building is modelled as an analytical substructure. A series of real-time hybrid simulations are performed to acquire response statistics under the design basis earthquake (DBE) and the maximum considered earthquake (MCE). The DBE has an intensity that is two-thirds that of the MCE, where the MCE has a 2% probability of exceedance in 50 years (BSSC 2004). Real-time hybrid simulations are conducted using the Real-Time Integrated Control System (Lehigh RTMD Users Guide 2009) of the Lehigh Network for Earthquake Engineering Simulation (NEES) Real-time Multi-Directional Earthquake Simulation Facility (RTMD) in conjunction with the finite element program HybridFEM (Karavasilis et al. 2009) that was specifically developed to enable nonlinear analytical substructure models to be created for real-time hybrid simulations.

17.2 Compressed Elastomer Damper

The compressed elastomer structural dampers used in the study are fabricated by bonding four pieces of an elastomer (butyl rubber blend) onto a longitudinal steel bar, as shown in Fig. 17.1a. The pieces of elastomer on this bar are then pre-compressed into a steel tube (Fig. 17.1b). Each prototype damper includes three tubes which are welded together (Fig. 17.1c). To enable the damper to be attached to the structure as shown in Fig. 17.1d, transverse bars with bolt holes are welded across the steel tubes and additional transverse attachment bars are welded across the narrow dimension of the longitudinal bars (Fig. 17.1c).

The interface between the elastomer and the steel tube is not mechanically bonded, which allows the elastomer to slip relative to the tube, producing friction when large deformations are imposed. The dampers are designed to slip before the elastomer tears or the bond to the longitudinal bar fails (Sweeney and Michael 2006). There is no slip limitation of the damper as long as the steel tubes are long enough so that the elastomer remains in the tube during the earthquake response.

Characterization tests of the prototype damper were conducted by Karavasilis et al. (2009b) at the RTMD facility (Lehigh RTMD Users Guide 2009). The experimental results showed that the dampers exhibit elastomeric behavior under small deformation (less than 15 mm) (Fig. 17.2a). When the deformation is larger

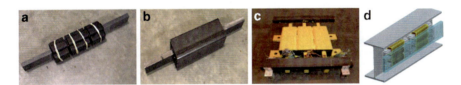

Fig. 17.1 Fabrication of compressed elastomer damper: (**a**) elastomeric material wrapped around longitudinal bar; (**b**) elastomeric material and bar compressed into the steel tube; (**c**) damper with additional transverse attachment bars in place and (**d**) installation to beam web (Sweeney and Michael 2006)

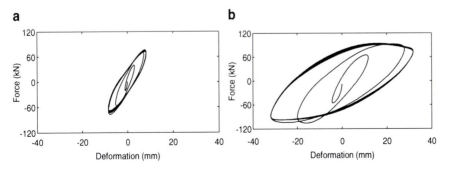

Fig. 17.2 Damper hysteresis from characterization tests (Karavasilis et al. 2009): (**a**) before slip, and (**b**) after slip

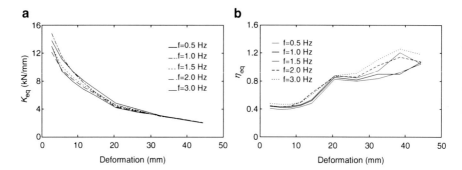

Fig. 17.3 Mechanical properties evaluated from characterization tests (Karavasilis et al. 2009): (**a**) equivalent stiffness, and (**b**) equivalent loss factor

than 15 mm, slip of the elastomer compressed inside the steel tube occurs and the dampers exhibit a combined elastomeric and frictional behavior (Fig. 17.2b). The damper assembly was tested with two prototype dampers (i.e. a total of six tubes) acting in parallel, referred to herein as a compressed elastomer damper.

The equivalent stiffness, K_{eq}, and the loss factor, η_{eq}, of the compressed elastomer damper were determined from the characterization test data. The K_{eq} and η_{eq} from the characterization tests are given in Fig. 17.3. Figure 17.3a shows that the stiffness decreases with increasing deformation and slightly increases as the frequency increases for a given deformation. Figure 17.3b shows that the energy dissipation, η_{eq}, is relatively constant for small amplitudes of deformation (less than 10 mm) and significantly increases after slip of the elastomer occurs at about 15 mm.

17.3 Steel MRF with Compressed Elastomer Dampers

17.3.1 Prototype Building

Figure 17.4a shows the plan view of the 2-story, 6-bay by 6-bay prototype office building used for the study. The building has four identical perimeter steel MRFs to resist lateral forces. Each MRF consists of four bays. The design study focuses on one typical perimeter MRF. This MRF is designed either as a conventional SMRF as defined in the International Building Code (2003), referred to herein as IBC 2003, or as an MRF with compressed elastomer dampers using the simplified design procedure (SDP) by Lee et al. (2005). In the latter case, dampers and diagonal braces are added to the two interior bays, as shown in Fig. 17.4b.

The nominal yield stress of the steel members of the MRF is 345 MPa. A smooth design response spectrum with parameters $S_{DS} = 1.0$ g, $S_{D1} = 0.6$ g, $T_0 = 0.12$ s and $T_s = 0.6$ s defined by IBC 2003 represents the DBE.

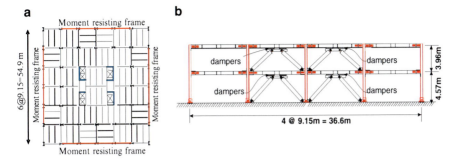

Fig. 17.4 Prototype building structure: (**a**) plan view, and (**b**) perimeter MRF with dampers and diagonal bracing

17.3.2 Design of Perimeter MRF as a Conventional SMRF Without Dampers

The perimeter MRF of Fig. 17.4b is initially designed as a conventional SMRF using the equivalent lateral force procedure from IBC 2003. This SMRF without dampers, referred to herein as UD100V, satisfies the member strength criteria and the drift limit of 2% of IBC 2003, where values of 8 and 5.5 are used for the response modification factor, R, and the amplification factor, C_d, respectively.

To study whether MRFs with compressed elastomer dampers can be designed to have a reduced strength compared to a conventional SMRF (without dampers) and achieve a prescribed level of seismic performance, the perimeter MRF was redesigned without dampers to have a design base shear equal to $0.50V$, where V is the design base shear of UD100V. This MRF design is referred to as UD50V, and is significantly lighter than UD100V. A comparison of weight is discussed later. UD50V is subsequently outfitted with compressed elastomer dampers, where the SDP by Lee et al. (2005) is used to design the dampers to limit the drift of the MRF to 1.65%. The SDP enables the design of the dampers to be integrated into the system design by specifying performance objectives that the combined MRF and damper system must achieve. Details of how the SDP is utilized are discussed later.

Table 17.1 summarizes the properties for the two MRF designs (UD100 and UD50V). The table lists the column section, beam sections, steel weight, fundamental period of vibration, T_1, and story stiffness (used later to design the dampers for the UD50V MRF). The last column of Table 17.1 provides estimates (based on the equal-displacement rule) of the expected maximum story drift, θ_{max}, under the DBE earthquake. Under the assumption of the equal-displacement rule, the UD100V frame was found to exceed the 2% story drift limit of IBC 2003, however, UD100V did satisfy the 2% story drift limit using the drift check procedure involving the use of C_d per IBC 2003.

Table 17.1 Properties of MRF designs

MRF	Column section	Beam sections	Steel weight (KN)	T_1 (s)	Story stiffness (kN/mm)	$\theta_{max}(\%)$
UD100V	W14 × 211	1st story: W24 × 84	200	1.08	1st story: 66574	2.40
		2nd story: W21 × 50			2nd story: 42018	
UD50V	W14 × 120	1st story: W24 × 55	124	1.48	1st story: 36007	3.23
		2nd story: W18 × 40			2nd story: 23894	

17.3.3 Design of Dampers for SMRF

The SDP idealizes the damper hysteresis loops as linear viscoelastic ellipses and the damper design variables are the equivalent damper stiffness and the loss factor. The thickness and the area of the elastomer assumed in the prototype MRF for the hybrid simulations presented in this paper are four times larger than the thickness and the area of the elastomer of the dampers tested by Karavasilis et al. (2009b). Such an elastomer thickness ensures that the dampers in the prototype MRF will remain undamaged (i.e., no slip) under the DBE, while slip is expected under the MCE. This design strategy ensures no need to replace the dampers for seismic events lower than or equal to the DBE. For earthquakes larger than the DBE (e.g., MCE), the dampers are treated as sacrificial elements that should be replaced after the seismic event, if needed. The properties of the large-scale compressed elastomer damper design were derived from the experimental data presented in Fig. 17.3 as follows. With t_{ref} and A_{ref} designated as the thickness and area, respectively, of the elastomer damper used in the characterization tests, the properties (thickness t and area A) of the damper designs for the prototype MRF are expressed in terms of the ratios t/t_{ref} and A/A_{ref}. Given the stiffness, $K_{eq}(u_{ref})$, and the loss factor, $\eta_{eq}(u_{ref})$, of the damper in the characterization tests, the stiffness and loss factor of the damper designs are: $K_d(u_d) = (A / A_{ref})\cdot(t_{ref} / t)\cdot K_{eq}(u_{ref})$ and $\eta(u_d)=\eta_{eq}(u_{ref})$, where u_d is the deformation imposed on the damper in the MRF, and u_{ref} is the deformation of the damper in the characterization tests, related to u_d through the expression $u_{ref} = u_d \cdot(t_{ref} / t)$. The expressions for $K_d(u_d)$ and $\eta(u_d)$ are derived by transforming the characterization test results for the damper from force – deformation ($F_{ref} - u_{ref}$) behavior to shear stress – shear strain ($\tau - \gamma$) behavior using the t_{ref} and A_{ref} dimensions (i.e., $\tau=F_{ref}/A_{ref}$ and $\gamma=u_{ref}/t_{ref}$), and then by transforming the shear stress – shear strain ($\tau - \gamma$) behavior to force -deformation ($F_d - u_d$) behavior of the damper designs using the t and A dimensions (i.e., $F_d=\tau A = F_{ref}\cdot A/A_{ref}$ and $u_d=\gamma t = u_{ref}\cdot t/t_{ref}$).

The SDP developed by Lee et al. (2005) is slightly modified herein to account for the strong dependence of $K_d(u_d)$ and $\eta_d(u_d)$ of the compressed elastomer dampers on deformation amplitude u_d. To achieve the target performance level (e.g., immediate occupancy under the DBE), detailed design criteria, such as story drift limits and limits on the internal forces of the members need to be established. For the study

17 Experimental Evaluation of the Seismic Performance of Steel Buildings... 329

Table 17.2 Design properties of UD50V MRF with dampers

α	Brace steel weight (kN)	B	T_1 (s)	η_d	ξ_t (%)	B	θ_{max} (%)	N_d Story 1st	2nd
10	17.2	1.0	1.04	0.60	15.00	1.35	1.65	8	5

herein, a value of $\theta_{max} = 1.65\%$ under the DBE is specified as the target performance objective. The modified SDP is then used, as explained below:

1. *Select an appropriate α value (ratio of total brace stiffness per story in the global direction to the MRF story stiffness).* This ratio should provide: (a) braces that are stiff enough so that the story drift produces damper deformation with minimal brace deformation; (b) braces do not buckle under the maximum forces transmitted by the dampers; and (c) only a small increase in the steel weight of the structure.
2. *Select an appropriate β value (ratio of total damper stiffness per story in the horizontal direction to the MRF story stiffness K_o).* The β value should provide a reasonable required number of dampers.
3. *Select an initial value of the damper loss factor, η_d.* With the η_d selected, the contribution of the dampers to the equivalent damping ratio of the MRF with the dampers, ξ_{eq}, is estimated based on the lateral force energy method (Sause et al. 1994). The damping reduction factor, B, is then obtained (BSSC 2004) as a function of the total damping ratio, ξ_t, which equals the sum of ξ_{eq} and the inherent damping ratio of the MRF building (assumed to be 2%).
4. *Response spectrum analysis.* The elastic response spectrum is reduced by the B factor, and the story drifts and damper deformation, u_d, are calculated based on a response spectrum analysis using the equal-displacement rule. In this analysis, dampers at each story are modelled with two linear springs (one spring at each interior bay), each having a horizontal stiffness equal to $K_o \cdot \beta/2$. With u_d known, the $\eta_d(u_d) = \eta_{eq}(u_{ref})$ is calculated from Fig. 17.3b using $u_{ref} = u_d \cdot (t_{ref}/t)$. Iterations of Steps 3 and 4 are performed until the value for η_d converges. If the story drifts after convergence do not satisfy the established performance criteria, Steps 2–4 are repeated, beginning by selecting a new value for β.
5. *Calculate required number of dampers.* With the u_d known, the damper design stiffness $K_{eq}(u_{ref})$ is determined from Fig. 17.3a, and $K_d(u_d) = (A/A_{ref}) \cdot (t_{ref}/t) \cdot K_{eq}(u_{ref})$. The required number of dampers, N_d, equals $(K_o \cdot \beta/K_d(u_d))$, rounded up to the nearest integer. If the number of dampers, N_d, is too large, a revised performance criteria and/or MRF design should be considered and Steps 1–5 are repeated.

Table 17.2 provides a summary of the damper design for the UD50V MRF where the performance criterion (as noted previously) is a design story drift of 1.65% under the DBE. It is observed that the MRF with eight compressed elastomer dampers in the first story and five compressed elastomer dampers in the second story exhibits a significantly better performance ($\theta_{max} = 1.65\%$) than that of the conventional steel SMRF UD100V ($\theta_{max} = 2.40\%$, see Table 17.1), where the design prediction of θ_{max} is based on the equal-displacement rule. Moreover, the UD50V MRF with dampers has a steel weight equal to 124 kN (beams and columns) + 17.2 kN (braces) = 141.2 kN, while the steel weight of the conventional steel SMRF UD100V is 200 kN. Thus, the UD50V MRF with dampers has a 30% reduction in steel weight compared to the conventional steel SMRF design, UD100V.

The damper imposes a limit on the peak damper force transmitted to the braces, the columns and foundation of the building by exhibiting a plastic (friction) behavior at higher drifts. This is a clear advantage over a conventional viscoelastic damper with uncontrolled peak damper force, since more economical designs can be achieved and braces able to safely support the dampers without buckling under high seismic intensity levels can be designed.

17.4 Real-Time Hybrid Simulations

17.4.1 Real-Time Integrated Control System Architecture

The performance of a perimeter MRF with compressed elastomer dampers is experimentally evaluated by conducting real-time hybrid simulations. As illustrated in Fig. 17.5, the experimental substructures are two individual compressed elastomer dampers with the remaining part of the building (MRF, braces, and gravity frames (shown as a lean-on column in Fig. 17.5)) modeled as an analytical substructure. Since the dampers at a story level are placed in parallel in the prototype MRF (Fig. 17.4b), they are subjected to the same velocity and displacement. Therefore, each of the damper setups in the laboratory represents all of the dampers in one story. In a real-time hybrid simulation the measured restoring force from a compressed elastomer damper is multiplied by the number of dampers N_d in a story to obtain the total restoring force of all the dampers at the story level in the MRF.

As discussed previously, the thickness and the area of the elastomer of the dampers that are used in UD50V MRF are considered to be four times larger than the thickness and the area of the elastomer of the dampers in the experimental substructure. Consequently, in the real-time hybrid simulation the command displacement of the dampers is scaled down by a factor of 4 and the measured restoring force is amplified by a factor of 4.

The nonlinear finite element program *HybridFEM* (Karavasilis et al. 2009) has been implemented into the real-time integrated control system at the NEES RTMD Facility (Lehigh RTMD Users Guide 2009). The architecture for the RTMD system is shown in Fig. 17.6. A digital servo controller (Real-time Control Workstation) with a 1,024 Hz clock speed (sampling time $\delta t = 1/1,024$ s) controls the motion of the servo-hydraulic actuators and is integrated with the Real-time Target Workstation, Simulation Workstation, and Data Acquisition Mainframe using a shared common RAM network (SCRAMNet). SCRAMNet has a communication rate of about 180 ns which enables the transfer of data among the integrated

17 Experimental Evaluation of the Seismic Performance of Steel Buildings...

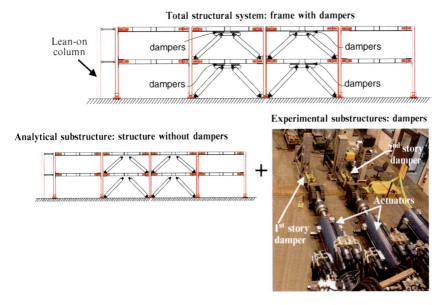

Fig. 17.5 Real-time hybrid simulation: analytical and experimental substructures forming the complete structural system

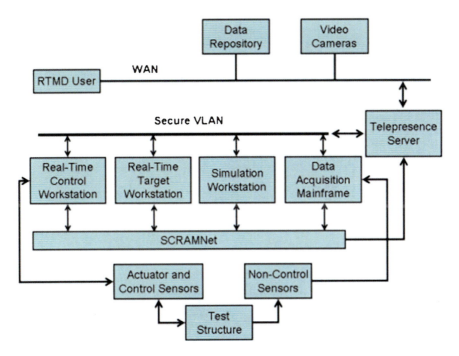

Fig. 17.6 RTMD control system architecture

workstations in real-time with minimal communication delay. *HybridFEM* has been developed in a manner that enables the analytical substructure modeling, servo-hydraulic control law, and actuator compensation scheme (discussed later) to be integrated into a single SIMULINK model on the Simulation Workstation and then downloaded onto the Target Workstation using Mathworks xPC Target Software (MATLAB 2007).

17.4.2 Analytical Substructure Modeling

The analytical substructure model of the MRF shown in Fig. 17.5 has a total of 122° of freedom and 71 elements. Inelastic behavior is modeled by means of a bilinear hysteretic lumped plasticity beam-column element with 3% hardening and an appropriate axial-moment yield surface. In order to overcome the shortcomings of the lumped plasticity modeling in predicting accurately the plastic rotation in the members of the structure, each physical member (i.e., beams and columns) was modeled with three beam-column elements in series, i.e., two elements were used to model the two plastic hinge regions at each end of the member with a length equal to 5% of the member length and one element with a length equal to the remaining 90% of the member length. For the steel MRF under consideration, this modeling approach was found to produce inelastic response close to the response obtained with a rigorous analysis using fiber beam-column elements (Karavasilis et al. 2009). Diaphragm action was assumed at every floor in the MRF due to the presence of a composite floor slab. The lean-on column was used to model P-Δ effects on the MRF from gravity loads carried by the gravity columns of the building that were in the tributary area of the perimeter MRF.

17.4.3 Experimental Substructure Setup

Figure 17.7 shows the experimental setup for the real-time hybrid simulations, which consists of the experimental substructures (two compressed elastomer dampers) and two servo-hydraulic actuators with supports, roller bearings and reaction frames.

The two actuators (see Fig. 17.5) have a load capacity of 2,300 and 1,700 kN with a maximum velocity of 840 and 1,140 mm/s, respectively, when three 1,514 l/min three-stage servo-valves are mounted on each actuator. The 2,300 and 1,700 kN actuators were attached to the experimental substructures associated with the first story damper and second story damper, respectively. The servo-controller for the actuator used in real-time hybrid simulations consists of a digital PID controller with a proportional gain of 20, integral time constant of 5.0 resulting in an integral gain of 4.0, differential gain of zero and a roll-off frequency of 39.8 Hz.

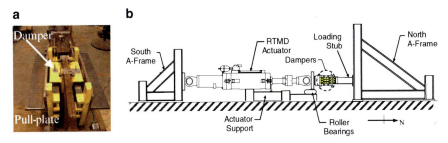

Fig. 17.7 Compressed elastomer dampers: (**a**) closeup of damper, and (**b**) details of test setup for each damper

17.4.4 Real-Time Integration of the Equations of Motion

For the MRF with the dampers of Fig. 17.5, the temporal discretized equations of motion at the $i+1^{th}$ time step can be expressed as

$$\mathbf{M} \cdot \ddot{\mathbf{x}}_{i+1} + \mathbf{C} \cdot \dot{\mathbf{x}}_{i+1} + \mathbf{r}^a_{i+1} + \mathbf{r}^e_{i+1} = \mathbf{F}_{i+1} \quad (17.1)$$

where $\ddot{\mathbf{x}}_{i+1}$ and $\dot{\mathbf{x}}_{i+1}$ are the acceleration and velocity vectors of the structure, respectively; $\mathbf{r}^a_{i+1}$ and $\mathbf{r}^e_{i+1}$ are the restoring force vectors of the analytical and experimental substructures, respectively; $\mathbf{M}$ and $\mathbf{C}$ are the mass and damping matrices of the structure, respectively; and $\mathbf{F}_{i+1}$ is the excitation force.

The CR unconditionally stable explicit integration algorithm (Chen and Ricles 2008a; Chen et al. 2009) is used to solve Eq. 17.1 for the structural displacement vector $\mathbf{x}_{i+1}$. According to the CR algorithm, the variations of the displacement and velocity vectors of the structure over the integration time step Δt are defined as

$$\dot{\mathbf{x}}_{i+1} = \dot{\mathbf{x}}_i + \Delta t \cdot \boldsymbol{\alpha}_1 \cdot \ddot{\mathbf{x}}_i \quad (17.2a)$$

$$\mathbf{x}_{i+1} = \mathbf{x}_i + \Delta t \cdot \dot{\mathbf{x}}_i + \Delta t^2 \cdot \boldsymbol{\alpha}_2 \cdot \ddot{\mathbf{x}}_i \quad (17.2b)$$

where $\mathbf{x}_i, \dot{\mathbf{x}}_i$ and $\ddot{\mathbf{x}}_i$ are the displacement, velocity and acceleration vectors of the structure at the ith time step, respectively; and $\boldsymbol{\alpha}_1$ and $\boldsymbol{\alpha}_2$ are matrices of integration parameters defined as

$$\boldsymbol{\alpha}_1 = \boldsymbol{\alpha}_2 = 4 \cdot \left(4 \cdot \mathbf{M} + 2 \cdot \Delta t \cdot (\mathbf{C} + \mathbf{C}_{eq}) + \Delta t^2 \cdot (\mathbf{K} + \mathbf{K}_{eq})\right)^{-1} \cdot \mathbf{M} \quad (17.3)$$

In Eq. 17.3 $\mathbf{K}$ is the initial stiffness matrix of the structure while $\mathbf{K}_{eq}$ and $\mathbf{C}_{eq}$ are matrices that contain terms associated with the equivalent stiffness k_{eq} and the damping c_{eq}, respectively, for the compressed elastomer dampers. For a sinusoidal deformation loading history of cyclic frequency ω and deformation amplitude u_d, k_{eq} and c_{eq} are equal to $K_d(u_d)$ and $\eta(u_d) \cdot k_{eq}/\omega$, respectively. In the real-time hybrid simulations presented herein, u_d and ω where assumed equal to the expected damper deformation from the SDP and the first mode cyclic frequency of the building, respectively. Preliminary real-time hybrid simulations showed that the response results where insensitive to small variations of the selected values of u_d and ω.

In a real-time hybrid simulation, Eqs. 17.2a and 17.2b are used to obtain the velocity $\dot{x}_{i+1}$ and displacement x_{i+1} vectors at the $i+1^{th}$ time step. The displacement vector x_{i+1} is decomposed into the analytical displacement vector x_{i+1}^a and the experimental (or command) displacement vector $\mathbf{x}_{i+1}^e$, which are imposed onto the analytical and experimental substructures, respectively, to obtain the restoring force vectors r_{i+1}^a and r_{i+1}^e. Strictly speaking, $\mathbf{x}_{i+1}^e$ contains deformations, i.e., differences in the displacements of the nodes at the ends of the experimental substructures. The analytical restoring force vector r_{i+1}^a is obtained with a standard nonlinear state-determination procedure for each beam-column element in the analytical substructure (Karavasilis et al. 2009), while the experimental restoring force vector r_{i+1}^e is obtained from the feedback forces measured using load cells that are placed in each compressed elastomer damper test setup. The equilibrium Eq. 17.1 is then employed to calculate the acceleration response vector $\ddot{x}_{i+1}$ at the $i+1^{th}$ time step, and the velocity $\dot{x}_{i+2}$ and displacement x_{i+2} vectors for the next $i+2^{th}$ time step are then readily available from Eqs. 17.2a and 17.2b. This process is repeated for each subsequent time step to obtain the response over the whole duration of the earthquake ground motion.

The integration time step Δt used for the hybrid tests is a multiple of the servo-hydraulic controller sampling time δt of 1/1,024 s, and equal to 10/1,024 s. This size of the time step was arrived at by performing a convergence study to ensure that value for Δt was sufficiently small enough that the integration algorithm produced accurate results. A linear ramp generator is used to apply the command displacement vector x_{i+1}^e through the hydraulic actuators to the experimental substructures at the servo-controller sampling rate, i.e., at a time step δt of 1/1,024 s. The interpolated command displacement vector is defined as

$$\mathbf{d}_{i+1}^{c(j)} = \frac{j}{n} \cdot (\mathbf{x}_{i+1}^e - \mathbf{x}_i^e) + \mathbf{x}_i^e \qquad (17.4)$$

In Eq. 17.4, $\mathbf{d}_{i+1}^{c(j)}$ is the command displacement vector at the jth substep within the $i+1^{th}$ time step; x_i^e is the command displacement vector at the ith time step; and j is the substep index for the interpolation within one single time step and ranges from 1 to n, where n is the integer ratio of $\Delta t/\delta t$ (i.e., equal to 10 for the tests presented herein).

As noted above, to proceed to the next $(i+2)^{th}$ time step the restoring force vectors r_{i+1}^a and r_{i+1}^e at the end of the $(i+1)^{th}$ time step must be obtained to calculate the displacement vector x_{i+2}. The available time to perform the state determination of the analytical substructure and form the r_{i+1}^a restoring force vector is equal to the integration time step Δt. For the analytical substructure of Fig. 17.5, r_{i+1}^a was obtained within the duration of Δt without creating any time delay issues. However, if the measured experimental restoring force vector r_{i+1}^e is fed back at the end of the time step after the actuators reach their corresponding command displacement $\mathbf{x}_{i+1}^e$,

17 Experimental Evaluation of the Seismic Performance of Steel Buildings... 335

a delay occurs while x_{i+2} is calculated and sent to the servo-controller, which reads the command displacement x^e_{i+2} one sampling time step δt later. To avoid this delay and ensure a smooth and continuous movement of the actuators, the experimental restoring force vector is extrapolated at the end of the $(n\text{-}1)^{\text{th}}$ substep within each time step (e.g., the $(i+1)^{\text{th}}$ time step) to become available before the actuators reach their command displacement x^e_{i+1} (Chen et al. 2009), where for each experimental substructure the restoring force contribution to r^e_{i+1} is:

$$r^e_{i+1} = r^{m(n-1)}_{i+1} + k_{eq} \cdot (x^e_{i+1} - d^{c(n-1)}_{i+1}) + c_{eq} \cdot (\dot{x}^e_{i+1} - \dot{x}^{m(n-1)}_{i+1}) \qquad (17.5)$$

In Eq. 17.5 $r^{m(n-1)}_{i+1}$ is the measured restoring force of the experimental substructure for the $(n-1)^{\text{th}}$ substep of the $(i+1)^{\text{th}}$ time step, $\dot{x}^e_{i+1}$ and $\dot{x}^{m(n-1)}_{i+1}$ are the target relative velocity between the nodes at the ends of the experimental substructure based on the CR integration algorithm (Eq. 17.2a) and the measured velocity in the damper for the $(n-1)^{\text{th}}$ substep of the $(i+1)^{\text{th}}$ time step, respectively, and, as noted above, c_{eq} and k_{eq} are the equivalent damping and equivalent stiffness of the elastomeric damper of the experimental substructure, respectively. The velocity of the experimental substructure is constant within the integration time step Δt due to the linear ramp generator. Therefore, in the extrapolation procedure the last term in Eq. 17.5 is included to minimize the error in the velocity-dependent restoring force of the experimental substructure (elastomeric damper) at the end of the time step by correcting for the difference between the target velocity $\dot{x}^e_{i+1}$ and the velocity produced by the linear ramp generator.

17.4.5 Actuator Delay Compensation

Due to inherent servo-hydraulic dynamics, the actuator has an inevitable time delay in response to the displacement command. This time delay is usually referred to as actuator delay and will result in a desynchronization between the measured restoring forces from the experimental substructure(s) and the integration algorithm in a real-time hybrid simulation. Studies on the effect of actuator delay (Chen and Ricles 2008b) show that actuator delay is equivalent to creating negative damping, which can destabilize a real-time hybrid simulation if not compensated properly.

To minimize the detrimental effect of actuator delay during the real-time hybrid simulations, the adaptive inverse compensation (AIC) method developed by Chen and Ricles (2011) was used to compensate for actuator delay during the simulations. The AIC method for a servo-hydraulic can be expressed using the following discrete transfer function that relates the compensated command displacement to the original command displacement for the actuator:

$$G_c(z) = \frac{(\alpha_{es} + \Delta\alpha) \cdot z - (\alpha_{es} + \Delta\alpha - 1)}{z} \qquad (17.6)$$

In Eq. 17.6 z is the complex variable in the discrete z-domain; $\Delta\alpha$ an evolutionary variable with an initial value of zero; and α_{es} the estimated actuator delay constant that is defined as

$$\alpha_{es} = \frac{estimated\ actuator\ delay}{\delta t} + 1 \qquad (17.7)$$

where the *estimated actuator delay* is the estimated duration for the servo-hydraulic to achieve the command displacement and δt the hydraulic servo-controller sampling time. The AIC method uses the initial estimated value for α_{es} for actuator delay compensation at the beginning of the hybrid simulation. The evolutionary variable $\Delta\alpha$ is used to adjust the initial estimated value for α_{es} to achieve accurate actuator control during a real-time hybrid simulation. The adaptation of the evolutionary variable $\Delta\alpha$ is based on a tracking indicator TI (Mercan 2007):

$$\Delta\alpha(t) = k_p \cdot TI(t) + k_i \cdot \int_0^t TI(\tau)d\tau \qquad (17.8)$$

In Eq. 17.8 k_p and k_i are proportional and integrative adaptive gains of the adaptive control law, respectively. The TI is based on the enclosed area of the hysteresis in the synchronization subspace plot shown in Fig. 17.8, where the actuator command displacement d^c is plotted against the actuator measured displacement d^m. The calculation of TI at each substep within a time step is formulated as (Mercan 2007)

$$TI_{i+1}^{(j)} = 0.5\left(A_{i+1}^{(j)} - TA_{i+1}^{(j)}\right) \qquad (17.9)$$

In Eq. 17.9, $A_{i+1}^{(j)}$ and $TA_{i+1}^{(j)}$ are the accumulated enclosed and complementary enclosed areas at the jth substep of the ramp generator at time step $i+1$, respectively, and are calculated as

$$A_{i+1}^{(j)} = A_{i+1}^{(j-1)} + dA_{i+1}^{(j)} = A_{i+1}^{(j-1)} + 0.5\left(d_{i+1}^{c(j)} + d_{i+1}^{c(j-1)}\right)\left(d_{i+1}^{m(j)} - d_{i+1}^{m(j-1)}\right) \qquad (17.10a)$$

$$TA_{i+1}^{(j)} = TA_{i+1}^{(j-1)} + dTA_{i+1}^{(j)} = TA_{i+1}^{(j-1)} + 0.5\left(d_{i+1}^{m(j)} + d_{i+1}^{m(j-1)}\right)\left(d_{i+1}^{c(j)} - d_{i+1}^{c(j-1)}\right) \qquad (17.10b)$$

The incremental values for the enclosed and complementary enclosed areas, $dA_{i+1}^{(j)}$ and $dTA_{i+1}^{(j)}$, respectively, are shown in Fig. 17.8 for the jth substep of time step $i+1$. At the beginning of the test, the enclosed and complementary areas have initial values of zero. The calculation of A and TA continues for every substep of each time step until the end of the real-time hybrid simulation. Chen and Ricles (2011) showed that the use of the AIC method for real-time hybrid simulation of structures with experimental substructures consisting of passive MR dampers resulted in good actuator tracking and test results in comparison with numerical simulation results for structural response.

For the real-time hybrid simulations a value of $\alpha_{es} = 30$ for the estimate of the actuator delay constant along with the values for the adaptive gains of $k_p = 0.4$ and $k_i = 0.04$

Fig. 17.8 Sychronization subspace with increment of enclosed area $dA_{i+1}^{(j)}$ and complementary enclosed area $dTA_{i+1}^{(j)}$ utilized in the determination of the tracking indicator *TI*

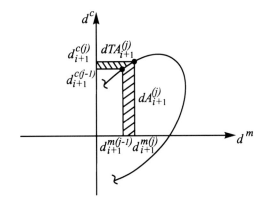

for both servo-hydraulic actuators were used. These values were established by conducting parametric studies of the servo-hydraulic systems at the RTMD (Lehigh RTMD Users Guide 2009).

17.5 Real-Time Hybrid Simulation Results

An ensemble of five earthquake ground motions recorded on stiff soil sites (without near-fault effects) are used in real-time hybrid simulation to evaluate the performance of the MRF with compressed elastomer dampers (Table 17.3). The ground motions were scaled to the DBE level using the scaling procedure of Somerville (1997). The amplitudes of these DBE ground motions were further scaled by 1.5 to represent MCE ground motions.

Time history results from real-time hybrid simulation are presented for the HSP090 record from the Loma Prieta 1989 earthquake scaled to the DBE and MCE intensities. A comparison between the measured, d^m, and command, d^c, actuator displacement for the two compressed elastomer dampers is presented in Fig. 17.9, where subspace synchronization plots of d^c versus d^m are plotted. Good agreement can be observed between d^m and d^c; where the plots for each actuator show no noticeable deviation between the measured and command actuator displacement. The root mean square (RMS) error between d^m and d^c was evaluated as

$$RMS = \sqrt{\frac{\sum_i \sum_{j=1}^{n}\left(d_i^{c(j)} - d_i^{m(j)}\right)^2}{\sum_i \sum_{j=1}^{n}(d_i^{c(j)})^2}} \qquad (17.11)$$

The RMS values were found equal to 3.6e-4 and 3.7e-4 for the actuators of the first story and second story damper under the DBE level, respectively. The corresponding

Table 17.3 Earthquake ground motions used in real-time hybrid simulations

Earthquake	Station	Component	Magnitude (M_w)	Distance (km)	PGA (g)	Scale factor	
						DBE	MCE
Loma Prieta 1989	Hollister – S&P	HSP090	6.93	27.67	0.18	1.99	2.99
Manjil 1990	Abbar	ABBAR–T	7.37	12.56	0.46	0.96	1.44
Northridge 1994	N Hollywood – Cw	CWC270	6.69	7.89	0.27	1.70	2.56
ChiChi 1999	TCU105	TCU105-E	7.62	17.18	0.12	2.45	3.67
ChiChi 1999	TCU049	TCU049-E	7.62	3.78	0.29	1.92	2.89

17 Experimental Evaluation of the Seismic Performance of Steel Buildings... 339

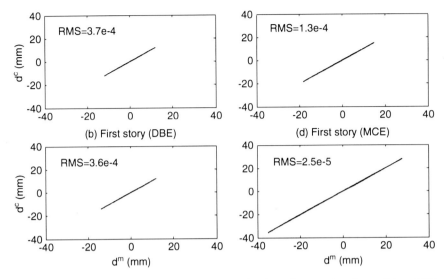

Fig. 17.9 Actuator displacement subspace synchronization subspace plots, Loma Prieta 1989 HSP090 ground motion

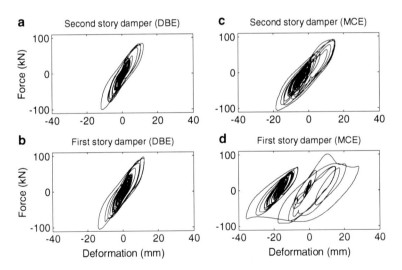

Fig. 17.10 Damper hysteresis from real-time hybrid simulation, Loma Prieta 1989 HSP090 ground motion

RMS values for the MCE level were found equal to 2.5e-5 and 1.3e-4. These values for the RMS and the results shown in Fig. 17.9 are representative of the accurate actuator control achieved in all of the hybrid simulations.

The hysteresis of the compressed elastomer dampers is presented in Fig. 17.10. Under the DBE the dampers in both stories exhibit an elastomeric behavior with

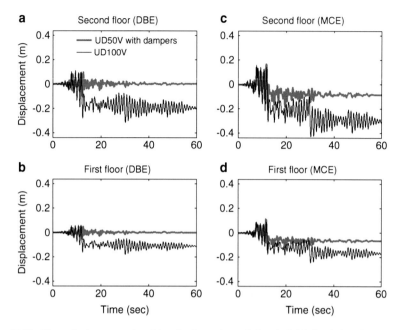

Fig. 17.11 Floor displacement time histories from the real-time hybrid simulation, Loma Prieta 1989 HSP090 ground motion

fairly rounded peaks. (i.e., Fig. 17.10a, b). Under the MCE the damper in the second story develops some minor slip (Fig. 17.10c), while the damper at the first story experiences an elastomeric-frictional behavior with slip (Fig. 17.10d) that results in permanent deformation, although the damper continues to develop energy dissipation.

Figure 17.11 shows the floor displacement time histories of the UD50V MRF with dampers. Also presented in Fig. 17.11 are the floor displacement time histories of the conventional UD100V SMRF from numerical analysis. The real-time hybrid simulations show that the lighter UD50V MRF with dampers experiences significantly lower transient and permanent story drifts than those of the conventional UD100V SMRF. Under the DBE earthquake the UD50V MRF with dampers has negligible permanent story drift since the dampers do not slip and have re-centering capability. Under the MCE the dampers act as sacrificial elements, which develop permanent deformation due to slip, however as discussed previously, the dampers can be replaced after the earthquake. Some modest yielding occurs in the beams and at the ground level of the columns of the UD50V MRF with dampers. As will be discussed below, the plastic (and associated permanent) deformations in the MRF with dampers are small. If the dampers were replaced or re-centered after the MCE, the residual drift of the MRF with dampers under the MCE could be significantly reduced.

The added benefit of real-time hybrid simulation is that it allows an unlimited number of ground motions to be applied to the structure and therefore, statistical

17 Experimental Evaluation of the Seismic Performance of Steel Buildings...

Table 17.4 Median response results from real-time hybrid simulations

Design		θ_{max} (%)		$\theta_{pl.max\_bm}$ (rad)		$\theta_{pl.max\_col}$ (rad)		v_{max} (m/s)		a_{max} (m/s)	
		DBE	MCE	DBE	MCE	DBE	MCE	DBE	MCE	DBE	MCE
UD100V	Story 1	2.60	2.90	0.008	0.014	0.010	0.015	0.78	1.00	5.32	6.60
conventional SMRF	Story 2	2.40	2.60	0.000	0.003	0.003	0.000	1.11	1.28	5.66	6.36
UD50V MRF	Story 1	1.35	2.50	0.002	0.006	0.006	0.002	0.61	0.90	4.18	5.70
with dampers	Story 2	1.40	1.80	0.000	0.000	0.000	0.000	0.77	1.10	5.16	6.50

experimental response results incorporating the ground motion variability can be obtained. In this paper, the seismic performance of the MRF with dampers is quantified in terms of various damage indices for both structural and non-structural components, and include the maximum story drift, θ_{max}; maximum plastic hinge rotation for beams, $\theta_{pl.max\_bm}$, and columns, $\theta_{pl.max\_col}$; peak floor absolute velocity, v_{max}; peak floor absolute acceleration, a_{max}; and the floor acceleration response spectra, $S_{a,flr}$. Table 17.4 presents median experimental response values for the θ_{max}, $\theta_{pl.max\_bm}$, $\theta_{pl.max\_col}$, v_{max} and a_{max} of the UD50V MRF with dampers from the real-time hybrid simulations. Also presented in Table 17.4 are the median values of the same response quantities of the conventional SMRF UD100V from the numerical analysis. Table 17.4 shows that the median θ_{max} value of 1.35% and 1.40% for the first and second stories, respectively, for UD50V MRF under the DBE is lower than the anticipated θ_{max} demand of 1.65% given in Table 17.2, while the median value of 2.60% and 2.40% for the first and second stories, respectively, the median θ_{max} for the UD100V SMRF is larger than the anticipated θ_{max} demand of 2.4% given in Table 17.1. Decreases in the median peak beam plastic hinge rotations in the UD50V MRF with dampers are approximately 75% and 57% for the DBE and MCE, respectively, compared to UD100V SMRF. For the columns, the median peak plastic hinge rotations in the UD50V MRF with dampers are approximately 80% and 33% less than that in the UD100V SMRF for the DBE and MCE, respectively. The median peak floor velocities of the UD50V MRF with dampers are 22% and 31% less at the first and second floors, respectively, than those of the UD100V SMRF for the DBE. For the MCE the median peak floor velocities of the UD50V MRF with dampers are 10% and 14% less at the first and second floors, respectively, than those of the UD100V SMRF The median peak floor accelerations of the UD50V MRF with dampers are 21% and 9% less at the first and second floors, respectively, than those of the UD100V SMRF for the DBE. Under the MCE, the UD50V MRF with dampers experiences a 14% reduction in the first floor median peak acceleration and a slightly higher second floor median peak acceleration than that of the UD100V SMRF.

Figure 17.12 shows the median floor acceleration response spectra S_a of UD50V MRF with dampers for the DBE and MCE levels. Also presented in Fig. 17.12 are S_a of the conventional UD100V SMRF from numerical analysis. The spectra show that the UD50V MRF with dampers performs significantly better

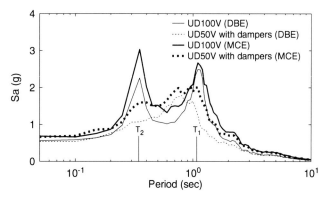

Fig. 17.12 Median second floor acceleration response spectra from real-time hybrid simulations

than conventional UD100V SMRF. It is evident that the resonance at the first and second modes of vibration of UD50V MRF with dampers is effectively damped (the period of vibration for the first and second modes are $T_1 = 1.04$ s and $T_2 = 0.35$ s, respectively).

17.6 Summary and Conclusions

An experimental program based on the use of real-time hybrid simulation to verify the performance-based seismic design of a steel perimeter MRF equipped with compressed elastomer dampers was presented. The experimental substructures for the simulation consisted of two individual compressed elastomer dampers, with the remainder of the MRF and associated tributary gravity columns and gravity loading of the building modeled as an analytical substructure. Real-time hybrid simulation allowed an ensemble of ground motions to be applied to the structure resulting in various levels of damage, without the need to repair the test specimens, since the damage was within the analytical substructure.

Statistical experimental response results incorporating the ground motion variability showed that a steel MRF with compressed elastomer dampers can be designed to perform better than conventional steel SMRFs under the DBE and MCE, even when the MRF with dampers is significantly lighter in weight than the conventional SMRF. In particular, the steel MRF with dampers was designed with a practical number of dampers (eight dampers in the first story and five dampers in the second story) and with a 30% lower steel weight than that of the conventional steel MRF. Real-time hybrid simulations showed that the MRF with dampers experiences significantly lower peak story drifts, peak plastic hinge rotations, lower peak absolute floor velocities and floor accelerations in addition to floor spectra accelerations than those of the conventional steel SMRF.

Acknowledgments This paper is based upon work conducted at Lehigh University and supported by grants from the Pennsylvania Department of Community and Economic Development through the Pennsylvania Infrastructure Technology Alliance, and by the National Science Foundation (NSF) under Grant No. CMS-04002490 within the George E. Brown, Jr. Network for Earthquake Engineering Simulation Consortium Operation. Any opinions, findings, and conclusions expressed in this paper are those of the authors and do not necessarily reflect the views of the sponsors.

References

BSSC (2004) NEHRP recommended provisions for seismic regulations for new buildings and other structures. Report FEMA 450, FEMA, Washington, DC

Chang KC, Lin YY (2004) Seismic response of a full-scale structure with added viscoelastic dampers. J Struct Eng 130(4):600–608

Chen C, Ricles JM (2008a) Development of direct integration algorithms for structural dynamics using discrete control theory. J Eng Mech 134(8):676–683

Chen C, Ricles JM (2008b) Stability analysis of SDOF real-time hybrid testing systems with explicit integration algorithms and actuator delay. Earthq Eng Struct Dyn 37(4):597–613

Chen C, Ricles JM (2011) Tracking error-based servo-hydraulic actuator adaptive compensation for real-time hybrid simulation. J Struct Eng 136(4):432–440

Chen C, Ricles JM, Marullo TM, Mercan O (2009) Real-time hybrid testing using the unconditionally stable explicit CR integration algorithm. Earthq Eng Struct Dyn 38(1):23–44

International Code Council (ICC) (2003) International building code. International Code Council, Falls Church

Karavasilis TL, Ricles JM, Marullo T, Chen C (2009) HybridFEM. A program for nonlinear dynamic time history analysis and real-time hybrid simulation of structures. ATLSS Engineering Research Center Report No 09–08, Lehigh University, Bethlehem, PA

Karavasilis TL, Sause R, Ricles JM (2011) Seismic design and evaluation of steel MRFs with compressed elastomer dampers. Earthq Eng Struct Dyn 40

Kasai K, Motoyui S, Ozaki H, Ishii M, Ito H, Kajiwara K, Hikino T (2009) Full-scale tests of passively-controlled 5-story steel building using E-Defense shake table. STESSA2009, Philadelphia

Lee KS, Fan CP, Sause R, Ricles JM (2005) Simplified design procedure for frame buildings with viscoelastic or elastomeric structural dampers. Earthq Eng Struct Dyn 34:1271–1284

Lehigh RTMD Users Guide (2009) http://www.nees.lehigh.edu/resources/users-guide, Lehigh University, Bethlehem, PA

MATLAB (2007) The Math Works, Inc., Natick, MA

Mercan O (2007) Analytical and experimental studies on large scale, real-time pseudodynamic testing. PhD dissertation. Department of Civil and Environmental Engineering, Lehigh University, Bethlehem, PA

Sause R, Hemingway GJ, Kasai K (1994) Simplified seismic response analysis of viscoelastic-damped frame structures. In: 5th U.S. National Conference on Earthquake Engineering, EERI, Chicago

Somerville P (1997) Development of ground motion time histories for phase 2 of the FEMA/SAC steel project. Report No. SAC/DB-97/04. FEMA, Sacramento

Sweeney SK, Michael R (2006) Collaborative product realization of an innovative structural damper and application. IMECE2006, ASME International Engineering Congress and Exposition, Chicago

Chapter 18
Experimental Investigation of the Seismic Behaviour of Precast Structures with Pinned Beam-to-Column Connections

Ioannis N. Psycharis, Haralambos P. Mouzakis, and Panayotis G. Carydis

Abstract Experimental results from three series of experiments that have been conducted at the Laboratory for Earthquake Engineering of the National Technical University of Athens, Greece, on assemblies or subassemblies of precast structures with dry, pinned beam-to-column connections are reported: the first series concerned shaking table tests on one-storey, practically full-scale 3D specimens composed of linear precast elements; in the second series, monotonic and cyclic pure shear tests were performed on pinned beam-to-column connections; and in the third series, shaking table tests were conducted on a part of a frame composed of a column and a beam. Various parameters were examined, as the diameter of the dowels, the number of dowels and the distance of the dowels from the beam's front. The effect of the dynamic loading on the behaviour of the connections was investigated by comparing the results of the dynamic tests with the corresponding results of the cyclic tests.

18.1 Introduction

The technology of prefabrication in building construction has been used for more than half of a century and is spreading considerably and successfully all over the world. The increasing lack of on-site labor work in the recent years on the one hand and the need for fast and economic constructions on the other hand, combined with

---

I.N. Psycharis (✉) • H.P. Mouzakis • P.G. Carydis
Laboratory for Earthquake Engineering, Department of Civil Engineering,
National Technical University of Athens, Iroon Polytechniou 5 Zografou,
15780 Athens, Greece
e-mail: ipsych@central.ntua.gr; harrismo@central.ntua.gr; pcarydis@central.ntua.gr

M.N. Fardis and Z.T. Rakicevic (eds.), *Role of Seismic Testing Facilities in Performance-Based Earthquake Engineering: SERIES Workshop*, Geotechnical, Geological and Earthquake Engineering 22, DOI 10.1007/978-94-007-1977-4_18,
© Springer Science+Business Media B.V. 2012

the need to minimize the influence of the weather conditions on the construction procedure, which affects the construction time and the quality of the structure, tend to impose more and more the prefabrication as one of the leading construction technologies.

In precast structures, of major importance is the type of the connections, their position into the structural system and the type of the structural system itself. Especially for seismic excitations, the response of the connections determines the behaviour of the entire system. On this subject, considerable research has been reported worldwide. For the design of precast framed structures, a state of the art paper on recent research and development was presented by Elliott (2000).

The most notable effort on the experimental investigation of the earthquake response of precast structures has been the PRESSS (Precast Seismic Structural Systems) project (Priestley 1996; Nakaki et al. 1999; Stanton 1998) carried out in the USA and Japan. The main objective of this project was to develop design guidelines for broader acceptance of precast concrete construction in seismic zones and to develop new concepts and technologies for precast systems to be applied in regions of high seismicity. Thus, it does address directly the main questions of traditional techniques of prefabrication as they usually appear around the world.

Significant research related to the behaviour of precast connections was reported by Imai et al. (1993) and Watanabe (2000), who have investigated semi–rigid connections within the European project COST C1 Action (COST C1 1999). A wide range of materials and geometries were studied; the results, however, showed significant diversion (Virdi and Ragupathy 1992a, b; Elliott 1998; De 1998; Elliott et al. 1998; Jolly and Guo 1998).

Most of the above-mentioned research has been focused on the seismic behaviour of moment resisting connections, since the international practice on prefabrication shows that, in many countries, emulative connections are used at beam-to-column joints. However, it is worth mentioning that extensive use of simple pinned beam-to-column connections of linear elements is common in southern Europe and elsewhere for single-storey or low-rise precast, mainly industrial buildings. Little investigation has been reported on such pinned connections; among others one should mention the research by Leong (2006), El Debs et al. (2006) and Rahman et al. (2008) on the strength and stiffness of the pinned dowel connections.

The dowel mechanism was investigated extensively by Vintzeleou and Tassios (1987), while Tsoukantas and Tassios (1989) conducted an analytical investigation on the shear resistance of connections between linear precast elements under monotonic and cyclic loading and proposed design values for the shear resistance of each connection and its corresponding shear slip for rough and smooth surfaces.

Recently, fib Bulletin 43 (International Federation for Structural Concrete (fib) 2008), drafted by Task Group 6.2 of fib Commission 6 "Prefabrication" was published, in which considerable information on practically every type of beam-to-column connection is given concerning its design and behaviour. However, these guidelines were based merely on the behaviour of the connections under monotonic loading and do not directly reflect the influence of strength and deformability of the connections on the overall behaviour of a precast structure.

Considering the limited research that has been conducted on structures composed by linear precast elements with pinned beam-to-column connections, the experimental research presented herein aims to investigate the behaviour of such structures under severe seismic excitation. To this end, various experimental setups were investigated: (1) shaking table tests on one-storey, practically full-scale 3D specimens composed of linear precast elements; (2) monotonic and cyclic pure shear tests on pinned beam-to-column connections; and (3) shaking table tests on subassemblies composed of a column and a beam.

All the tests were performed at the Laboratory for Earthquake Engineering of the National Technical University of Athens, Greece within the framework of two wider analytical and experimental research programmes: the FP5 project "Precast EC8: Seismic behaviour of precast concrete structures with respect to Eurocode 8 (Co-Normative Research)", which was coordinated by Professor G. Toniolo of the Polytechnic of Milan, Italy and the FP7 project "SAFECAST: Performance of innovative mechanical connections in precast building structures under seismic conditions", which is coordinated by Dr. Antonella Colombo, ASSOBETON, Italy. Research centres from Italy, Greece, Portugal, Slovenia, Turkey and China participated in the projects.

18.2 First Series of Tests: Seismic Behaviour of Single-Storey Precast Structures

The experimental investigation of the seismic response of single-storey precast structures was performed in the framework of the FP5 project "Precast EC8: Seismic behaviour of precast concrete structures with respect to Eurocode 8 (Co-Normative Research)."

18.2.1 Description of the Specimens

The shaking table tests were performed on four specimens (#2 to #5) that were designed to comply with typical precast single-storey frame systems, used extensively for industrial and residential buildings in Greece and other countries in Europe. Such precast frame systems consist of a series of portal frames, each one composed of cantilever columns fixed at their bottom into socket foundations and roof beams (in one or both directions) simply connected to the columns by means of steel dowels (pinned connections). The roof is formed by precast roof elements (usually T-shape girders) simply supported at their ends on the roof beams.

Due to size restrictions imposed by the dimensions of the shaking table (4.00 m × 4.00 m), each specimen was composed of only two portal frames (four columns) and had much smaller plan dimensions compared to typical real structures

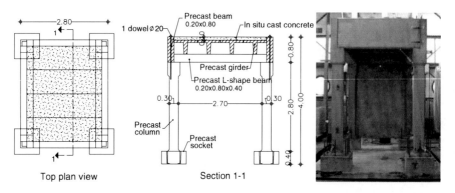

Fig. 18.1 First series of tests: geometry of specimens #2 and #3 (The photo is of specimen #3)

of this type. The beam-to-column connections were constructed in physical scale, in order to avoid scaling effects during their inelastic response. Two types of connections were checked: In specimens #2 and #3, the beam-to-column connection was provided by one steel dowel Ø20 at each beam end, grouted and bolted on top. The connection was "enhanced" by in situ cast concrete, poured in a nest provided through flanges extruding upward from the column's upper faces. This type of connection cannot be considered as an emulative one, since the in situ poured concrete was practically unreinforced and could not establish a monolithic connection between beams and columns. In specimens #4 and #5, the beam-to-column connections were purely dry, provided by two steel dowels Ø32 at each beam end, again grouted and bolted on top. Rubber pads were added at the seating of the beams on the columns.

The roof was not the same in all specimens: in specimens #2 and #3 (Fig. 18.1), the roof was stiff, composed by precast T-shape girders and a lightly reinforced topping layer, 0.10 m thick; in specimens #4 and #5, the roof was more flexible, composed of precast girders of smaller height that were interconnected by cast in situ concrete poured on 0.30 m wide strips along the joints.

The column-to-foundation connections were provided by traditional precast sockets. The gap between the columns and the sockets was filled with non-shrinking concrete. In specimens #4 and #5, it was observed during testing that the relatively large gap between the column faces and the foundation sockets led to partial fixation at the bottom of the columns, especially for strong base excitations, for which the infill concrete cracked severely. This partial fixation of the columns resulted in smaller stiffness, similar to the one expected after the formation of plastic hinges at the base of the columns in a real structure.

The cross sections of the beams were in physical scale but the cross sections of the columns were smaller than in real structures (0.30 m × 0.30 m), in order to decrease the overall stiffness. Additional mass of 4 Mgr was added on top of each specimen in order to simulate the inertia forces that would be developed at a larger structure. Thus, it can be said that the specimens corresponded to small-size, single-storey buildings.

18 Experimental Investigation of the Seismic Behaviour of Precast Structures... 349

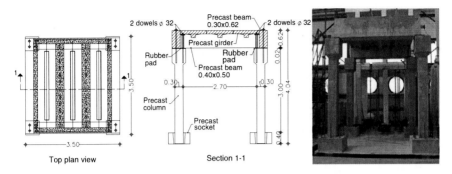

Fig. 18.2 First series of tests: geometry of specimens #4 and #5 (The photo is of specimen #4)

Specimens #2 and #3 were identical, apart from the presence of a precast panel wall that was mounted on one side of specimen #3. Specimens #4 and #5 were also identical, apart from the missing secondary beams in specimen #4, in which only beams in one direction were mounted (Fig. 18.2).

All specimens were designed according to Eurocode 2 (European Committee for Standardization (CEN) 2004a) and Eurocode 8 (European Committee for Standardization (CEN) 2004b) for a behaviour factor $q = 3.50$. Pinned connections between beams and columns were considered in the structural model. The concrete grade in all specimens was C25/30 and the steel grade was B500C. More details concerning the specimens and the results can be found in (Psycharis et al. 2006).

18.2.2 *Experimental Results*

Each specimen was subjected to a sequence of seismic excitations using the two horizontal components of the Edessa record of the Griva, Greece, 1990 earthquake ($M = 5$). The records show long duration of the strong shaking, about 6.0 s, which is governed by almost harmonic pulses of period about 0.6 s. For each specimen, several tests were performed with the base motion being step-wise amplified up to the point that significant damage was caused to the specimens, or when the limits of the shaking table were reached. The base excitation that corresponded to the last test was significantly larger than the design earthquake in all cases.

The instrumentation setup included accelerometers and displacement transducers properly placed for the measuring of accelerations, absolute and relative displacements and rotations at critical regions of the specimens, as the sliding between beams and columns, the rotation at the joints and the rotation at the base of the columns.

Quite different behaviour was observed for the two precast systems that were examined. For specimens #2 and #3, the tests stopped for maximum base excitation equal to 0.50 g, which caused severe damage. At this level of base motion, large

Fig. 18.3 First series of tests, specimens #2 and #3: typical damage at the joints

relative displacements were produced between beams and columns, which could not be accommodated by the small gap between beam ends and flanges of the columns. The infill concrete at the joints was disorganized and it rather worsened the behaviour by imposing random kinematic constraints that produced larger permanent deformations. The observed damage includes (Fig. 18.3): (i) residual opening of the joints due to permanent dislocations and torsional rotations of the beams, showing that the dowels bent during the seismic response; (ii) inclination of the columns as a result of the joints opening; (iii) damage at the beam and the column edges due to the impact between them (in these specimens no bearing pads were used for the seating of the beams on the columns); (iv) damage around the top of the dowels, showing that high stresses were developed at these places; (v) vertical cracks in the column flanges, probably caused by the impacts with the beams; (vi) severe cracks (practically total damage) in the infill concrete that was poured in the nest formed by the column flanges, showing that it was not able to provide any improvement in the seismic response of the joint. It is worth mentioning that small cracks were also observed at the bottom of the columns, indicating the formation of plastic hinges during the strong shaking.

Specimens #4 and #5 showed a much better behaviour and suffered minor damage only (small dislocations of the beams and some cracks at the beam edges, caused by the inevitable impact to the column during large rotations), even though they were subjected to very strong base excitations, up to 0.90 and 0.80 g respectively. For PGA = 0.50 g, when most of the damage of specimens #2 and #3 occurred, no permanent dislocations were observed. Their good behaviour must be attributed to the stronger beam-to-column connection, which, however, was able to accommodate the large deformations induced during the experiments.

The peak displacements that were recorded at the top of specimen #5 (slab and the top of columns) and the related slip at the joints are plotted in Fig. 18.4a for various levels of the peak base acceleration (PGA). It can be seen that, for PGA = 0.80 g, the slip at the joint exceeded 22 mm during the earthquake. The maximum drift was

18 Experimental Investigation of the Seismic Behaviour of Precast Structures...

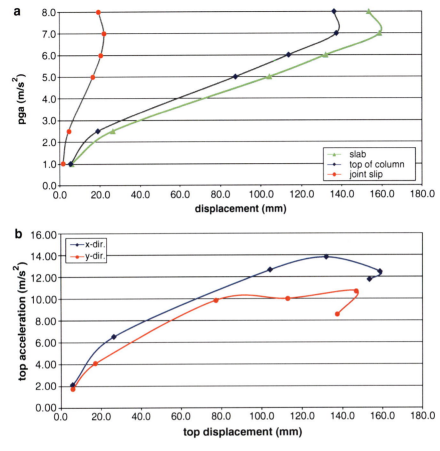

Fig. 18.4 First series of tests, specimen #5: (**a**) peak displacement and slip at the joints vs base peak acceleration (PGA); (**b**) roof peak acceleration vs roof peak displacement in x– and y–directions

recorded for specimen #4 (5.5% for PGA = 0.90 g), but it was mainly caused by the rotation at the base of the columns due to the severe cracking of the infill concrete of the foundation sockets; the resulted partial fixation at the base of the columns decreased the top accelerations and increased the top displacements. This behaviour simulates the response of structures, in which plastic hinges have been developed at the base of the columns. In Fig. 18.4b, the relation between the roof acceleration and the roof displacement in x– and y–directions are shown for specimen #5. These are capacity curves, similar to the ones that would be obtained by an incremental dynamic analysis (Vamvatsikos and Cornell 2002). It should be noted that the maximum capacity shown in these curves was determined by the bending capacity at the base of the columns and not by the shear capacity of the joints.

18.3 Second Series of Tests: Monotonic and Cyclic Pure Shear Tests on Pinned Beam-to-Column Connections

The experimental investigation of the behaviour of pinned beam-to-column connections under pure shear monotonic and cyclic loading was performed in the framework of the FP7 project "SAFECAST: Performance of innovative mechanical connections in precast building structures under seismic conditions."

The tested typologies represent the common solution in precast construction practice in Europe and elsewhere, mainly for single-storey industrial buildings. In total, 20 experiments (Table 18.1) were performed and the investigation was focused on the effect of:

- The diameter of the dowels. Most tests were performed on connections with 2 Ø25 dowels, the failure of which could be achieved within the limits of the hydraulic actuator. However, tests with 1 Ø25, 1 Ø32 and 2 Ø16 dowels were also performed. Connections with Ø16 dowels are rather unusual in practice, but were tested in order to investigate the response of connections with low strength and high deformability.

Table 18.1 Experimental programme of 2nd series of tests

Specimen – test	Dowels	Distance d of dowels from edge (m)	Type of test	Remarks
2D25-d10-PSH	2 Ø25	0.10	Push	
2D25-d10-PLL			Pull	
2D25-d10-CY-1			Cyclic	Start in push-direction
2D25-d10-CY-2			Cyclic	Start in pull-direction
2D25-d10-CY-PL			Cyclic	Start in pull-dir./with anchored steel plate
2D25-d15-PLL		0.15	Pull	
2D25-d15-CY			Cyclic	Start in pull-direction
2D25-d20-PLL		0.20	Pull	
2D25-d20-CY			Cyclic	Start in pull-direction
1D25-d10-PSH	1 Ø25	0.10	Push	
1D25-d10-PLL			Pull	
1D25-d10-CY			Cyclic	Start in pull-direction
1D25-d10-CY-F			Cyclic	Start in pull-direction/fibre reinforced concrete
2D16-d10-PSH	2 Ø16	0.10	Push	
2D16-d10-PLL			Pull	
2D16-d10-CY-1			Cyclic	Start in push-direction
2D16-d10-CY-2			Cyclic	Start in pull-direction
1D32-d20-PSH	1 Ø32	0.20	Push	
1D32-d20-PLL			Pull	
1D32-d20-CY			Cyclic	Start in pull-direction

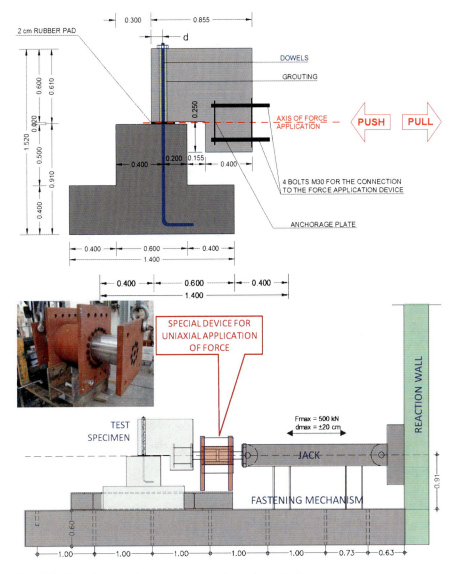

Fig. 18.5 Second series of tests: specimens and experimental setup

- The number of dowels. Most tests were performed on connections with two dowels, but in some tests connections with one dowel were used in order to investigate how the behaviour is related to the number of dowels.
- The distance d of the dowels from the beam edge (Fig. 18.5). In many cases in real structures, due to geometrical restrictions, the dowels are placed close to the beam's front face with the concrete cover being around 10 cm. The small cover of the dowels at the beam side results in smaller resistance of the connection in the pull direction than the one in the push direction.

18.3.1 Description of the Specimens

The connection that was tested and the experimental setup are presented in Fig. 18.5. The specimens consisted of a part of the beam and a part of the column, connected by steel dowels. They were constructed of high strength concrete C30/37 and the steel grade was B500C for both the dowels and the reinforcement. For the assembly of the specimens, waiting ducts of Ø65 were provided in the beams, which were grouted after construction with non-shrinking concrete. The dowels were bolted on top. The experiments were performed 24 h after grouting, when the strength of the grout had reached a value around 20 MPa. For the tests, a 500 kN capacity MTS hydraulic actuator, with displacement range ±200 mm, was used. No extra vertical load was applied.

The force was acting at the level of the joint, through a special device that ensured that it was applied in the longitudinal axis only and that no moment was induced to the joint.

The monotonic tests were executed in the pull or in the push direction up to the failure of the joint or when the maximum capacity of the actuator (500 kN) was reached. The cyclic tests were performed with displacement control at a rate of 0.2 mm/s. Three cycles were performed at each displacement level. The displacement step was initially equal to $0.5dy$ and later, towards the end of the test, equal to dy. The yield displacement, dy, was determined from the monotonic pull or monotonic push tests; in case of difference between the pull and the push direction, the minimum value was considered. The tests were stopped when significant strength degradation (over 20%) was observed.

18.3.2 Experimental Results

Indicative results of the experimental investigation are presented in the following. More details can be found in SAFECAST: Performance of innovative mechanical connections in precast building structures under seismic conditions (2011a).

The monotonic pull and monotonic push diagrams for the specimens with 2 Ø25 dowels and distance of the dowels from the beam edge $d = 10$ cm are presented in Fig. 18.6. In the push direction, the test stopped when the displacement was 50 mm and the shear force was 466 kN. The dowels did not break, but spalling of the concrete occurred in both the beam and the column in the direction normal to the loading (Fig. 18.7a). Also, significant splitting occurred at the concrete on top of the beam, around the dowels (Fig. 18.7b). Compared to the theoretical strength that was calculated from the formula by Vintzeleou and Tassios (1987) (equal to 199 kN), the achieved force was more than double.

In the monotonic pull test, the achieved force was much less, because spalling of the concrete occurred early in the experiment at the front face of the beam (Fig. 18.8a). After that point, a local plateau was observed at the force (around 117 kN, Fig. 18.6); however, the force started increasing again when the displacement

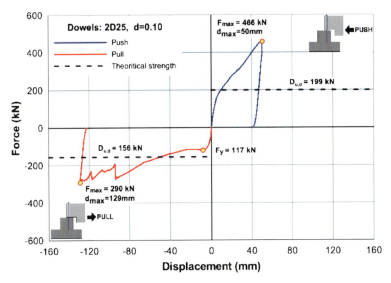

Fig. 18.6 Second series of tests: force–displacement diagrams for monotonic push (*blue line*) and monotonic pull (*red line*) tests of specimens 2D25-d10

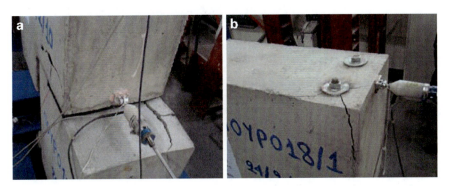

Fig. 18.7 Second series of tests: damage at the end of the monotonic push test 2D25-d10-PSH

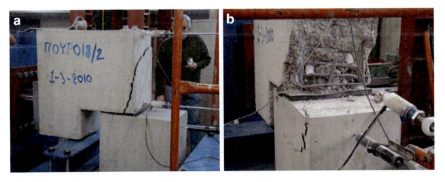

Fig. 18.8 Second series of tests: damage during the monotonic pull test 2D25-d10-PLL: (**a**) at early stage of the test; (**b**) at the end of the test

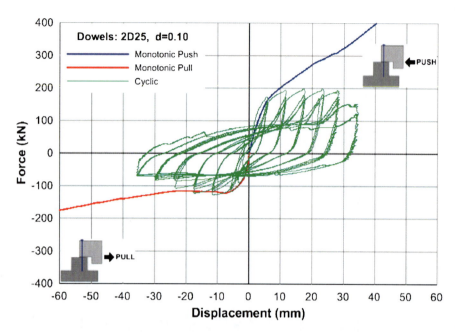

Fig. 18.9 Second series of tests: force–displacement diagrams for the cyclic tests of specimens 2D25-d10 and comparison with the results of monotonic tests

was 25 mm, when the dowels touched the horizontal hooks and thus their displacement was restricted. Compared to the theoretical strength (156 kN), the maximum attained force was almost double, but the first plateau occurred at smaller force than the theoretical value. The damaged beam at the end of the test is shown in Fig. 18.8b. Both dowels broke during the test.

The results of the corresponding cyclic tests are shown in Fig. 18.9. As expected, the response was not symmetric in the push and in the pull direction, because the resistance in the pull direction was reduced due to the small distance of the dowels from the beam edge ($d = 10$ cm). In both push and pull directions, the cyclic response follows the curves of the monotonic loading up to a point a little further yield. Compared to the theoretical prediction, larger resistance was observed even in the pull direction. The observed damage (Fig. 18.10) was similar to the one that occurred during the monotonic pull test.

For large displacements, the dowels suffered significant tension, which caused their residual elongation. During the cyclic tests, this permanent elongation forced the dowels to slide along the ducts, in order to accommodate the difference in their length between max/min displacement (maximum length) and zero displacement (minimum length). The sliding of the dowels along the ducts was verified from the upwards movement of the bolts on top of the beam. The plots that are presented herein correspond to smoothed curves, which were obtained by eliminating the wiggles due to friction sliding.

Fig. 18.10 Second series of tests: typical damage observed at cyclic tests 2D25-d10-CY-1&2: (**a**) at early stage of the tests; (**b**) at the end of the tests

The dowels broke during most of the cyclic tests. However, it must be mentioned that failure of the dowels does not necessarily lead to loss of resistance. This happens because dowels broke at points inside the beam or the column. Thus, a portion of the dowel extruded from the column or the beam, respectively, inside the opposite element, and this part of the dowel continued to resist the horizontal movement. In some cases, this resistance was larger than the one before the failure of the dowel. Due to this phenomenon, it was not always evident during the experimental procedure whether the dowels had failed or not. Also, in some cases, it was observed that the dowels broke at a second point inside the opposite element.

As mentioned above, the small cover of the dowels ($d = 10$ cm) led to significantly reduced resistance in the pull direction. The response was much improved for increased values of d, as shown in Fig. 18.11, where the results of the monotonic pull tests for $d = 10$, 15 and 20 cm are compared. The specimens with $d = 15$ and 20 cm showed almost double resistance in comparison to the strength of the specimen with $d = 10$ cm. Similar was the improvement in the behavior in the cyclic test, where the specimens with larger distance d showed significantly increased resistance in the pull direction (almost symmetrical diagram) and much wider hysteretic loops (less pinching). The observed damage was also reduced, with much less spalling occurring at the beam.

In Fig. 18.12, comparison of the obtained results for the specimens with 2 Ø25 and 1 Ø25 dowels is shown for cyclic loading. For the specimen with one dowel, spalling of the cover concrete of the dowel at the beam side also occurred, similarly to what happened to the two-dowel specimens. As shown in Fig. 18.12, the resistance of the specimen with one dowel was practically one half of the one of the specimen with two dowels, though with much less pinching. The same ratio of one half was observed in the resistance for monotonic push and monotonic pull loadings.

Several conclusions can be drawn by similar comparisons of the experimental results for various cases. For example, the cross section of 2 Ø16 and 1 Ø25 is almost the same. For this reason, for monotonic push loading, the response of the specimen

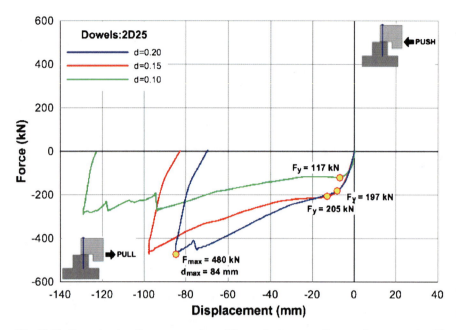

Fig. 18.11 Second series of tests: comparison of force–displacement diagrams for monotonic pull for the specimens with 2 Ø25 dowels and various values of the distance d of the dowels from the beam edge

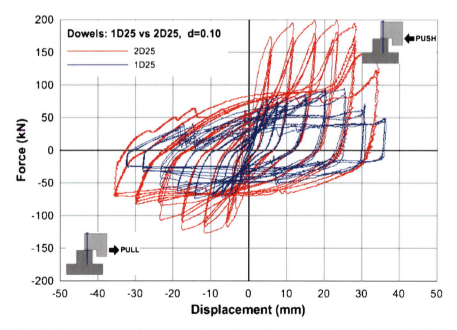

Fig. 18.12 Second series of tests: comparison of force–displacement diagrams for cyclic loading for the specimens with 2 Ø25 and 1 Ø25 dowels

with 2 dowels Ø16 was almost the same with the one of the specimen with 1 dowel Ø25. However, in the monotonic pull direction, the response was better for the specimen with 2 Ø16 dowels, because the ratio of the distance d over the diameter of the dowels was larger, thus, less spalling occurred. For cyclic loading, the behaviour of the two specimens was similar in the pull direction, but the specimen with 1 Ø25 dowel responded with significantly larger resistance in the push direction.

The specimens with 1 dowel Ø32 were constructed with distance $d = 20$ cm for both the beam and the column. In monotonic loading, the yield force was the same in both push and pull directions, but smaller hardening was observed in the pull direction. Spalling of the concrete at both the beam and the column occurred during the monotonic pull test. The cyclic test produced an almost symmetric response. Compared to the theoretical prediction, larger resistance was observed.

18.3.3 Main Features of the Response

The main conclusions that were derived from the experimental investigation of pinned beam-to-column connections in pure shear can be summarized as follows:

- The distance, d, of the dowels from the beam edge is important: for small values of d, the behaviour is asymmetric, with the resistance in the pull direction being significantly lower than the one in the push direction. The behaviour can be improved by increasing the distance d or by applying mechanical devices, as an anchored steel plate in front of the dowels, in order to reduce the spalling of the concrete.
- In cyclic tests, the resistance is significantly lower than the maximum strength that can be achieved by monotonic loading. In general, the cyclic curves follow the monotonic ones up to a point a little further after yield, where the maximum resistance is achieved.
- Failure of the dowels is not necessarily followed by strength drop. This happens because dowels usually brake inside the beam or the column and a portion of the dowels extrudes inside the opposite element.

18.4 Third Series of Tests: Shaking Table Tests on Frames with Pinned Beam-to-Column Connections

The experimental investigation of the seismic response of frames with pinned beam-to-column connections was performed in the framework of the FP7 project "SAFECAST: Performance of innovative mechanical connections in precast building structures under seismic conditions."

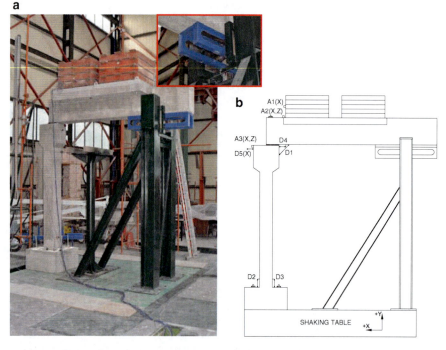

Fig. 18.13 Third series of tests: (**a**) general experimental setup and special device for the sliding pinned support at the free-end of the beam; (**b**) instrumentation setup

The purpose of the research, which is still on-going, was to investigate the effect of the dynamic loading on the behaviour of the connections by comparing the results of the dynamic tests with the corresponding cyclic ones that were obtained during the second series of tests.

Additionally, the influence of the vertical component of the seismic excitation on the behaviour of the connections was investigated with the shaking table tests. To this end, similar tests were performed with and without vertical excitation.

18.4.1 Description of the Specimens

The dynamic tests were performed using specimens that consist of a beam attached to a column by means of dry-pinned connections with dowels, similar to the ones tested under static monotonic and cyclic tests. Due to the limited plan dimensions of the shaking table and the limitations in force and overturning moment, one half of the corresponding one bay frame was modeled. For the support of the beam, a special device was constructed, which was providing free sliding and free rotation at the free end of the beam. The experimental configuration and the instrumentation setup are shown in Fig. 18.13.

With this configuration, the joints were subjected to rotation (through the bending of the column) and sliding (through the shear deformation of the connection) simultaneously during the shaking. The amount of the rotation and the value of the shear force that could be achieved during each test depended on the flexural stiffness of the column and its strength. Three types of columns were examined:

- Stiff and strong column (SS) with dimensions of cross section 60 cm×40 cm, heavily reinforced. In this case, the column responded elastically during the ground shaking, the effective period of the system was relative small and large shear forces were developed at the joint, capable of damaging it severely or producing its failure. The rotations at the joint were small and, practically, the joint behaved as in the pure shear tests under cyclic loading (second series of tests).
- Flexible and strong column (FS) with dimensions of cross section 30 cm×40 cm, heavily reinforced. In this case, the column responded elastically during the ground shaking, although with significant bending. The effective period of the system was around 1.0 s, but we managed to induce large shear forces at the joint, capable of producing its damage, by properly selecting the base motion. At the same time, significant rotations were developed at the joint. In this way, the response of the joint under simultaneous shear and rotation could be investigated.
- Flexible and weak column (FW) with dimensions of cross section 30 cm×40 cm, lightly reinforced. In this case, a plastic hinge was formed at the base of the column and the effective period of the system was large. In this case, large rotations were induced at the joint, but the shear forces that could be developed were small due to the yielding of the column, not capable of producing any severe damage. With these tests, the response of the joint to large rotations could be investigated.

In all cases, the beam had an orthogonal cross section 40 cm×60 cm, similar to the one used for the static monotonic and cyclic tests. At the middle of its span, the beam had a T–shaped cross section, in order to facilitate the placement and the fastening of the additional mass, equal to 10 Mgr, that was added on top. The distance of the dowels from the beam edge was $d = 10$ cm in all cases. The beam was sitting on the column through an elastomeric pad of 2 cm thickness. As in the static tests, the dowels were grouted and bolted on top. The specimens were made of concrete of grade C30/37 with reinforcement B500c. They were designed according to Eurocode 2 (European Committee for Standardization (CEN) 2004a) and Eurocode 8 (European Committee for Standardization (CEN) 2004b). The analyses were performed considering the beam-to-column connection as pinned, while the column was assumed fixed at its base.

18.4.2 *Experimental Results*

Each specimen was tested on the shaking table under uniaxial excitation in the horizontal x-direction (longitudinal direction of beam) or biaxial excitation in the horizontal x- and the vertical y-direction. For each specimen, several experiments

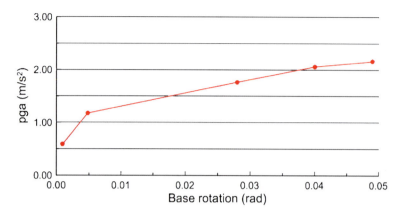

Fig. 18.14 Third series of tests: rotation at the base of the column versus PGA for FW specimens

were performed with the base acceleration increasing step-wise in successive experiments, up to the point that significant damage occurred to the specimen or when the limits of the shaking table were reached. In the latter case, the critical quantity was the peak ground displacement (max 100 mm).

Before the execution of the shaking table tests, the dynamic properties of each specimen were measured through sine logarithmic sweep excitation of low amplitude. This procedure was also repeated at various stages within the experimental procedure, in order to determine the change in the dynamic properties of the specimens after initiation of damage.

In the following, indicative results are presented for tests FS and FW. More details can be found in SAFECAST: Performance of innovative mechanical connections in precast building structures under seismic conditions (2011b).

18.4.2.1 Tests of the FW Specimens

Most of the experiments were performed using as base excitation the Jensen Filter Plant (JFA) record of the Northridge, 1994 earthquake. The component JFA-292 (PGA = 0.59 g in the original record) was applied in the horizontal x-direction. In case of biaxial excitation, the vertical component JFA-UP of the record (PGA = 0.40 g in the original record) was added in the y-direction.

As mentioned above, the column of specimens FW was designed for small earthquake loads and ductile behaviour, therefore plastic hinge was developed at its base during the experiments. Indeed, small cracks appeared at the base of the column even for weak base motion. In subsequent tests with stronger excitations, the cracks at the base of the column enlarged and extended, showing the development of a plastic hinge. This is also evident from the rotation versus PGA diagram depicted in Fig. 18.14.

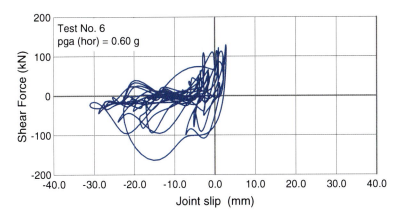

Fig. 18.15 Third series of tests: hysteretic response of the joint for the FS specimen subjected to the Petrovac base motion amplified to PGA = 0.60 g

As expected, the response of the joint was almost elastic due to the small forces that were developed. The maximum top horizontal acceleration was around 0.15 g and was recorded for PGA = 0.18 and 0.21 g. By comparing the hysteretic behaviour of the joint with the corresponding results obtained for cyclic loading, the following were observed: (i) the maximum force attained during the dynamic excitation was practically equal to the corresponding maximum force of the cyclic response in both the push and the pull directions; and (ii) the stiffness of the joint was much larger in the dynamic excitation than it was in the cyclic loading; thus, the maximum force was attained at much less displacement.

When the damaged specimen was subjected to simultaneous horizontal and vertical acceleration, the response was very similar to the one for horizontal excitation only, which shows that the vertical base motion did not have any practical effect on the behaviour of the damaged structure.

18.4.2.2 Tests of the FS specimens

These experiments were performed using the Petrovac record of the Montenegro 1979 earthquake as the base motion. The NS component (PGA = 0.45 g in the original record) was applied in the horizontal x-direction. In cases of biaxial excitation, the vertical component of the record (PGA = 0.21 g in the original record) was added in the y-direction.

In this case, the strong columns of the specimens did not suffer any damage, while significant damage was observed at the joint, mainly spalling of the concrete in front of the dowels, due to the large forces that were developed. The hysteretic response of the joint during the test with the strongest base motion (PGA = 0.60 g) is shown in Fig. 18.15. It is seen that the joint experienced quite large displacements in the pull direction (around 30 mm), evidently due to the damage that occurred in the front side of the beam, while the displacement in the push direction was much smaller (around 3 mm). By comparing this behaviour with the corresponding response of the joint under cyclic loading (Fig. 18.9), it can be concluded that the joint behaved similarly to the cyclic response in the pull direction, with the maximum force being around 150 kN. However, the displacement was quite smaller in the push direction in which the joint behaved practically elastically with the maximum attained force being equal to one half of the resistance; in this regime, the average stiffness of the joint was larger than that for the cyclic loading, similarly to what was observed in tests FW.

Acknowledgements The research presented herein was conducted in the framework of the FP5 project "Precast EC8: Seismic behaviour of precast concrete structures with respect to Eurocode 8 (Co-Normative Research)", Grant Agreement number G6RD-CT-2002-00857, Growth program and the FP7 project "SAFECAST: Performance of innovative mechanical connections in precast building structures under seismic conditions", Grant Agreement number 218417, Research for SME Associations. The financial support provided by the Commission of the European Communities through these projects is greatly appreciated. Special thanks are due to Professor S. Tsoukantas of NTUA for his valuable advice and suggestions. Georgia Kremmyda and Lucia Karapitta, graduate students at NTUA, made a significant contribution to the execution of the experiments and the processing of the data. The help of civil engineer Tryfon Topintzis in designing the specimens was also important. Last but not least, we would like to thank the Greek companies PROET S.A., Interbeton S.A. and Chalyvourgiki S.A. for supplying the specimens or the materials for their construction.

References

COST C1 (1999) Control of the semi–rigid behaviour of civil engineering structural connections. Final report, European Commission, EUR 19244

De Chefbedien (1998) A Precast beam-to-column head connections. In: Proceedings of the COST C1 international conference on control of the semi-rigid behaviour of civil engineering structural connections, Liege, pp 35–44

El Debs MK, El Debs ALH, Miotto AM (2006) Experimental analysis of beam–to–column connection with semi–rigid behaviour of precast concrete structures. In: Proceedings of the 2nd fib congress, Naples

Elliott KS (1998) Semi-rigid connections in precast concrete structures and bridges. In: Proceedings of the COST C1 international conference on control of the semi-rigid behaviour of civil engineering structural connections, Liege, pp 3–12

Elliott KS (2000) Research and development in precast concrete framed structures. Prog Struct Eng Mater 2(4):405–428

Elliott KS, Davies G, Mahdi AA, Gorgun H, Virdi K, Ragupathy P (1998) Precast concrete semi-rigid beam-to-column connections in skeletal frames. In: Proceedings of the COST C1 international conference on control of the semi-rigid behaviour of civil engineering structural connections, Liege, pp 45–54

European Committee for Standardization (CEN) (2004a) Eurocode 2: design of concrete structures – Part 1–1: general rules and rules for buildings. EN 1992-1-1, Brussels

European Committee for Standardization (CEN) (2004b) Eurocode 8: design of structures for earthquake resistance – Part 1: general rules, seismic actions and rules for buildings. EN 1998-1, Brussels

18 Experimental Investigation of the Seismic Behaviour of Precast Structures... 365

Imai H, Castro JJ, Yanez R, Yamaguchi T (1993) A new precast system for frame structures and its structural characteristics. In: Dhir RK, Jones MR (eds) Concrete 2000. E & N Spon, London, pp 655–670

International Federation for Structural Concrete (2008) Structural connections for precast concrete buildings. fib Bulletin 43

Jolly CK, Guo M (1998) Application of numerical analysis to connection in precast concrete frames. In: Proceedings of the COST C1 international conference on control of the semi-rigid behaviour of civil engineering structural connections, Liege, pp 65–74

Leong DCP (2006) Testing of pinned beam–to–column connections of precast concrete frames. Master thesis, University Technology, Kuala Lampur

Nakaki SW, Standon J, Sritharan S (1999) An overview of the PRESSS five story precast test building. PCI J 44(2):26–39

Priestley MJN (1996) The PRESSS program – current status and proposed plans for phase III. PCI J 41(2):22–40

Psycharis IN, Mouzakis HP, Carydis PG (2006) Experimental investigation of the seismic behaviour of prefabricated RC structures. In: 2nd international fib Congress, Naples

Rahman AB Abd, Ghazali AR, Hamid Z Abd (2008) Comparative study of monolithic and precast concrete beam–to–column connections. MCR J 2(1):42–55

SAFECAST: Performance of innovative mechanical connections in precast building structures under seismic conditions (2011a) Deliverable 2.1: experimental behaviour of existing connections. Commission of the European Communities, Brussels

SAFECAST: Performance of innovative mechanical connections in precast building structures under seismic conditions (2011b) Deliverable 2.2: quantification of the effects of dynamic loads. Commission of the European Communities, Brussels

Stanton JF (1998) The PRESSS program in the USA and Japan – Seismic testing of precast concrete structures. In: Proceedings of the COST C1 international conference on control of the semi-rigid behaviour of civil engineering structural connections, Liege, pp 13–24

Tsoukantas SG, Tassios TP (1989) Shear resistance of connections between reinforced concrete linear precast elements. ACI 86(3):242–249

Vamvatsikos D, Cornell CA (2002) Incremental dynamic analysis. Earthq Eng Struct Dyn 31:491–514

Vintzeleou EN, Tassios TP (1987) Behavior of dowels under cyclic deformations. ACI Struct J 84(1):18–30

Virdi K, Ragupathy R (1992a) Tests on precast concrete subframes with semi-rigid joints. In: Proceedings of the 1st COST C1 state of the art workshop on semi-rigid behaviour of civil engineering structural connections, Strasbourg, pp 120–131

Virdi K, Ragupathy R (1992b) Analysis of precast concrete frames with semi-rigid joints. In: Proceedings of the 1st COST C1 state of the art workshop on semi-rigid behaviour of civil engineering structural connections, Strasbourg, pp 296–307

Watanabe F (2000) Seismic design for prefabricated and prestressed concrete moment resisting frames. In: Proceedings of the PCI/FHWA/FIB international symposium on high performance concrete, Orlando, pp 820–829

Chapter 19
Experimental Investigation of the Progressive Collapse of a Steel Special Moment-Resisting Frame and a Post-tensioned Energy-Dissipating Frame

Antonios Tsitos and Gilberto Mosqueda

Abstract Quasi-static "push-down" experiments were performed to evaluate the progressive collapse resistance of steel buildings considering multi-hazard extreme loading. A 1:3 scale three-story, two-bay conventional special moment resisting frame (SMRF) and a post-tensioned energy-dissipating frame (PTED) of similar geometry, designed and previously tested for seismic performance on a shaking table, were adapted for quasi static collapse testing. The experiments simulated the structural response of a prototype building after the sudden failure of a base column. An objective of the tests was to evaluate the effectiveness of earthquake resistant design details in enhancing the resistance to progressive collapse. An effort was made to document the sequence of damage in the frames and to correlate observed damage events with changes in the global resisting strength. Significant vertical displacement capacity and the ability of the steel components to redistribute loads after the failure of a single column were demonstrated by both tests. The resistance of the SMRF specimen against progressive collapse depends on the ultimate deformation capacity of the beam-to-column connections including panel zones, while the vertical load carrying capacity of the PTED frame depends primarily on the performance and ultimate strength of the tendons. Numerical simulations, using simplified models of the frames, investigate the capability of structural analysis software to reproduce the experimentally measured response.

A. Tsitos (✉)
Department of Civil Engineering, University of Patras, Rion 26500, Greece
e-mail: atsitos@upatras.gr

G. Mosqueda
Department of Civil, Structural & Environmental Engineering, University at Buffalo – The State University of New York, Buffalo, NY 14260, USA
e-mail: mosqueda@buffalo.edu

M.N. Fardis and Z.T. Rakicevic (eds.), *Role of Seismic Testing Facilities in Performance-Based Earthquake Engineering: SERIES Workshop*, Geotechnical, Geological and Earthquake Engineering 22, DOI 10.1007/978-94-007-1977-4_19,
© Springer Science+Business Media B.V. 2012

19.1 Introduction

Recent terrorist attacks against government and military buildings worldwide and most notably the September 11, 2001 events at the World Trade Center in New York City have motivated the structural engineering community to focus significant efforts towards better understanding the sequence of events leading to progressive collapse in building structures. The ultimate goal is to establish rational and reliable methods for the assessment and the enhancement of structural resistance to extreme accidental events, as well as numerical tools for the analysis of structures in an extreme loading and deformation context. Although design methodologies (such as the alternative path method and the tie force method) and analysis procedures to enhance resistance to progressive collapse are already proposed in guideline documents issued by U.S. authorities (U.S. General Services Administration 2003; Department of Defense 2005), there is a scarcity of experimental data to support the numerical modeling of building structures under extreme loads, particularly to the point of failure.

Experimental investigations of large scale building models or sub-assemblies are scarce in the literature. The resistance of different types of steel frame connections has been evaluated (Karns et al. 2006, 2007). The connections were initially subjected to blast loads and then pushed using monotonic loading, in order to determine their post-blast integrity for the purpose of mitigating progressive collapse. Recently, a series of quasi static collapse tests was performed on 3/8 scale models of concrete frame beams (Sasani and Kropelnicki 2008). In-situ full scale tests were performed on existing concrete buildings (Sasani and Sagiroglu 2008), on steel buildings (Sezen 2007, personal communication) and on masonry buildings (Zapata and Weggel 2008). All of the above followed the standard approach of the sudden loss of one or more exterior columns ("missing column scenario"), simulating an extreme accidental load, such as bomb blast or impact of a heavy vehicle. For the cases involving real buildings, the structures were able to redistribute the loads without the propagation of failure to additional members. Columns were not removed until the point of failure, thus the margin of safety against collapse could not be directly determined.

The purpose of this study is to experimentally investigate the progress of damage through collapse in steel frames when subjected to a missing column scenario and determine the mechanisms and capacity of the frames to transfer the loads to neighboring columns. The idea of investigating the influence of seismic detailing of a building on its progressive collapse resistance in a multi-hazard framework was presented in a case study of the 1995 terrorist attack on the Alfred P. Murrah Federal Building in Oklahoma City (Hayes et al. 2005). A key finding of that study was that strengthening the perimeter elements of the building using current seismic detailing techniques could have greatly reduced damage to the structure. In this perspective, a series of tests were conducted, following the missing column scenario, on two 1/3 scale three-story, two-bay steel frames: (i) a special moment resisting frame (SMRF) and (ii) a post-tensioned energy dissipating frame (PTED). The test results are summarized herein (a more detailed analysis is presented by Tsitos (2009)).

The concept of the PTED frame for seismic design has been developed and experimentally investigated (Christopoulos et al. 2002). The PTED frame (see Fig. 19.1a for a typical configuration in buildings) has beams that are not

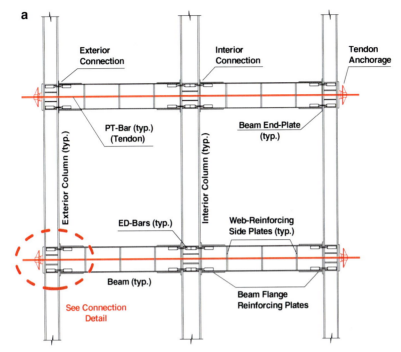
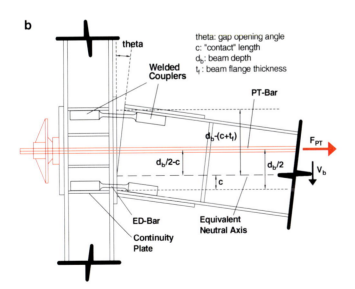

Fig. 19.1 The concept of the PTED frame: (**a**) typical steel frame with PTED connections and (**b**) detail of an exterior connection at a deformed configuration

welded or bolted to the column flanges and relies instead on the post-tensioning force provided at each floor by two tendons located at mid-depth of the beam for shear force transfer and for re-centering when the frame is subjected to lateral seismic loads. Four symmetrically placed energy dissipating (ED) bars located at each connection provide energy dissipation under cyclic loading as gaps at the beam-column interface open and close due to the rocking of the beam relative to the column. The ED bars are properly restrained so that they can yield both in tension and compression without buckling. The deformed state of a typical exterior PTED beam-column connection is presented in Fig. 19.1b. The beams of the PTED frame are reinforced with transverse web reinforcing plates as well as reinforced beam flanges at the member ends, to mitigate buckling phenomena that may arise as a result of the post-tensioning compressive action effect in combination with stresses from cyclic loading.

The SMRF and the PTED frame of the present study were previously tested for seismic performance on the shake table at the University at Buffalo Structural Engineering and Earthquake Simulation Laboratory (SEESL). The results of the seismic tests are available in the literature (Wang 2007; Wang and Filiatrault 2008). The seismic study concluded that the PTED frame had desirable re-centering capability after the maximum considered seismic event and was very effective in terms of limiting accelerations and damage to easily replaceable components, namely the ED bars. The same frame with minimal seismic damage was adapted for this study, in order to conduct collapse experiments simulating the missing column scenario.

19.2 Experimental Setup

For the quasi-static "push-down" collapse testing, the two frames were installed on the 24 in. (60.9 cm) thick reinforced concrete Reaction Wall and Strong Floor at SEESL. The loading was applied by means of a MTS 244.51S servo-controlled actuator with a stroke of 40 in. (101.6 cm) and a force capacity of 220 kips (978.6 kN). Due to the quasi-static loading protocol, the actuator was used in its "static" configuration, using a low capacity servo-valve (SEESL 2008). A 2 in. (5.1 cm) thick steel plate, attached to the strong floor, served as a base of the whole experimental setup. The specimens were equipped with thin steel sliding plates at the columns. Additionally, a support and sliding mechanism, consisting of eight steel pedestals with TEFLON sliding pads, was designed, in order to restrict the out-of-plane motion of the frames, while allowing the development of very large in-plane deformations with relatively low friction. Geometric properties of the frames, the loading method and a general view of the experimental setup are illustrated in Fig. 19.2 for the SMRF specimen and in Fig. 19.3 for the PTED specimen. Due to geometric and loading constraints the specimens were installed horizontally, i.e. the frame plane was parallel to the strong floor.

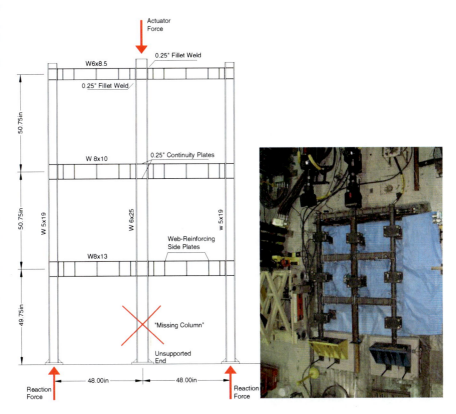

Fig. 19.2 Schematic of the steel moment resisting frame (SMRF) with boundary and loading conditions (*left*) and a general view of the experimental setup prior to testing (*right*)

Standard wide flange "W" sections were used, in order to maintain geometric similitude to the full scale structure. The DYWIDAG Mono-Strand post-tensioning system was used for connecting the beams to the columns of the PTED frame. Two tendons were used at the mid-depth of the beam of each floor with initial post-tensioning forces of approximately 20 kips (89 kN), 15 kips (66.7 kN) and 12.5 kips (55.6 kN) per tendon for the 1st, 2nd and 3rd floor, respectively. The tendons were equipped with custom-made load cells installed adjacent to their anchorage mechanism. The beams of the PTED frame incorporated web flange reinforcing plates and reinforced flange segments at both ends, to prevent the beams from buckling under compression, when subjected to seismic cyclic loads. It should be noted that due to the requirement of direct comparison of the two frames' dynamic behavior during the previous shake table testing, the same beam web reinforcements were used in the SMRF specimen, although it was not required in principle by the design.

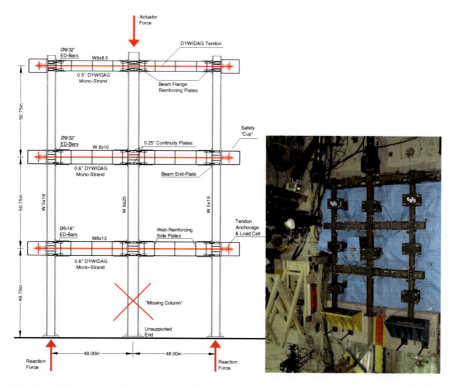

Fig. 19.3 Geometry of the post-tensioned energy-dissipating frame (PTED) with boundary and loading conditions (*left*) and a general view of the experimental setup prior to testing (*right*)

The exterior columns of the frames were connected to specially designed reaction blocks, consisting of a rectangular reinforced concrete base and an overlying steel block shaped as a triangular prism. Each reaction block was firmly attached to the strong floor with four post tensioned DYWIDAG Thread-Bars loaded to 100 kips (444.8 kN) each. The central (interior) column of each frame was left unsupported, in an attempt to simulate the "missing column" scenario, and was allowed to move within the plane defined by the frame.

Special steel safety "cups" were designed, in order to arrest the components of the tendons' anchorages, in the event of sudden failure of the tendons of the PTED frame during testing. The installed specimens were fully instrumented with uni-axial strain gauges attached to the steel members, displacement transducers (string potentiometers) and a 3D imaging Krypton system, operating with infrared LED's (SEESL 2008). A schematic of the instrumentation used in the case of the PTED specimen is presented in Fig. 19.4: an array of 76 strain gauges, 17 cable position transducers, 27 linear potentiometers and 7 load cells were installed, coming to a total of 127 instrumentation channels. All instrumentation channels were connected to a Pacific Instruments 6000 Data Acquisition System.

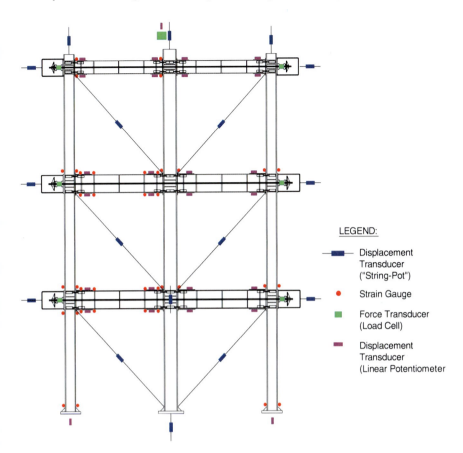

Fig. 19.4 Schematic of the instrumentation layout for the PTED specimen

19.3 Experimental Procedures and Results

Low amplitude identification tests (low-level displacement "ramp"- and "sawtooth"-shaped histories) were performed on the specimen for equipment check and calibration of system performance (e.g. calculation of "elastic" stiffness and comparison with analytic predictions) as well as for the assessment of the actual condition of the two specimens after the completion of the shake table testing program. These preliminary tests were followed by a final "push to collapse", using a slow displacement ramp with a loading rate ranging of 0.2 in./min (0.51 cm/min). The data acquisition rate was set at 10 Hz, which was considered more than sufficient, given the quasi-static nature of the experiments

The most representative results of the "push-down" tests are the plots of the force applied by the actuator versus the displacement of the central column of the SMRF and PTED specimens. These plots are presented in the top halves of Figs. 19.5 and 19.8

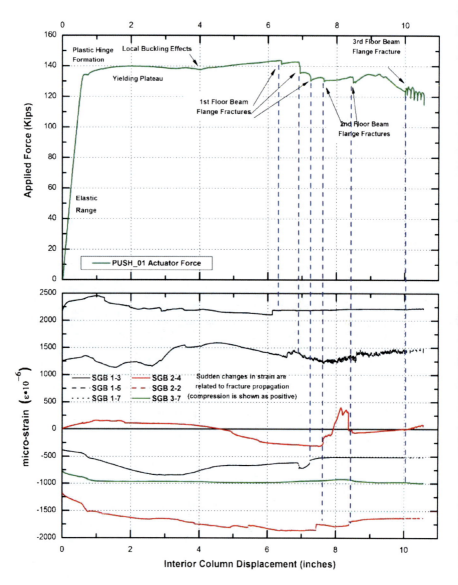

Fig. 19.5 Combined plot of the global force vs. interior column vertical displacement response of the SMRF specimen (*top*) and measured strain vs. interior column vertical displacement of selected uni-axial strain gauges (*bottom*) correlating the global behavior with localized fracture events in the beams

for the SMRF and PTED specimens respectively. It should be noted that the applied displacement was controlled automatically up to 10 in. (25.4 cm) during the SMRF test or 13.75 in. (34.9 cm) during the PTED test and manually – at increments of 0.2 in. (0.51 cm) – thereafter. Control issues associated with the manual approach

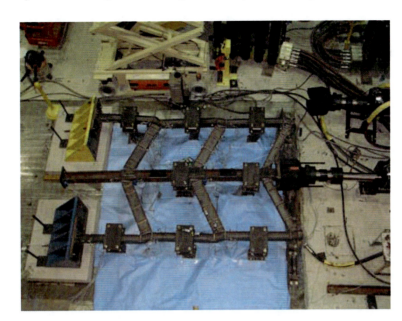

Fig. 19.6 General view of the deformed state of the SMRF specimen after the push-down test

are the cause of the oscillations observed in the plots after the aforementioned displacement levels, likely due to stress relaxation of the yielded material in between load steps when the displacement is held constant (the tests were performed in displacement-control mode).

The SMRF specimen exhibited essentially linear elastic behavior up to a displacement level of 0.4 in. (1.0 cm) that corresponded to an applied force level of 135 kips (600.5 kN) (see top half of Fig. 19.5). Plastic hinges were formed at the ends of the beams as the applied displacement increased. The plastic hinge formation was accompanied by significant plastic shear deformation of the beam-column connection panel zones at the exterior columns which eventually led to out-of-plane buckling of the panel zones. This was primarily due to the fact that the exterior connections were not restrained against rotation (e.g. by adjacent frame bays or slab action) and as a result no significant global hardening behavior was observed. The deformed state of the SMRF specimen at the end of the destructive push-down test is presented in the photograph in Fig. 19.6. Apart from the expected damage due to plastic hinge formation and fractures at the ends of the beams, the most outstanding feature is the deformation of the exterior columns due to the aforementioned inelastic deformation of the connection panel zones. Additionally, local buckling of the beam flanges was observed up to a level of displacement of 6 in. (15.2 cm) where the maximum resistance of 145 kips (645 kN) was obtained, after which a series of fractures occurred that could be traced with the help of strain gauge signals to local events at specific members (see bottom half of Fig. 19.5). The gradual deterioration of the SMRF specimen continued thereafter up to the

Fig. 19.7 Close-up view of a destroyed interior beam-column connection of the SMRF specimen after the push-down test

end of the test. A detail of the final damaged state of an interior beam-column connection is shown in Fig. 19.7: the initial stage of plastic hinge formation was followed by buckling of the top flange of the beams and by initiation and propagation of cracks from the bottom flange into the web.

The push-down testing of the PTED specimen was performed similarly, but with distinctly different observations. Due to the non-linear elastic behavior (gap opening at the beams' ends) of the PTED frame at low deformation levels, the elastic region could not be well-defined near the origin of the global curve (top half of Fig. 19.8). The resistance of the frame increased monotonically to a maximum value of 121 kips (538.2 kN) at a displacement of 3.5 in. (8.9 cm) with a slight stiffness softening. Up to that point, only minor slippage of the beams with respect to the columns and yielding of the ED bars and of the flanges of the central non-reinforced sections of beams had been observed. With increasing displacement, a series of tendon ruptures, failures of the energy dissipating bars (not very important for the global response) and buckling at the beams of the 2nd and 3rd floor led to a rapid decrease of the frame resistance. The residual "strength" of the PTED frame after the application of 19 in. (48.3 cm) of displacement was insignificant; less than 15 kips (66.7 kN). It is noted that the buckling of the post-tensioned beams at the left bay of the 2nd and 3rd floors occurred at the end of the reinforced zones, far from the interface with the column flange.

By comparing the top plot on Fig. 19.8 (global behavior) with plots of the tendon forces vs. the interior column's displacement in the bottom half of Fig. 19.8, it can

19 Experimental Investigation of the Progressive Collapse of a Steel Special... 377

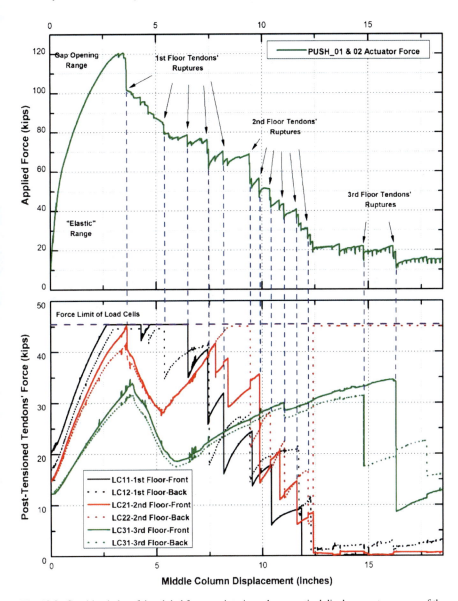

Fig. 19.8 Combined plot of the global force vs. interior column vertical displacement response of the PTED specimen (*top*) and measured post-tensioning force vs. interior column E-W displacement for all six tendons of the PTED (*bottom*) correlating the global degradation behavior with single rupture events in the Mono-Strand wire bundles

be observed that all major drops in the strength resistance of the frame correspond to failures of single wires of the front tendons (each DYWIDAG Mono-Strand tendon consists of seven bundled wires, all encased in a protective PVC pipe). Ruptures of single wires are represented by vertical force drops and can help trace damage

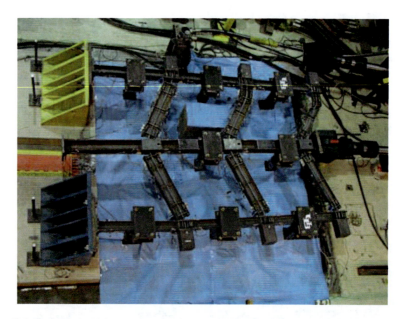

Fig. 19.9 General view of the "collapsed" state of the PTED specimen after the push-down test

events observed in the global plot back to the approximate area where the damaging event occurred. In addition to the tendons' tensile fractures, the local minima of the force traces for the 2nd and 3rd floors (in the vicinity of 5 in. (12.7 cm) global displacement) are explained by the simultaneous apparition of heavy buckling in the respective beams. The photograph in Fig. 19.9 illustrates the final state of the PTED specimen upon the completion of the push-down test. In contrast to the SMRF specimen, the columns of the PTED frame survived the test practically intact, with the sole exception of plastic hinge formation at the top and bottom of the 1st floor. A close-up view of a disintegrated interior PTED connection is presented in Fig. 19.10. The ED bars have ruptured in tension and the tendon is in a severely degraded state, although this is not directly visible due to the PVC encasing.

Both push-down tests demonstrated that a single column failure is not sufficient to cause progressive collapse of a frame designed with this level of seismic detailing. The most severe dead plus live load combination for the specific frame geometry would be of the order of 30 kips (133.4 kN) (scaled), which is three times less than the maximum measured capacity of the PTED. This means that a dynamic amplification factor (as the real phenomenon is not quasi-static) of at least 3 can be accommodated without danger for the structure.

A direct comparison if the global resistance of the two specimens would favour the SMRF, both in terms of maximum resistance and deformation ductility. Nonetheless, such a comparison would not be very meaningful in this case, because, as already explained, the design of the SMRF specimen was biased in the sense that it tried to match the elastic dynamic characteristics of the PTED during the previously

Fig. 19.10 Detailed view of a disintegrated interior beam-column connection of the PTED specimen after the pushdown test

performed shake table simulations. The steel sections used for the members, the welding details and the additional reinforcing plates of the beams' webs led to an overall "stocky" special moment-resisting frame which would not be typical in standard construction practice.

19.4 Numerical Simulations with Simplified Models

Selected results of numerical simulations of the push-down tests are presented in Fig. 19.11 for the SMRF specimen and in Fig. 19.12 for the PTED specimen. The numerical model of the SMRF specimen was constructed with the open-source research software OpenSEES (http://opensees.berekeley.edu). A distributed plasticity, fiber section approach was followed with definition of degradation at the material level using a branch of appropriate "negative" stiffness at the material constitutive law. The ultimate resistance and the deformation capacity of the specimen were accurately simulated up to the point of first fracture. With this type of modeling the degradation phase was depicted at an average sense (i.e. without abrupt drops corresponding to fractures at specific locations).

The numerical model of the PTED specimen was constructed using SAP2000 with a macro-model approach in which the PTED connection was modeled by an assembly of beam and rigid elements connected by non-linear springs and "gap"

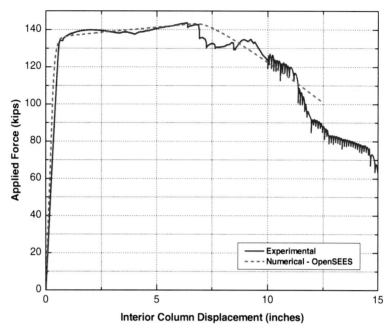

Fig. 19.11 Numerical vs. experimental push-down curve for the SMRF specimen

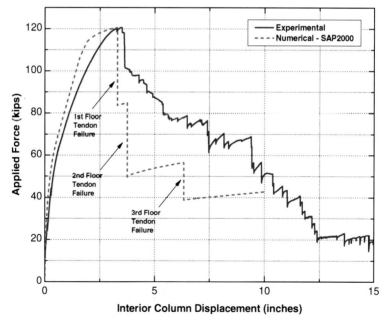

Fig. 19.12 Numerical vs. experimental push-down curve for the PTED specimen

elements (Christopoulos 2002). The model explicitly accounted for the PTED beams' depth and the gap opening. The ED bars were modeled as bi-linear springs with a smooth transition zone. The tendons (PT bars) were modeled as linear elements with axial stiffness and ability to deform when loaded perpendicularly to their axis. Concentrated plasticity approach was used both for the beams and the tendons. A more comprehensive description of the model is given by Tsitos (2009). The model captures the observed behavior with sufficient accuracy, particularly the peak strength. The strength degradation is more gradual in the actual experiment because it was caused by a sequence of failure of individual strands in tendons. In contrast, the complete tendon at a floor level failed instantaneously since all strands were assigned the same nominal properties.

19.5 Conclusions

As part of an experimental study to investigate the resistance of steel buildings against progressive collapse, a conventional SMRF and a PTED frame were subjected to quasi-static push-down tests following a missing-column scenario.

The frames behaved very well in terms of being able to resist loads three times larger than the most severe dead and live load combination without significant damage. In the case of the sudden loss of a central column, both frames would have been able to support the remaining building gravity load and also have significant remaining deformation capacity. In this multi-hazard extreme loading case, the seismic design and detailing seems to have provided the strength and ductility needed to prevent a progressive catastrophic failure of the tested structures.

The experimental data recorded from the test provided useful insight into the types and propagation of damage that would be expected when a conventional moment frame and post-tensioned frame are subjected to extreme displacements, such as those expected in a collapse situation. In the case of the moment frame, it is important to account for all "sources" of deformation capacity and not rely only on plastic hinges at the ends of members, when trying to predict the margin of safety against progressive collapse. The critical role of the tendons' strength and failure mode is clearly reflected on the global behavior of the PTED frame and underlines the need for using adequate models of this crucial part of post-tensioned frames, when attempting numerical simulations.

The numerical modeling of a progressive collapse scenario needs to be further researched and coded, with the ultimate objective of providing a simple yet reliable tool to the professional structural engineering community. Special attention should be paid to the modeling of the degradation phase of the structural components, as it is governing the post-peak force-displacement behavior.

Acknowledgments This research was supported in part by the National Science Foundation under award ECC-9701471 to the Multidisciplinary Center for Earthquake Engineering Research (MCEER). The first author was financially supported by a Fellowship from the "Alexander S. Onassis" Public

Benefit Foundation. The opinions and conclusions expressed in this paper are those of the authors and do not necessarily reflect the views of the sponsors. The authors express their gratitude to the SEESL technicians and to DYWIDAG-Systems International for providing the post-tensioning systems used in the PTED frame and the experimental setup.

References

Christopoulos C (2002) Self-centering post-tensioned energy dissipating (PTED) steel frames for seismic regions. PhD dissertation, Department of Structural Engineering, University of California, San Diego

Christopoulos C, Filiatrault A, Uang CM, Folz B (2002) Post-tensioned energy dissipating connections for moment resisting steel frames. ASCE J Struct Eng 128(9):1111–1120

Department of Defense (2005) Design of buildings to resist progressive collapse. Unified Facilities Criteria (UFC, 4-023-03). U.S. Department of Defense, Washington, DC

Hayes JR Jr, Woodson SC, Pekelnicky RG, Poland CD, Corley WG, Sozen M (2005) Can strengthening for earthquake improve blast and progressive collapse resistance? ASCE J Struct Eng 131(8):1157–1177

Karns JE, Houghton DL, Hall BE, Kim J, Lee K (2006) Blast testing of steel frame assemblies to assess the implications of connection behavior on progressive collapse. In: Proceedings of ASCE 2006 Structures Congress, St. Louis

Karns JE, Houghton DL, Hall BE, Kim J, Lee K (2007) Analytical verification of blast testing of steel frame moment connection assemblies. In: Proceedings of ASCE 2007 structures congress, Long Beach

Sasani M, Kropelnicki J (2008) Progressive collapse analysis of an RC structure. Struct Des Tall Spec Build 17(4):757–771

Sasani M, Sagiroglu S (2008) Progressive collapse resistance of Hotel San Diego. ASCE J Struct Eng 134(3):478–488

SEESL (2008) Online lab manual of the structural engineering and earthquake simulation laboratory. http://seesl.buffalo.edu/docs/labmanual/html/. Accessed 1 Nov 2009

Tsitos A (2009) Experimental investigation of the progressive collapse of steel frames. PhD dissertation, Department of Civil, Structural & Environmental Engineering, University at Buffalo, The State University of New York, Buffalo

U.S. General Services Administration (GSA) (2003) Progressive collapse analysis and design guidelines for new federal office buildings and major modernization projects. U.S. General Services Administration, Washington, DC

Wang D (2007) Numerical and experimental studies of self-centering post-tensioned frames. PhD dissertation, State University of New York at Buffalo, Buffalo

Wang D, Filiatrault A (2008) Shake table testing of a self-centering post-tensioned steel frame. In: 14th world conference on earthquake engineering, Beijing

Zapata BJ, Weggel DC (2008) Collapse study of an unreinforced masonry bearing wall building subjected to internal blast loading. ASCE J Perform Constr Facil 22(2):92–100

Index

A
Accreditation, 265–284
Actuation technology, 119–132

B
Bearing tester, 65–80
Bridge
    cable-stayed, 25–26
    long-span, 32–34
    piers, 36–39

C
Cable dynamics, 25–26
Calibration, 287–302
Centrifuge, 83–98
Certification, 265–284
Computer vision, 159–176
Connections, 345–364
Corrosion, 36–39

D
Damage measures, 1–19
Damper, 323–343
Distributed
    computing, 177–196
    test control, 177–196
    testing, 199–218
Dowels, 345–364

E
Earthquake protection, 221–243

Elastomer, 323–343
Energy dissipation, 83–98
Engineering demand parameters, 1–19
ESB box, 135–156
Eurocode 8, 1–19
Experimental constraints, 43–60

F
Field vision systems, 1–19

G
Geotechnical earthquake engineering, 83–98,
    135–156, 247–262

H
High definition cameras, 159–176
High performance computing, 177–196,
    199–218
Historic buildings, 221–243
Homogenisation, 247–262
Hybrid actuators, 119–132

I
Instrumentation management,
    287–302
Interaction
    human-structure, 34–36
    kinematic, 29–32
    soil-structure, 29–32,
    135–156
ISO/IEC 17025, 287–302

M.N. Fardis and Z.T. Rakicevic (eds.), *Role of Seismic Testing Facilities
in Performance-Based Earthquake Engineering: SERIES Workshop*, Geotechnical,
Geological and Earthquake Engineering 22, DOI 10.1007/978-94-007-1977-4,
© Springer Science+Business Media B.V. 2012

L
Laminar box, 135–156

M
Markers detection, 159–176
Model(-ing)
    analytical, 22
    large numerical, 199–218
    model container, 135–156
    numerical, 22
    physical, 22, 43–60
Moment-resisting frames, 367–382
Moment-rotation, 83–98
Monuments, 221–243
Multi-support excitation, 32–34

N
Non-linear structural dynamics, 26–29
Nonstructural building equipment, 305–321

O
Optical measurements, 159–176

P
Parametric excitation, 25–26
Pedestrian-induced vibrations, 34–36
Performance based
    earthquake engineering, 1–19
    engineering (PBE), 265–284
    seismic design, 323–343
Piles, 29–32
Pinned connections, 345–364
Post-tensioned frames, 367–382
Precast structures, 345–364
Prefabrication, 345–364
Progressive collapse, 367–382

Q
Qualification, 265–284, 287–302, 305–321
Quality management system, 265–284

R
Reaction walls, 1–19, 65–80
Reinforced concrete deterioration, 36–39
Reinforced soils, 247–262
Repeatability, 265–284, 287–302
Reproducibility, 265–284, 287–302

Resonant frequency, 83–98
Rocking, 83–98

S
Second gradient media, 247–262
Seismic
    experimental activities, 99–116
    experimental facilities, 99–116
    experimental needs, 99–116
    qualification, 43–60
    response, 83–98, 345–364
    strengthening, 221–243
    testing facilities, 99–116, 287–302
    testing methods, 99–116
    testing needs, 99–116
Self-centring structures, 26–29
Servo-systems, 119–132
Shaking table, 22–24, 43–60, 65–80, 135–156, 221–243, 247–262, 345–364
Shallow foundation, 83–98
Similitude relations, 43–60
Simulation
    finite element simulation, 177–196
    hybrid simulation, 199–218
    parallel simulation, 199–218
    real-time hybrid simulation, 323–343
Steel frames, 367–382
Steel moment resisting frame, 323–343
Strong walls, 22–24
Substructure
    algorithm, 199–218
    testing, 199–218
Systems analysis, 305–321

T
Test(-ing)
    benchmark testing, 287–302
    cyclic, 345–364
    distributed, 199–218
    dynamic, 119–132
    large scale, 1–19
    monotonic, 345–364
    proof, 43–60
    pseudodynamic, 1–19
    seismic performance, 305–321
    seismic testing facilities, 99–116, 287–302
    seismic testing methods, 99–116
    seismic testing needs, 99–116
    structural, 99–116
    substructure, 199–218
    validation, 83–98